AD

THE QUA

"*The Quantum Revelation* is that unique book that plumbs the depth of the current scientific revolution and at the same time speaks from the heart to the heart of everyone. It is what I call the new paradigm—embodied in a readable, enjoyable, and above all supremely informative volume!"

—ERVIN LASZLO
Founder and President of the Club of Budapest, Founder and Director of the Laszlo Institute of New Paradigm Research, and author of *What Is Reality?* and many other books

"In *The Quantum Revelation,* Paul Levy reveals a picture of our world that is shockingly different than the one we were taught. It is a view in which mind matters, and in which consciousness is fundamental. The implications of these insights for how we live our life and for our very survival as a species are being sensed by an increasing number of individuals. No intelligent, thinking person in today's world can afford to be uninformed about these developments. For your sake—and that of future generations—put *The Quantum Revelation* on the top of your stack."

—LARRY DOSSEY, MD
Author of *ONE MIND: How Our Individual Mind Is Part of a Greater Consciousness and Why It Matters*

"*The Quantum Revelation* is a deep and engrossing contemplation of the quantum world that creates a spiritual awakening within the reader, a must-read for deep contemplation of a spiritual path."

—BARBARA HAND CLOW
Author of *Alchemy of Nine Dimensions, Revelations of the Ruby Crystal,* and many other books; founder of Bear and Co.

"Paul Levy's *The Quantum Revelation* brilliantly reveals how science is capable of enhancing our understanding of spirituality. In creatively pointing out the correlation between psyche and matter, Paul is helping to illumine the dreamlike nature of the universe in a way that can help all of us to further awaken. Bravo!"

—DR. ARNY MINDELL
Author of *Quantum Mind: The Edge between Physics and Psychology,* and founder of process-oriented psychology

"Perhaps the greatest 'spiritual teacher' of the twentieth century was Albert Einstein because he grasped a fundamental truth of existence: that science and spirituality are not separate. In *The Quantum Revelation*, Paul Levy brilliantly elucidates the reality that separation does not exist on any level and that humanity's principal task in the throes of the global crisis is to deeply reconnect with oneself, with each other, and with Earth. The question, 'Is there a higher power?' is now embarrassingly obsolete. Levy masterfully demonstrates that in fact, there is no one and nothing else."

—CAROLYN BAKER, PHD
Author of *Dark Gold: The Human Shadow and the Global Crisis* and *Love in the Age of Ecological Apocalypse: Cultivating the Relationships We Need to Thrive*

"*The Quantum Revelation* lives up to its title in spades. Every page offers a breathtaking, provocative revelation that shows how the quantum world, far from inhabiting an esoteric niche of our reality, constitutes its very fabric. Learned, bold, piercingly insightful, and ever summoning us into the larger human story, this book helps us come to terms with the true nature of our reality, that we might awaken to the mystery and agency of our own true natures. To assimilate the message of this book is to help forge the future we all so deeply need to create, as individuals and as a collective force on this planet."

—PHILIP SHEPHERD
Author of *Radical Wholeness* and *New Self New World*

"For years we've seen a flood of books telling us everything we need to know about quantum physics, linking it to religion, consciousness, evolution, even sex. You might think that by now it would all seem elementary—but you'd be wrong. Paul Levy's *The Quantum Revelation* shows that there are still mysteries to discover in the strange world of subatomic particles and their even stranger relation to consciousness and the reality we share. This seems to be, as Levy tells us, a 'participatory' one, in which we are active players engaged in bringing the world into being, a truth here indeed brought home with the force of revelation. Heisenberg may have been uncertain about some things, but I bet observed and observer alike will appreciate this book."

—GARY LACHMAN
Author of *Beyond the Robot: The Life and Work of Colin Wilson*

THE QUANTUM REVELATION

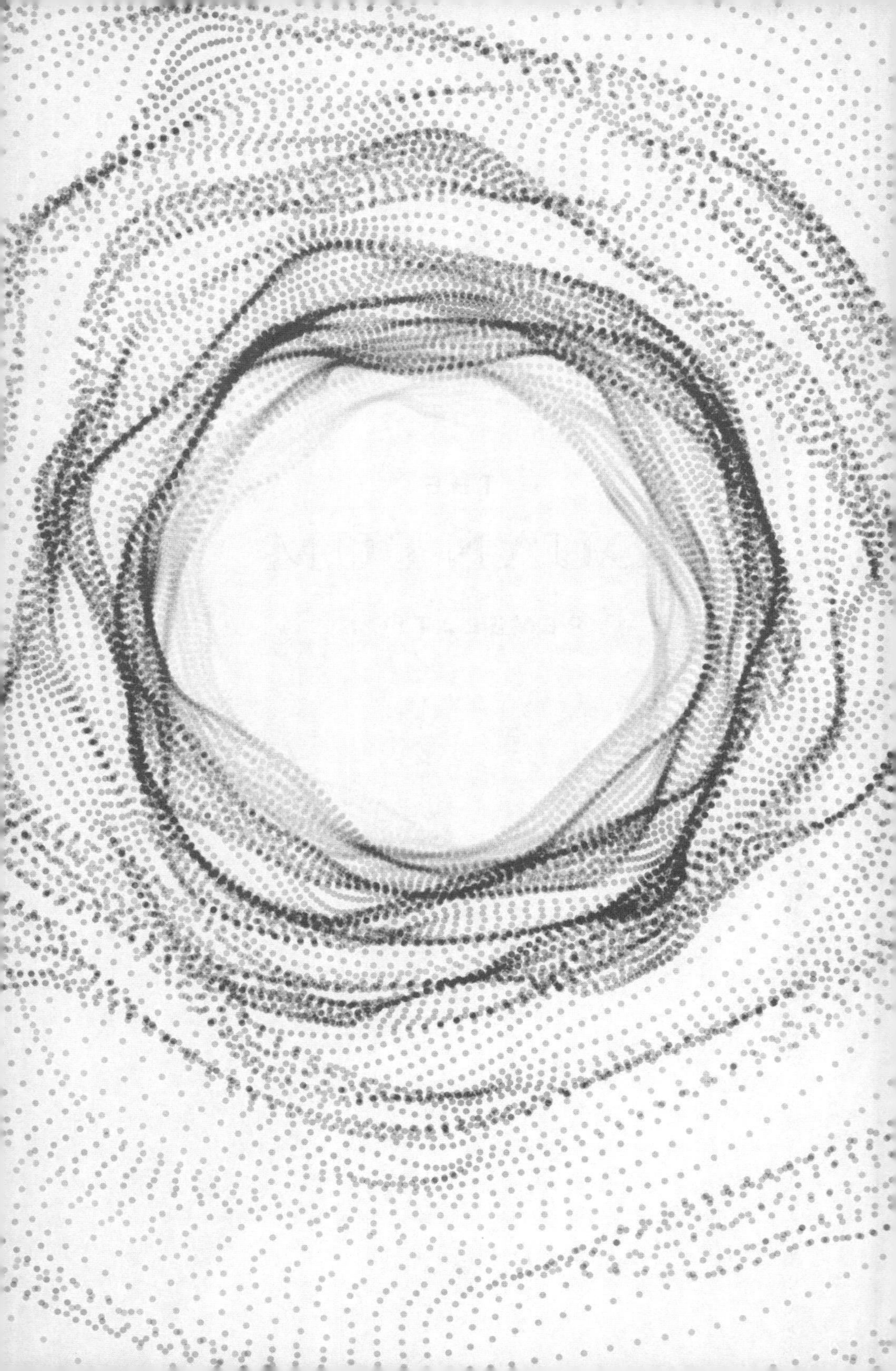

PAUL LEVY

THE QUANTUM REVELATION

A Radical Synthesis of Science and Spirituality

SELECTBOOKS
NEW YORK

Photo of John Archibald Wheeler on the book's dedication page is printed
with permmision of UT Office of Public Affairs Records, di_10047,
The Dolph Briscoe Center for American History,
The University of Texas at Austin

This edition published by SelectBooks, Inc.

Designed by Pauline Neuwirth.

For information address SelectBooks, Inc., New York, New York.

FIRST EDITION

ISBN 978-1-59079-448-7

Library of Congress Cataloging-in-Publication Data

Names: Levy, Paul, 1956- author.
Title: The quantum revelation : a modern-day spiritual treasure / Paul Levy.
Description: first [edition]. | New York : SelectBooks, Inc., 2018. |
Includes bibliographical references.
Identifiers: LCCN 2017041102 | ISBN 9781590794487 (pbk. : alk. paper)
Subjects: LCSH: Quantum theory--Religious aspects. | Quantum
theory--Philosophy.
Classification: LCC BL265.P4 L48 2018 | DDC 201/.653--dc23 LC record
available at https://lccn.loc.gov/2017041102

Manufactured in the United States of America

10 9 8 7 6 5 4 3 2 1

PERMISSIONS

Quotations printed in chapters on the following pages of *Quantum Revelation* were originally published in *At Home in the Universe*, 1994, John Archibald Wheeler, © American Institute of Physics, with permission of Springer Nature: "A Septet of Sybils: Aids in the Search for Truth," pp. 3, 6, 7, 8, 9, 15, 16, 18; Genesis and Observership," pp. 23, 26, 35, 39, 42, 43, 44, 45; "Our Universe: The Known and the Unknown," p. 64; "The Morale of Research People," p. 75; "Delayed-Choice Experiments and the Bohr-Einstein Dialogue," pp. 113, 114, 115, 120, 123, 124, 126, 128, 130; "Hermann Weyl and the Unity of Knowledge," pp. 184, 186, 187, 188; "Hideki Yukawa as Uniquely Ecumenical," p. 197; "Science and Survival," p. 226; "Beyond the Black Hole," pp. 272, 294; "It from Bit," pp. 296, 299, 300, 301, 303, and 307.

Quotations printed on pages 64, 84, 90, 103, 119, 148, 153, 167, 168, 173, 182, 223, 227, 235, 256, 262, 263, 270, 287. 292, 296, 297, 307, 329, 330, 334, 335, 336, 337, 338, 339, 341, 342, 343, 344, 345, 346, 347, 351, and 353 were originally published in *Geons, Black Holes and Quantum Foam: A Life in Physics* by John Archibald Wheeler with Kenneth Ford, Copyright © 1998 by John Archibald Wheeler and Kenneth Ford, Used by permission of W. W. Norton & Company, Inc.

Quotations by C. G. Jung on pages 169, 215, 223, 229, 280, 327, 328, 354, 384, 441, 513, 515, 516, 517 and 518 from *The Structure and Dynamics of the Psyche*, CW 8, C. G. Jung, Edited by Sir Herbert Read, Michael Fordham, Gerhard Adler, William McGuire, © 1969, are reproduced by permission of Princeton University Press.

Quotations in *The Structure and Dynamics of the Psyche*, CW 8, C. G. Jung, Edited by Sir Herbert Read, Michael Fordham, Gerhard Adler, William McGuire, © 1969, Routledge, reproduced by permission of Taylor & Francis Books UK.

Quotations by E. Schrödinger on pages 6, 7, 11, 13, 17, 19, 20, 21, 27, 48, 49, 51, 52, 53 and 57 from *Science and Humanism: Physics in Our Time*, E. Schrödinger, © 1951 are printed with permission provided courtesy of Cambridge University Press.

Quotations from Philip K. Dick on pages 75, 76, 131, 294, 376, 479, 482, 483, 588, 705, 706, 707, 708, 709, 718, and 886 from *The Exegesis of Philip K. Dick*, Philip K. Dick, Edited by Pamela Jackson and Jonathan Lethem, © 2011 are printed with permission provided courtesy of Houghton Mifflin Harcourt.

JOHN ARCHIBALD WHEELER

This book is dedicated to John Archibald Wheeler, as well as to all of the scientists, theologians, natural philosophers, creative artists, shamans, dreamers, and truth-seekers of every sort who, like Wheeler, have devoted their lives to shining light on who we really are.

CONTENTS

FOREWORD

by JEAN HOUSTON

Let me say it up front: This is one of the most fascinating, evocative, and important books that I have ever read. Paul Levy brings his massive curiosity to what is perhaps the most curious and massive discovery of our time. Quantum physics is the name of the game, and pursuing its mysteries prepares one to become a game changer in both one's life, one's understanding of reality, and how one agrees to participate in the biocosmic field that is intrinsic to our nature.

This remarkable offering provides one of the clearest understandings of this scientific phenomenon (which involves all of us) that I have read. It is quantum physics from a deeply human point of view, as we follow Levy's heroic journey into the dark forest of the quantum nature of the universe. I recall that in the *Divine Comedy* Dante began his grand tour of the deep and dark dimensions of reality in the dark forest (*selva oscura*). What he found there in the thirteenth century were the mythic worlds that seemed to underlay the domains of existence at that time. Levy does much the same for our time, our science, and our vernacular.

What I write now should really be an epilogue rather than a foreword to this book, for I design my quantum-inspired experiments from the kind of understandings that Levy brings so brilliantly to this book. As a lifelong pioneer into the strange and beautiful country of the human potential, I have discovered that quantum physics, when applied to latent human capacities, opens up a realm of experience and exploration that, up until now, has remained more mythical than real. With quantum physics we become "mything" links, living lives that allow the extraordinary to become the ordinary.

The discovery of the quantum nature of our universe, and thus of ourselves, is so major an event that its profound implications cannot be overstated. Quantum theory demands a radical re-visioning of the role of consciousness as the underlying organizing principle of the universe. With this understanding, quantum physics is introducing us to ways of seeing that profoundly impact human thinking, feeling, sensing, knowing, and being. In different states of consciousness we can be brought to subtle levels where the mind interacts with the universe itself in what the Buddhists call interdependent co-arising. The self, when understanding that we do not just live in the universe but the universe lives in us, becomes, in some sense, identical to the quantum mind and therefore has many more capacities than those operating in local consciousness.

Working with these concepts—both spiritual and scientific—I find that we can enable students to be, to do, and to create in ways that are suggestive of higher levels of human accomplishment. What follows, then, are some of the experiential processes I have led, which take as their basis the quantum nature of our possibilities. These include introducing more fluid categories of space and time, for instance where one is able to experience subjective time. This is where a short amount of clock time is felt to be much longer and thus one is able to experience adventures, write books, finish projects, go voyaging in the seas of the unconscious, even learn or rehearse things that would normally take a much longer time to do. Similarly, from the quantum perspective of the simultaneity of past, present, and future, we are able to change the story of minor past events until it becomes a realistic part of one's memory.

But even our imagination is transcended by the universe in its quantum aspect and we find we are no longer simply imaginative but imaginal. We are in the quantum holofield of consciousness wherein all potential patterns reside. We are no longer caught in our own habits and expectations. We seemingly access the blueprints, the guidance, forms, and patterns of what, up to now, we have only imagined. Granted, this requires a metanoia, a big time-shift in one's belief systems, for as the quantum paradigm shows, belief structures reality. This book is in itself a powerful stimulus to believe and to live a radically different belief and therefore a new way of being in this world and time. As Paul Levy writes:

> Quantum physics heralds a change so momentous that it can—and already has—transformed the course of human history. . . . This great change is already underway and yet there remains a long way to go for the full transformational impact of the discoveries of quantum physics to be assimilated by humanity, i.e., for its insights to transform the nature of the collective common sense of human beings. . . . We find ourselves in the role of midwives, helping to birth a quantum understanding of the world. What quantum physics is revealing to us is so radical, with implications so far-reaching, that to call it merely revolutionary would not do it justice. The conceptual revolution of quantum theory has literally turned physics on its head. What it is revealing about our universe is turning right side up what had been inverted and upside down regarding our understanding of the nature of reality.
>
> The advent of quantum physics can be thought of as a revolution in the realm of ideas. Quantum physics is introducing us to a radically new way of seeing, conceptualizing, and understanding that profoundly impacts human thinking, feeling, sensing, knowing, and being. As if the universe itself were giving us a cosmic physics lesson, what quantum physics is revealing to us requires a completely new way of picturing and thinking about the universe, our place in it, as well as who we are.

I take his words as a stimulus to my own, and particularly to my experiments as to how quantum physics can change how we re-conceptualize and allow ourselves to discover potentials that, until recently, have seemed more mythical than real. What follows is arguably more art than exact science, but its consequences are profound as well as suggestive of the fact that we are organism-environments who are symbiotic and indeed self-similar with the quantum continuum within which we live and abide and have our being. A fuller exploration can be found in my chapter in *What Is Consciousness?*, a work I cowrote with Ervin Laszlo and Larry Dossey.

• • •

EVERYDAY LIFE IN THE QUANTUM UNIVERSE

The implications, indeed the revelations of existing in a quantum belief system so vividly expressed by Paul Levy, find support in a number of areas: identity, time, parallel selves, and the paradox of being human.

With regard to identity, the quantum perspective holds that we are not encapsulated bags of skin dragging around a dreary little ego. Rather, we contain many selves and personas, which, when developed, give us skills and capacities that make us multimodal and thus able to bring fresh perspectives to the contents of our life and work. This latency, which can be ignored one's entire life, is to be found within one's quantum nature and can be called upon whenever needed. This is the case in my own experience: while I am almost phobic about writing, in my persona as a cook I have neither fear nor resistance and happily bring my cooking skills to the mindset of the writer, with my "dishes" now resulting in over thirty published books on many subjects. This feast of creativity I pass on to my students, who then discover new competencies and abilities that grow from their alternate personalities. In addition to these, there resides a still higher level of consciousness and capacity. I call this the quantum blueprint, the fullest expression of the self. It can be considered to be the optimal template, that state which gives them access to higher states of functioning and skillful means. It is their emergent evolutionary nature that is filled with the codes of their higher possibilities.

With guided imagery, participants are led to experience the optimal template as integral to their nature. I might say to them, "As the optimal template moves in you, know that you are becoming a superb catalyst, a carrier of new genesis as the world is getting ready to move. Your mind is growing so you can think in many ways—in words, in images, thoughtways that border on genius, in fact can become genius. Much that may have been lost in you from your childhood and adolescence, as well as those remarkable skills and qualities that were latent in you, are now becoming real . . . courage, passion for the possible, rigor, diligence, a wave tide of joy and belief, the creative life, your spirit inspirited by the God, the Cosmic consciousness, the Supreme Beloved who calls you into being. You are loved, nurtured, empowered, called forth to your highest destiny for this life in this time." With this template in place, they gain the sensibility and openness to explore life in the quantum universe.

One of the most intriguing ways to explore our quantum nature is to enter the mysteries of time. Most people have forgotten that we are multi-temporal by nature and can experience short periods of time as being much longer. Thus, for example, we can learn an alternate temporal process. This involves experiencing subjective time, which differs greatly from linear objective time, also called clock time. With subjective time participants report experiences of writing books and music, rehearsing and improving skills, and pursuing activities that would normally take a long time objectively, but occur subjectively in a few minutes.

Since time in the quantum universe is simultaneous, we have the ability to recreate minor events in the past so that the present and future are enhanced by this change. After gaining confidence in this process, participants can go further into the omnidirection of time. Quantum physics tells us the times of past, present, and future are occurring simultaneously in ways that are difficult for us to understand. With this in mind, we can change a minor happening in the past so that it affects one's present and future in positive new ways. This requires us to enter into an altered state of consciousness wherein we select a minor past incident that we would like to re-create (never a major traumatic event) and dramatize it in positive ways so that the remembered event is now what is in the forefront of one's memory while the so-called historic event becomes a dim dream. Whether this is just putting in an overlay of new memory in the brain or actually shifting the past event itself, I cannot say. But certainly the effects have been profound. In some cases, people who were involved in the original event start to remember the event in its changed variations.

The point is that we can remember and even re-create the past and the future. Memories can be changed on this simultaneous all-at-once continuum in which the universe—including all its times, experiences, and dimensions—can be changed, transformed, rewritten, and reexperienced because the universe (which includes our time and memory) is regenerating itself every nanosecond. Since we are conscious participants in the living universe, we can enter the Akashic fields of memory and shift elements of our own history.

Living cosmically can also effect opportunities in alternate spatial places and dimensions. The new science of quantum singularity gives us the freedom to explore parallel selves, which are both similar to us as well as dissimilar in health, knowledge, profession, and lifestyle. In this regard, we take people to alternative or parallel selves where they enter into friendly and cooperative relationships and then with the agreement of the parallel self, bask in the energy field of their health. One of my students, Jennifer, had a near fatal form of Lyme disease. I guided her to the parallel place in the universe where there lived a Jennifer in a superb state of health and well-being. With mutual agreement their fields connected and our Jennifer felt herself being re-tuned to a better state of health. Upon returning to her regular place in reality, she began to improve. As of now, she is in a state of health similar to that of her parallel self. Subsequently, Jennifer was able to take her comatose dying brother to a healthy parallel self where he became immersed in the field of this healthy self, shared the other's "frequencies" and upon "returning" woke up from his coma and proceeded to grow healthier each day. He then went on to live a normal life.

We can easily dismiss this kind of phenomenon as strong suggestion in an altered state of consciousness that then affects health and well-being. However, after having seen dozens of these kinds of learnings, changes, and shifts with several dozen students, you cannot close your accounts with reality and therefore come to surmise that something much more interesting is happening. If there is any truth to this matter, it is no wonder that we want to keep the lid on this multitemporal unconscious! And yet perhaps these different worlds or times of our experience are not self-contained but bleed through in other states of consciousness—dreams, reveries, creative inspirations, and spiritual and other potent experiences.

To read *The Quantum Revelation* is to know that we are players in a great game called Paradox. And what is the paradox? It is that we can be a child sitting on a window ledge knowing everything and we can be the universe knowing the child at the same time. Quantum physics suggests that we are both infinite and finite beings. As finite beings we are God-stuff incorporated in space and time. As infinite quantum beings we are the living universe in an eternal yet spirited form of itself. As this infinite, self-expressing aspect of cosmic consciousness and as a form of the living universe, we find ourselves capable of creating and sustaining an individual finite self—that is you, the human being that is the microcosm or, if you will, the fractal of the infinite self.

The human selfing game, in which we are divided into many selves, may be what infinity does for fun. Not realizing this, we live in a state of galloping ambiguity, caught in a limited time vehicle and yearning for our greater self. Then when we make the rare excursion into our greater being, becoming our cosmic selves, we suddenly yearn like Dorothy in Oz to get back home to the farm in Kansas. Why is this? To continue the metaphor, to live in Kansas, however joyous and rewarding, is to chronically confront our limitations of our body and mind, and those of others, to which we are accustomed, whereas entering into infinite life is rather difficult to navigate and transcends all understanding.

What do we do, where do we go, who do we meet as infinite beings? I believe that to live in a state of both/and is to become who and what we were patterned to be. As we cannot shrink the infinite to fit into the finite—because if we do so we just end up with a fundamentalist God—we can extend through conscious work on ourselves the capacity to expand and thus to enter into partnership with the infinite.

Then, and this may be the goal of the Paradox game, we do indeed discover that we are an infinite self creating and sustaining our individual human self. Do you see the stupendous import of this statement? To me, it is a mind-cracking, soul-buffeting, life-enlarging realization. Once understood and internalized, it adds tremendous power to our freedom to be, our enormous capacity to grow, evolve, and re-create ourselves, and our ability to live simultaneously as finite and infinite beings. The infinite self directs the development and unfolding of the finite self, and the finite self offers joy, entertainment, and knowledge to the infinite self. This is the paradox of partnership resolved. The game is to overcome the illusion of separation. This is the revelation of quantum physics that Paul Levy brings to life in a most splendid and captivating way.

Jean Houston is a visionary thinker, teacher, and philosopher who pioneered the human potential movement.

PREFACE

It is easy for me to feel that the process of writing this book about quantum physics has been an act of ceremonial magic. I say this because the very experience of thinking and writing about quantum theory has had a profound effect on me, changing something deep within me. Perhaps it hasn't so much changed me, as brought something within me more to the forefront of my consciousness that had previously been in the background. I can say that for me quantum physics has brought the deep nature of my "self," the part of me that is an aspect of the universal self, into focus in a new way. At the very least, it has given me a useful framework for contextualizing my inner experiences of the world that both inspires and makes sense to me. In learning about quantum physics, my life has stopped "making sense" in one way, while making deeper sense—and infusing my life with meaning—in another. It is these new quantum physics-induced insights that I aim to share with the readers of this book, hopefully evoking similar revelations in the reader's mind.

The sense I have from my own personal encounter with the quantum is that upon being introduced to the quantum level of reality, no one will remain the same; this has certainly been true for me. The more I understand what quantum theory is a theory of, the more I realize that modern day physicists have stumbled upon, via exploring the outside world, a correlate to something that I, and I imagine many other people, experience as our inner, subjective experience. This is to say I began to notice the correlation and mirrored interreflection of the reality that quantum theory describes with my own mind. The ultimate subject of quantum physics turns out to be isomorphic with what most interests me—the nature of the human mind. The universe's quantum nature that physics has discovered is, in a very real sense, our own quantum nature. The world of the quantum is a magic mirror reflecting back to us our own essential nature. Our very mind stuff, the medium and substanceless substance[1] of our present field of awareness, is itself not separate from, but rather an indivisible expression of the quantum field.

As I continually deepen my inquiry into the realm of the quantum, it becomes ever more apparent that there is a precise, elegant, and exact parallel, down to the most minute details, between the laws of the subatomic realm and the laws of the human mind. This parallel between physics and psychology should come as no big surprise, for it is the human mind that has created quantum physics in the first place. We should therefore expect to find the genius—for discovering reality as well as deceiving ourselves regarding it—within the mind's own creations.

In writing this book I feel as if I am continually distilling the essence of the quantum gnosis into a magic potion of immense potency, but this elixir is in the form of words on a page. My quantum nature evidently wanted to learn about itself, so it inspired me to write a book about quantum physics in order to teach me about the subject. As such, this book is the record of my own ongoing investigations into quantum physics. *The Quantum Revelation: A Radical Synthesis of Science and Spirituality* first started out as one article, which as soon as it was finished quickly turned into a second one complementing the first. I've designed the book around these two original articles, which have amplified themselves and grown enormously—becoming quantum entangled, so to speak—integrating themselves into each other to eventually become this book, while still keeping their distinct structure as two halves that fit together, forming interrelated aspects of an integrated whole.

In this book I invoke the wisdom of the greatest physicists of the last century, people such as Albert Einstein, Niels Bohr, John Wheeler, Erwin Schrödinger, Wolfgang Pauli, David Bohm, Richard Feynman, and Werner Heisenberg (who, it should be noted, is a very controversial figure), as well as some non-physicists who have in their own unique way contributed to articulating a parallel vision of the deeper nature of reality—pioneers such as the psychologist C. G. Jung, His Holiness the Dalai Lama, and the science fiction author Philip K. Dick. In the course of writing this book I began feeling as if I were becoming part of an ongoing dialogue that they were—and still are—having about the nature of our universe, all taking place inside of my own head. Feeling as though I were having some sort of wild dream, these great thinkers have become "dream figures" inside of my own mind, and my task was to somehow reconcile what they were saying with my own experience. In dialoguing with them, I felt like I was practicing active imagination.[2] I can only imagine what it was like for them.

As someone who is by nature curious and interested in the nature of things, it has been truly fascinating to become familiar with physicists' thoughts on the matter of "matter." In addition, it was really interesting to learn how physicists think, particularly how they think about thinking. As a lifelong student of the human psyche, I've found myself interested in the field of physics in my own unique way, such that I am in one sense more interested in the physicists' unconscious reactions to the physics they are inventing than in the physics itself. This is not quite true because the physics *is* very interesting, but I found myself unable to separate out the physicists from the physics, as if they were conjoined in a quantum system that was revealing itself through their interplay, thereby reflecting the very same kind of seamless inseparability that their type of physics requires.

After reading one of the original articles I had written on the subject, one physicist—a mentor of mine—reflected to me that it was to my advantage to be a non-physicist, as I have more of an open, unprejudiced mind, typical of an "outsider," which enables me to see and contemplate with fresh eyes the strange goings-on in the field. It can feel a bit weird to be writing about a field that I'm not a practitioner of, but then again, the field of physics has gotten sufficiently weird that maybe my writing about it isn't that weird after all. As human beings we all have a quantum nature, a pure and coherent nature which bestows upon us the right—and responsibility—to consciously realize it as best as we are able. The benefits of realizing our quantum nature are as boundless as the potentials intrinsic to the quantum itself.

Oftentimes great discoveries take years to catch up with and be assimilated by the collective into a shared, transformed worldview.

The more I study quantum physics, the more I become convinced that it is unquestionably one of the greatest discoveries of the human mind in all of our history. Quantum physics does not exist separate from the human mind that has discovered (or invented) it. It is in fact an emergent phenomenon, a semantic structure arising within and inseparable from the human mind itself. As such, quantum physics is nothing short of a reflection and modern map of the human mind. To say this differently: turning the mind back upon itself, quantum physics is a revelation of the mind to itself, through the instrument of science. In all of history we have never quite encountered anything like this.

I am of the opinion that the quantum revelation ranks with the Buddha's discovery of the nature of the enlightened mind. It is the same realization expressed through the art of science (at least in my imagination). A true revelation always manifests itself outside the sphere of our own will; it is never our own activity, but is *as if* something seemingly outside of ourselves reveals itself to us. As the revelation that is quantum physics is taken into and integrated within ourselves, it will change everything, particularly ourselves, with no "philosopher's stone" left unturned. So as you read this book, proceed at your own risk—don't say you haven't been forewarned.

Quantum physics is truly revelatory, and as its theory points out, we ourselves play a crucial part in its living revelation. We and quantum physics are part of one and the same quantum system, a system not composed of separate parts, but rather, one that is always seamlessly unified and whole. Studying the revelations of quantum theory is a form of yoga, the deeper meaning of which has to do with "yoking," i.e., joining with something greater than and beyond our ordinary conception of who we are—our quantum nature. Quantum physics is the bridge that elegantly connects two heretofore separate parts of ourselves in one stunning stroke, unifying the physical with the spiritual dimension of experience.

We should be screaming from the rooftops about so earthshaking a discovery. This is a true game changer, a priceless treasure that could not be more relevant and helpful for all of us, one tailor-made for the time we live in. To say this differently: it is as if we, the human species, have come up with a revolutionary new conception of the universe and our place in it—in other words a "saving idea"—with its own quantum style of grace that can literally transform our lives. The quantum is an idea that is thirsting to be freely shared; getting more widely disseminated will only increase its potency, or so I imagine. Once the idea of the quantum and what it is revealing to us gains enough momentum, thereby reaching a critical mass in the human psyche, all bets are off as to our potential future.

The quantum is readily available, existing all around us and within us; we just have to develop the eyes to see it and the inner capacity to feel its subtle and coherent presence. It is as if, in its infinite compassion, the universe—via quantum theory—is offering us a boundless gift beyond measure, an ever-abundant gift that keeps on giving. Encountering the quantum is like discovering a treasure that has been hidden within our own mind that changes everything. The quantum is a sacred treasure whose liberating power can transform the whole universe. To receive its benefits, we simply have to recognize what is being revealed to us.

ACKNOWLEDGMENTS

It is said that it takes a village to raise a child. Even though writing is an isolated venture, I think the same can be said about writing this book. It has taken a whole community to birth it. This book would never have come to fruition in the way it has without the Awakening in the Dream community in Portland, Oregon, that has formed around my work. This group of people has truly been like a family to me. I'd like to personally acknowledge what has become known as "The Dreaming Eight," an amazing group of creative dreamers who magically helped to dream this book into being: John Thomas, Gene Latimer, Donna Zerner, Bob Welch, Kimberli Matin, Larry Berry, Neil Levenson, and Lucretia Hatfield. A special note of thanks goes to Larry Berry for his many years of unending selfless service. He has always been there in whatever way was needed. I'd also like to express much gratitude to Mark Hartley.

I want to thank all of my many teachers—whether they be in the form of people, books, dreams, or nature herself—that always seem to synchronistically show up at exactly the right time. I'd like to acknowledge the great help I've received from a thoroughly quantum-physicized friend of mine who, for his own unique reasons, wishes to remain anonymous—you know who you are. I can't sufficiently express in words my heartfelt appreciation for my beloved teacher of the Dharma, Khenpo Tsewang Dongyal Rinpoche.

I'd also like to express my appreciation for Kendra Crossen Burroughs, who generously offered her incredible gifts to help make this book as good as it could be. And a big thank-you to John Rose for his help when things got down to the wire. I also want to thank my publisher, SelectBooks, and particularly Kenzi and Nancy Sugihara for their belief in this book. And deep appreciation goes to my editor, Molly Stern.

I'd be remiss if I didn't give a nod to the nonlocal field—one of the key things I'm pointing at in all of my work—for its active participation and collaboration in helping me write this book. I am truly unable to express my gratitude to everyone—be they local, nonlocal, fully embodied, or no longer in a body—who has helped to midwife this book into living form.

INTRODUCTION

For the last few years, all I've wanted to do is to read about quantum physics.[3] I've been studying quantum physics off and on for decades, but have never gone as far down the rabbit hole as I have this time. It feels like I've gone through the looking glass to the point of no return. The more I contemplate what quantum physics is telling us, the more my mind gets blown into phantasmal traces of "nonexistent" subatomic particles. Studying quantum theory is like ingesting a mind-altering, timed-release psychedelic that keeps coming on with greater intensity the more I wrap my mind around what this world-changing revolutionary theory is telling us about the nature of reality. Taking in what quantum physics is revealing to us about our universe is "psycho-activating" beyond belief, in that it activates the psyche, inspires the imagination, and synchronistically dissolves the boundary between mind and matter. To say that quantum physics is the greatest scientific discovery of all time is no exaggeration; its profound revelations and implications cannot be overstated. We literally have to create a new form of language—not to mention a new way to think—to do it justice.

In discovering the quantum, physics has indisputably encountered consciousness. There is simply no denying this fact. The highly respected mathematical physicist John von Neumann said, as early as 1932, that consciousness exists in—and has entered the equations of—quantum mechanics, yet no one knows exactly where to find it. Quantum theory demands a radical re-visioning of the role that consciousness plays in the deep structure and ongoing unfolding of reality. Quantum physics unequivocally points out that the study of the universe and the study of consciousness are inseparably linked, and that ultimate progress in the one will be impossible without progress in the other. The change that began with the discovery of the quantum realm wasn't solely a transformation of the worldview of science, but is potentially an expression of and vehicle for the evolutionary mutation of human consciousness itself.

To create context for its discovery, at the beginning of the twentieth century the prevailing opinion among many physicists was that there was nothing new to be discovered in physics except for more precise measurements. A unique development in human history, the discovery of the quantum nature of our universe, has brought about a seismic, tectonic shift in the very foundation of physics and the roots of our scientific worldview. The change in the concept of reality emerging in quantum theory is not simply a continuation of the past, but rather a radical break from it. The gap between the new version of reality that quantum reality reveals and our old, habitual ways of thinking about reality are wider than the abyss of the Grand Canyon, the two sides of which are at least on the same level. With the emergence of quantum physics we are encountering an entirely new universe that is of a totally different order than the one we've been used to.

Quantum physics heralds a change so momentous that it can—and already has—transformed the course of human history. The discovery of the quantum has inaugurated what Niels Bohr conceived of as being a new epoch.[4] This great change is already underway and yet there remains a long way to go for the full transformational impact of the discoveries of quantum physics to be assimilated by humanity, i.e., for its insights to transform the nature of the collective common sense of human beings. All great theories spend a long time gestating and being born; it can be dangerous to understand new things too quickly. We find ourselves in the role of midwives, helping to birth a quantum understanding of the world. What quantum physics is revealing to us is so radical, with implications so far-reaching, that to call it merely revolutionary would not do it justice. The conceptual revolution of quantum theory has literally turned physics on its head. What it is revealing about our universe is turning right side up what had been inverted and upside down regarding our understanding of the nature of reality.

The advent of quantum physics can be thought of as a revolution in the realm of ideas. Quantum physics is introducing us to a radically new way of seeing, conceptualizing, and understanding that profoundly impacts human thinking, feeling, sensing, knowing, and being. As if the universe itself were giving us a cosmic physics lesson, what quantum physics is revealing to us requires a completely new way of picturing and thinking about the universe, our place in it, as well as who we are. Quantum theory is teaching us that implicit in our very thinking are certain flaws and misperceptions that, unseen and taken for granted, unnecessarily constrain and limit our ability to apprehend the nature of nature, including our own. The founders of quantum physics—oftentimes referred to as "genius physicists"[5]—people such as Max Planck, Albert Einstein, Niels Bohr, Werner Heisenberg, Wolfgang Pauli, and Erwin Schrödinger, famously argued that quantum physics is first and foremost a new way of thinking. Indeed, the most far-reaching impact "of that uniquely twentieth century mode of thought, quantum physics,"[6] will be found within the human mind.

The discoveries of quantum physics require a novel response in us which, when more fully understood and integrated, will irrevocably change us—both individually and as a species—in the very core of our being. Regarding the implications of quantum physics, John Bell, one of the most important physicists of the latter half of the twentieth century, is of the opinion that "the new way of seeing things will involve an imaginative leap that will astonish us."[7] It is hard to imagine something truly astonishing that we wouldn't tend to initially rule out as preposterous. This new way of seeing things, this imaginative leap, is truly an evolutionary up-leveling, a real quantum jump in consciousness, in which quantum physics is inviting each of us to partake.

Quantum physics is the most subversive of all the sciences, having created a "reality crisis" in the field of physics such that the very idea of "reality" has been undermined, relegated to being a questionable, ambiguous, twilight concept. The very "reality" that pre-quantum physics had been studying has been demonstrated by quantum physics to not even exist! Speaking about the new view of reality emerging from quantum physics, theoretical physicist Robert Oppenheimer relates that "there was terror as well as exaltation in their new insight."[8] It was as if physicists were experiencing the terror of realizing that there was no solid ground to stand on, that they were being invited to face the existential abyss, and it was their very discipline itself that was sending out the invitation. To quote physicists Bryce DeWitt and Neill Graham:

> No development of modern science has had a more profound impact on human thinking than the advent of quantum theory. Wrenched out of centuries-old thought patterns, physicists of a generation ago found themselves compelled to embrace a new metaphysics. The distress which this reorientation caused continues to the present day. Basically physicists have suffered a severe loss: their hold on reality.[9]

The greatest experts of quantum physics, if it's even possible to speak of "experts" in a field that, according to Nobel prize-winning physicist Richard Feynman, "nobody understands,"[10] literally do not know "what" they are talking about.[11] There is not a physicist alive who knows what quantum physics is a science of. Speaking about quantum physics, Nobel laureate Murray Gell-Mann is often quoted as saying, "We have learned to live with the fact that nobody can understand it."[12] One of the self-professed reasons that the novelist D. H. Lawrence enjoyed studying quantum physics was *because* he didn't understand it. One of the effects of quantum physics is to turn the mind back on itself in a way that our classically-conditioned mind finds impossible to understand.

Physicists who study their own theory have, in their attempts at grasping its implications, lost their grip on reality, finding absolutely nothing to hold onto. Quantum physics has pulled the rug out from under us only to reveal no floor below, no place on which to take a stand as the notion of a solid, objectively existing world evaporates like dewdrops in the morning sunlight. To quote Einstein, "If [quantum theory] is correct, it signifies the end of physics."[13] It should be pointed out that from every indication, based on decades of experiments of every kind imaginable, there is not a shred of evidence that points to quantum physics being incorrect. We can draw our own conclusions.

The discovery of the quantum has created a real crisis in the field of physics, not to mention within the minds of physicists. To clarify, there is no crisis within quantum physics itself. Quantum theory is internally consistent and its predictions are unassailably accurate; the crisis has to do with the fact that the logical foundations of classical science are violated in the quantum realm. The trouble is border trouble; it arises along the quantum-classical frontier. To quote science writer Timothy Ferris, "These border skirmishes raise questions sufficiently baffling as to constitute the scientific equivalent of a Zen koan. Quantum weirdness is so counterintuitive that to comprehend it is to become not enlightened but confused."[14]

The very laws of nature that physicists were discovering in the new world of the quantum were as different from what they were used to as the strange world encountered by Alice after she fell down the rabbit hole. To quote Nobel laureate Leon M. Lederman and coauthor Christopher T. Hill from their book *Quantum Physics for Poets,* the founders of the quantum realm had discovered "a new kind of 'dream logic' reality."[15] To quote psychologist and physicist Arnold Mindell, "Quantum theory is indeed absurd in that it is more dreamlike than real—and its dreamlikeness renders accurate results."[16] Synthesizing the practical and esoteric, quantum theory's accurate results reveal that this universe we are inhabiting seems to be of the nature of a shared collective dream that all of us are collaboratively dreaming up into materialization each and every moment.

Speaking about the quantum, Mindell continues, "It is the world in which the shaman moves, the world each of us meets every second of our lives, the realm we enter every night in dreams."[17] Not accessible solely for physicists, shamans, or the privileged few, each and every one of us interfaces with the world of the quantum "every second of our lives," both in our waking life and our dreams at night. From the agreed-upon consensus reality point of view of classical physics (whose perspective is the very opposite of the universe being a collective dream), the strange goings-on in the world of the quantum makes no sense at all. In 1912 Einstein wrote to a friend, "The more success the quantum theory has, the sillier it looks."[18] Silly, yes, but revolutionizing the world in its silliness.

Quantum theory claims to be, and in practice certainly is, the foundational theory and primary principle upon which our present scientific understanding of the universe is built. If we dig deeply enough into any natural phenomenon—physical, chemical, biological, or cosmological—we will invariably encounter the world of the quantum. Some physicists who have had glimpses of understanding what quantum physics is revealing describe it as one of the greatest achievements of the human race. California Institute of Technology physicist John Preskill has called quantum theory "the crowning intellectual achievement of the last century."[19] Quantum theory is not just one of many theories in physics; it is the one theory that has profoundly affected every other branch of physics. In Einstein's words, the discovery of quantum theory "set science a fresh task: that of finding a new conceptual basis for all of physics."[20]

Quantum physics is not merely wild and abstract speculations of "out there" theoretical physicists who live in ivory towers; it has down-to-earth practicality. There is hardly an aspect of contemporary society or of our own individual lives that has not already been fundamentally transformed by the ideas and applications of quantum physics. One third of our economy involves products based on quantum mechanics—things such as computers and the internet, lasers, MRIs, TVs, DVDs, CDs, microwaves, electron microscopes, mobile phones, transistors, silicon chips, semiconductors, quartz and digital watches, superconductors, and nuclear energy. Yet even with the huge impact quantum physics has already had on each of our lives, this effect is infinitesimally small compared to what will occur when more of us recognize and internalize the implications of what it is revealing to us about the nature of reality as well as of ourselves. These revelations will literally transform our most fundamental sense of human identity, of what it means to be human. The amazing technologies that quantum physics has helped us to currently develop have been likened to "low-hanging fruit" compared to what's yet to come, whether in revolutionary technologies or even more far-reaching technologies of mind.

The discoveries of quantum physics, practically speaking, have given us the capacity to both increase the quality of our lives and/or to potentially ravage the environment on an unprecedented scale, even to obliterate our species altogether. Though imbued with its own measure of divine creativity, our species is also the greatest destroyer of life that our planet has ever known. One of the great dangers of our modern age is that our spiritual and moral qualities—and the resultant expansion of consciousness—lag far behind our advances in technology. There is, however, a revelation about the nature of our minds that is hidden within the very source of our technological advances that could potentially save our species from continuing to shortsightedly hurtle along its current ill-fated trajectory. To quote Henry Stapp, widely considered to be the current dean of quantum theorists, "Yet along with this fatal power it [science] has provided a further offering which, though subtle in character and still hardly felt in the minds of men, may ultimately be its most valuable contribution to human civilization, and the key to human survival."[21] We are presented with a crucial question: Do we use the discoveries of quantum physics to become conscious cocreators with the universe, working for the betterment of our species, or do we use them to destroy ourselves? Quantum theory reflects back to us that the choice is truly ours.

Quantum physics works like a charm. It is like a higher-dimensional talisman, a physics of possibilities. The precise accuracy of its mathematical formalism and methodology is beyond debate; none of its predictions has ever been shown to be wrong. It is literally the most successful scientific theory of all time, so much so that if a new theoretical model contradicts quantum mechanics, Feynman unabashedly counsels us to "abandon it." To quote Gell-Mann, "Quantum mechanics is not a theory, but rather a framework within which we believe any correct theory must fit."[22] The revelations that compose quantum theory are like mana from heaven—a veritable gift from above—but in this case the mana comes from the interaction of nature with our own minds. "Natural science," writes Heisenberg, "does not simply describe and explain nature, it is part of the interplay between nature and ourselves."[23] As if a psychic nutrient secreted by the universe, the heavenly quantum mana, arises within and from the mind for the benefit of the mind itself to realize its own nature.

It is as if physics has discovered a wonderful magic wand that works every time, but the amazing thing is that no one knows why. I have never in all of my life come across a field in which all of the supposed "experts" disagree with each other about the meaning of their own theory. The deep philosophical question that begs to be answered is: What does quantum physics mean? When the alleged experts can't agree, we can feel free to choose our preferred expert, or to explore and speculate on our own. Instead of superimposing my own interpretation, my intention is to allow quantum physics itself to lead the reader to his or her own interpretation.

I am certainly not a physicist. In fact, the more I contemplate the deeper philosophical underpinnings of quantum physics, the more I can't help but wonder what nature herself is revealing to us through her new physics. But as someone who has no academic training in physics, I am writing as an "outsider" and nonprofessional regarding the field. I have no authority to comment on the nuts-and-bolts physics of things, or the technical mathematical models that are involved in the professional process of "doing physics," which I literally know next to nothing about. On the contrary, I am simply giving voice from my perspective as a deeply curious person who is sincerely trying to make sense of what it means when physics tells us that the world we live in is quantum through and through.

To quote Feynman, "Nature isn't classical, dammit, and if you want to make a simulation of nature, you'd better make it quantum mechanics."[24] Practically speaking, quantum reality is not terribly remote or inaccessible; it is everywhere—around us, inside us, and everywhere in between. To quote physicist David Bohm, "Quantum mechanics has been experienced by everybody far more than classical mechanics."[25] As a citizen in the recently recognized quantum world, I am writing as an "innocent bystander," except that quantum physics invariably implicates me as participating in what I'm writing about.

Regarding quantum physics, most of us have little or no idea what we have been missing. We have been ill-informed and left out in the dark regarding these over-the-top discoveries that have everything to do with the ultimate nature of the reality in which we live our everyday lives. At first glance, quantum physics can seem completely inscrutable and incredibly intimidating. However overwhelming it appears, the fundamental essence of quantum physics is actually quite simple, and contains the deepest relevance for all of us. In 1908 Einstein wrote in a letter, "This quantum problem is so uncommonly important and difficult that it should be everyone's concern."[26] Speaking about the public's ignorance regarding the momentous discoveries in the new physics, Nobel Prize-winning physicist Isidor Isaac Rabi simply says, "It's a great pity."[27] With reference to quantum physics, what we don't know *can* hurt us. In our modern age, scientific literacy has become a political and moral necessity. In the following inquiry, we should prepare ourselves to be astonished.

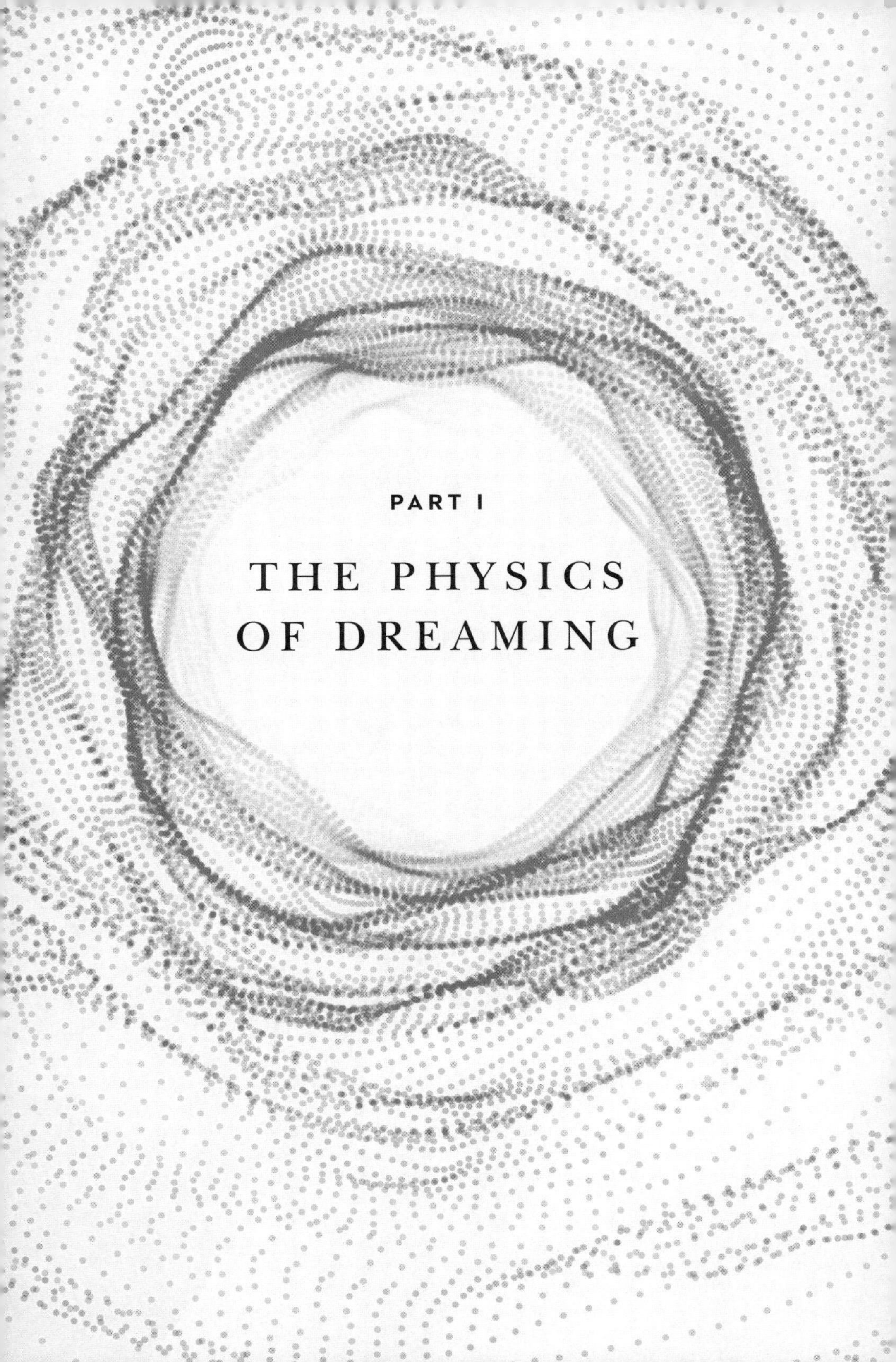

PART I

THE PHYSICS OF DREAMING

• CHAPTER ONE •

PARTICIPATORY UNIVERSE

Renowned theoretical physicist John Archibald Wheeler, a colleague of both Albert Einstein and Niels Bohr, is considered to be one of the greatest physicists of the twentieth century. A professor emeritus at Princeton, Wheeler has been called a "sage of modern physics" as well as "the last of the greats." Drawn to explore the very limits of science, Wheeler was unafraid to face the big issues of his field. His list of accomplishments in physics is staggering; the whole universe—big and small—was the playground for his poetic, creative imagination. In 2006, two years before Wheeler's death, cosmologist Max Tegmark remarked that Wheeler was "the last Titan, the only physics superhero still standing."[28] After discussing physics with Wheeler, Einstein himself said, "What Wheeler told me left a big impression on me. . . . I don't think I'll live to find out who is correct. . . . A possibility would be a combination of his ideas and mine."[29] Acting as a tour guide, Wheeler will lead us in our investigation into the quantum world. "The quantum," to quote Wheeler, is "the most revolutionary principle in all of science, and the strangest."[30]

A scientist-philosopher who was simply fascinated by how nature works, Wheeler—he himself was seen as a "wonder of nature" by his colleagues—became the poet laureate of existence itself. Physicist Freeman Dyson says that Wheeler's style is "inseparable from the substance of his thinking, as poetry is inseparable from science in his mind."[31] Wheeler was a pied piper among physicists. Due to his fondness for speculating on what directions future science might take, he was considered to be the Delphic oracle of physics. A mentor to Richard Feynman, he was an inspiring teacher for many of the greatest and most innovative physicists of our current day. His goal was to plant ideas deep in the minds of his students which, like timed-release capsules, might find some way to flower five, ten, or fifty years later.

To say Wheeler was an out-of-the-box creative thinker would be an understatement; for Wheeler the box that he was "out of" was a higher-dimensional hypercube which existed in the realm of the imagination. Wheeler was a rigorous scientist as well as a visionary whose musings went far beyond the orthodoxy of his time, often astounding and/or exasperating narrow specialists. A speculative dreamer with the soul of a surrealist poet, he has been described as someone who "dreams with open eyes," and "a twentieth century Leonardo da Vinci." As the leading light of theoretical physics, many of Wheeler's fellow scientists are convinced that his "earthshaking ideas" and prophetic insights into the foundation of modern-day physics will spur a revolution in our perception of the universe. "Wheeler's message," to quote Dyson, "is a call for a radical revolution."[32] John Wheeler's vision for the future, to quote physicist Paul C. W. Davies, "beckons not merely an advance in science, but an entirely new type of science."[33] Truly a legend in the physics community, Wheeler's impact on the field of modern-day physics is hard to overstate.

As Wheeler has pointed out, the majority of developments in science have come out of asking the right questions. The questions we ask are determined by our way of thinking. What we think about and how we perceive the world seems as if it subtly affects reality at a very deep and basic quantum level, thereby informing, influencing, and in some mysterious way modifying the underlying fabric out of which third-dimensional reality emerges. What we wonder about and how we wonder about it subtly alters the way in which reality presents itself to us. Wheeler comments, "One of the biggest problems is how to state the problem. It's an old saying that the minute you can state a problem correctly you understand 90 percent of the problem."[34] *We ourselves create the reality of human experience with the questions we ask and the procedures that we undertake to find the answers to them.*

To quote Werner Heisenberg, "What we observe is not nature itself, but nature exposed to our method of questioning."[35] It is easy to assume that when we ask questions of nature, of the world seemingly outside of ourselves, that there is an actual reality existing independently of what can be said about it. Parsifal-like, we have to ask the right question, but finding the right question often takes a good deal of imagination. Wheeler writes in his personal journal that "it may be more important to look for the right questions than to look for the right answers."[36] The asking of one question prevents and excludes the asking of another.

To quote Wheeler, "The question is what is the question?"[37] The issue of how to ask the question and when it is asked plays an important role in what answer we get. There is a mysterious relation, or to quote Wheeler, a "deep and hidden" relation, between question and answer. Unlike many "experts," Wheeler had a fondness for examining questions without the need to pretend that he knew the answers. Einstein, who considered his greatest gift his insatiable curiosity,[38] reminds us that the most important thing is never to stop questioning. Wheeler always told his students that "no question is stupid enough not to be an interesting question."[39] He felt that inquiring into the nature of our situation is what we are here for. If we have not gotten the universe to answer, maybe it is because we didn't realize it was alive and hence did not question it. What is the universe revealing to us? Wheeler comments, "No question? No answer!"[40]

THE UNIVERSE WILL NEVER BE THE SAME

Classical physics, the physics that existed before the discovery of quantum physics, was about uncovering what were thought to be the preexistent laws of a separately existing universe that objectively existed independent of observation. The idea of an objective world, pursuing its course in space and time independently of any observing subject has, in Heisenberg's words, "been the guiding image of modern science."[41] Quantum physics, however, has obliterated the classical notion of an independently existing world forever. According to quantum theory, the idea of a world independent of our observation has a conventional meaning in everyday life, but ultimately is incorrect. To quote Wheeler, "Nothing is more important about quantum physics than this: it has destroyed the concept of the world as 'sitting out there.' The universe will never afterwards be the same."[42] Quantum physics forever shattered the idea of there being an objectively existing world—it has proven that there is no such thing!

In Wheeler's words, "the mind-bending strangeness of quantum theory"[43] is revealing that there is no world "out there" apart from our observation of it. Our observations are part and parcel of what we observe. Our perception of the universe is a part of the universe happening through us that has an instantaneous effect on the universe we are observing. The act of observing changes the observed—this is known as "the observer effect." It is ironic that physics, long considered the most "objective" of all the sciences, in pursuing its dedicated quest to understand the deep nature of the material universe, has dispelled the very notion of an objective universe.

One of the insights of quantum physics is that "purely objective" science turns out to be impossible. Speaking about quantum physics' realization that the observer affects the observed, the great doctor of the soul C. G. Jung comments, "the result being that reality forfeits something of its objective character and that a subjective element attaches to the physicist's picture of the world."[44] In other words, the subjective component of our knowledge must necessarily be taken into account. The world that we experience is showing itself to us in the specific way it does as an instantaneous reflex (reflex-ion/reflection) of the way we are observing it. This means that our inner attitudes, thoughts, beliefs, and assumptions, all subjective states of mind, are playing a central role in the evocation of the particular form that the world appears to us moment by moment. To understand our world as fully as possible, we need to unify the objective/scientific and the subjective/mental spheres of knowledge, moving from a science of objectivity to a science of intersubjectivity.

The unification of the objective and the subjective areas of experience should conserve the richness of each as well as maintaining their relative independence. Speaking "from the perspective of life," Nobel laureate Wolfgang Pauli is of the opinion that we are not treating matter "properly" if we are "totally neglecting the inner state of the 'observer.'"[45] In a similar spirit, wondering why science has historically solely focused on understanding the nature of the seemingly objective world—at the expense of marginalizing our own subjective experience of the outside world—His Holiness the Dalai Lama simply asks, "Why is there a fixation on the external?"[46] Good question. The very fact that this question is rarely asked is a sign that this fixation is so widespread that it is simply accepted as normal without noticing how one-sided it is.

Quantum physics brings into the foreground the relation between a subject's experience and the rest of the universe. It makes no sense to think of ourselves as a self-enclosed, encapsulated, independent agent existing separate from the universe. Previous to the advent of quantum physics, physicists pretended that they were not involved in their own experiments by maintaining the illusion of a disembodied objectivity. The psyche of the observer, however, is an integral part of the process being observed. Quantum theory has opened up the door to a profoundly new vision of the cosmos, where the observer, the observed, and the act of observation are inseparably united.[47]

We, as observers, are participating in what philosophy professor Tu Weiming calls a "joint venture" with the world out there. Our interaction with the seemingly outside world, mediated through our experimental instruments, results in information being added to both our minds and the universe at large. Niels Bohr, one of the founding fathers and principal interpreters of quantum physics, pointed out that, just like within a dream, in our lives we are simultaneously actors and spectators; we are both observers and the observed, subject and object, dreamers and the dream.[48] It's as if in our sharing a dream space together, we are collectively dreaming up our universe while simultaneously being dreamed up by it.

Quantum physics has shown that the idea of safely standing behind a slab of plate glass while passively observing the universe is impossible, as our observation of even something as miniscule as an electron necessitates the shattering of the glass and reaching into, so to speak, the electron's subatomic world, which changes the electron, ourselves, and the whole universe. Wheeler refers to the idea of the universe being "out there," existing separately from ourselves, as an "old idea."[49] It is an outdated idea whose expiration date has been reached.

In the quantum world, the act of observation produces a change in the system that is different in kind, not just in degree. In quantum physics, there appears to be an eerie connection between the physical state of a system and conscious awareness of it by some observing being. The act of observation is epistemological, in that it adds to our knowledge about nature—but it is also at the same time a metaphysical event, in that it changes things in nature from being indeterminate to being determinate.

It is impossible to gain information without changing the state of the system being measured. The more information extracted in the measurement, the greater the perturbation. We invariably bring about a different world by the very act of trying to determine the state of the world. A simple way to envision this is to imagine a blind person trying to understand what a snowflake is. The blind person can touch the snowflake (which will melt it from their body heat) or put it in their mouth and taste it (which will dissolve it), but through whichever means they try to apprehend the snowflake they inevitably change it. Another example is when we use a thermometer to measure temperature—this process, however slightly, heats up or cools down the thing being measured.

Speaking about reality, physicist Vlatko Vedral gets right to the point when he says, "Rather than passively observing it, we create reality."[50] In quantum physics, the physicists themselves—in collaboration with the universe—produce the results of their experiments. In quantum physics, we are no longer passive witnesses of the universe, but rather we unavoidably find ourselves in the new role of active participants who inform, give shape to, and in some mysterious sense "create" the very universe we are interacting with. Making this point, Wheeler says, "Useful as it is under everyday circumstances to say that the world exists 'out there' independent of us, that view can no longer be upheld. There is a strange sense in which this is a 'participatory universe.'"[51]

Though the words sound the same, this is not some overly simplistic and naïve New Age philosophy which proclaims that "we create reality" (which oftentimes conflates and confuses the ego—the false sense of self—with the true or deeper self that is the creator of our experience). Human consciousness has always inescapably participated in how the world around us manifests, but many people are *unconsciously* participating in this process. The revelations of quantum physics are potentially helping us to begin to become conscious of our participation in dreaming up our world. To quote Wheeler, "In order to make sense out of the mysteries ahead, we'll find ourselves forced to recognize the participatory character of the universe in a much deeper way than we now see."[52]

A perfect illustration is when we are absorbed in a dream and have forgotten that we ourselves have something to do with creating it. Wheeler was fond of bringing up the words of the poet Antonio Machado: "Traveller, there is no path. Paths are made by walking."[53] Becoming conscious of a process that we have always been unconsciously involved in not only opens up the possibility of a radical transformation of the human experience, but could also open up previously unimagined frontiers of human freedom that could utterly remake our world.

TWO HOLES THAT CHANGED THE WORLD

The central mystery expressing the nonclassical "dream logic" of quantum theory is contained within one experiment, in Feynman's words, "the experiments with the two holes"—the "double-slit" (also known as the "two-slit") experiment. To quote Feynman, it "has in it the heart of quantum mechanics. In reality, it contains the *only* mystery."[54] This famous experiment, which has been repeated innumerable times, can't be explained in any classical way, and contains encoded within it all of the peculiarities and paradoxes of quantum mechanics.[55] Feynman felt that all of quantum physics can be gleaned from carefully thinking through the implications of this single experiment. To quote Feynman, "Any other situation in quantum mechanics, it turns out, can always be explained by saying, 'You remember the case of the experiment with the two holes? It's the same thing.'"[56] The essential features (or "weirdness," as many physicists put it) of this experiment are found to characterize all quantum systems throughout the universe; we shouldn't forget that the universe itself is one massive quantum system.

The trade magazine *Physics World* voted the double-slit experiment the single most beautiful experiment of all time. In the double-slit experiment, we start by shining a beam of light through an opaque wall with two slits in it onto a photographic plate (a screen) that detects the arrival of the light. As the beam passes through both slits, it spreads out and, as we would expect, behaves like waves interfering with each other, creating an interference pattern on the screen. If we close one of the slits, the interference pattern (which is thought to be the result of light waves interacting with each other) disappears, and we can tell by the pattern that the light makes on the screen that it behaves as particles. This makes sense, for as the light passes through the one open slit, it has nothing to interfere with to create an interference pattern. So far, so good.

The source of light is then dimmed to the point of emitting only one photon (a discrete, indivisible particle of light; this experiment works just as well with electrons, atoms, and countless other subatomic particles) at a time through the two open slits onto the screen behind it. Each single photon leaves as a localized, discrete, and indivisible particle and arrives at the screen as a particle (as is evidenced by a tiny flash of light at a definite point). The next photon is emitted only after the previous photon has been detected on the screen. In between the emission and detection of each photon, however, something highly mysterious is going on.

When both slits are open, we would expect that each individual photon should go through one or the other of the two open slits. Over time, as the number of photons builds up, this would result in two stripes on the photographic plate, one behind each slit. This is not what happens, however. The photon seems able to manifest as several probabilistic counterparts of itself and explores all possible pathways open to it simultaneously. As each individual photon is sent through the slits and onto the photographic plate, over time an interference pattern builds up.[57] Being indivisible, a photon should be able to go through only one of the two slits, thus making an interference pattern impossible. It shouldn't matter if the other slit is open or closed.

For an interference pattern to happen, it is as if the single photon would have to be in two places at once, traveling through both slits at the same time. In the physics version of "one hand clapping," the photon is interfering with itself. It is seemingly impossible for the photon, which is indivisible, to divide itself up and then comes back together in its unified state on the other side of the slits. "Quantum mechanics," Wheeler writes, "says that the cloud of probability that *is* the photon until it is detected can take both routes at once!"[58]

When both slits are open, an interference pattern builds up; when one slit is closed, the interference pattern disappears. The questions naturally arise: how does any given photon know when the second slit is open and when it is not? In passing through one part of the experiment, how can it know what is happening in some other part? The results of the experiment indicate that whether or not the second slit is open informs the photon regarding what sort of pattern to build up on the photographic plate. How do these individual and seemingly separate photons conspire with each other to build up an interference pattern over time? And how does a single photon "know" where to place itself in the overall pattern? Why doesn't every photon follow the same trajectory and end up in the same spot on the other side of the slits? All good questions, I imagine Wheeler saying.

A detector is then placed at one of the slits to see which slit the photon actually goes through. Whenever this is done—trying to catch the photon in the act—the photon always manifests as a particle that goes through one slit or the other and the interference pattern disappears. In other words, when we look, the photon always manifests as a normal everyday particle. When we don't look, the photon manifests its wavelike aspect. So not only does the photon know whether both slits are open or not, it also knows whether or not we are watching it, and adjusts its behavior accordingly. It is as if the photon knows beforehand that we are lying in wait ready to observe it and, as a result of this, chooses to maintain its particle persona. Appearing self-conscious, the photon doesn't seem to want to be seen in its quantum weirdness, as if not wanting to be caught in the act of going both ways at once. If the detector at the slit is turned off, we then have no knowledge of the route the photon has taken, its secret is safe, and it resumes its mysterious wavelike behavior and the interference pattern comes back.

The photon not only knows which slit is open or closed, but seemingly possesses information about the whole experimental setup, including the observer, which in principle suggests that it knows the quantum state of the entire universe. Astrophysicist Massimo Teodorani writes, "You may also think of a kind of 'intelligence' of particles, since when a particle passes through both slits simultaneously it seems to have a perfect consciousness of the past and the future in order to create the correct figure of interference."[59] This suggests that the quantum world is truly sentient, as well as holistic; each of its parts are in touch with the whole.

There is no clearer example of the interaction of the observer with the observed than the double-slit experiment. Consciousness interfered in the experiment in such a way so as to have a direct effect at the quantum level. It is important to understand that Wheeler's idea of a participatory universe was not based on wild theoretical speculation, but on the most rigorously tested scientific experiments imaginable. Quantum theory is so counterintuitive that he could never have dreamt it—even in his wildest dreams—without the constant guidance provided by experiments. The double-slit experiment was the very first experiment in which consciousness literally entered the physics lab in a way that was both impossible to ignore and demanding to be accounted for; there were countless other types of more elaborate experiments to follow.

The double-slit experiment helped to reveal the dual, seemingly schizophrenic nature of the quantum world. It was the experiment that first showed that how light manifested—as a wave or a particle—depended upon how it was observed. This is the prototypical experiment that revealed how consciousness not only was an inextricable part of the universe, but actually affected the very universe of which it was aware through the act of awareness itself. "Consciousness," as Bohr reminds us, "is inseparably connected with life."[60] The double-slit experiment was the first major clue that "shed light" on the fact that whether light manifested as a wave or a particle depended on the questions asked (i.e., how the experiment is arranged), an insight which has enormous implications for understanding the nature of the universe at large. As Wheeler reminds us, the questions we ask are crucial.

The double-slit experiment also revealed something astounding: The subatomic particles under investigation behave *as if* they know they are being looked at. It is *as if* while we are observing these quantum entities, they are observing us observing them. It is *as if* they are continually monitoring and processing their environment—just like us. This is to say that the inhabitants of the microworld appear to have sentience. Not only do these quantum entities act as if they know they are being watched, they appear to precognitively know how they are going to be watched in the future and are able to retroactively change their behavior accordingly. This experiment indicates that observations made after these quantum entities have "done their thing" (whatever their "thing" is) seems, from all of the evidence, to have an effect on their behavior after the seeming fact.[61] Though manifesting in time, these quantum entities appear to be not "bound" by time, but seem to exist both inside and outside of time simultaneously, casting our previously held notion of linear time into question.

In addition, these quantum entities act *as if* they can interfere with themselves (in the same way that we can) and appear to have the superhero power of being able to be in two places at the same time (most of us are still working on this). This is to say they are not localized and can't be pinned down in the normal way, both in terms of space and time. Quantum entities act as if they are party to some bizarre, mysterious form of nonlocal telepathic communication that transcends the normal parameters of space and time. To say they are psychic is an understatement; in some way they are inseparable from, connected to, and manifestations of the psyche itself. The double-slit experiment was the archetypal process that pulled the curtain back and revealed that consciousness is an integral part of our universe.

Quantum entities are truly holistic in that they seem to have an aspect that is localized in space while at the same time having a nonlocal aspect in touch with the whole. Based on their behavior, these inhabitants of the microworld act as if they know they are part of—and are merely playing a role in—a more extensive entity called a probability wave. Seemingly aware of more of the world than their immediate locality, these quantum entities, based on all the evidence, appear to know and have the ability to instantaneously synchronize themselves with the quantum state of the entire universe. The implications, which are still being unpacked and explored in new and ingenious experiments,[62] are beyond huge.

The double-slit experiment gave an incarnate birth to the field of quantum physics. It was as if the physical world (i.e., matter) took the form of this experiment so as to transmit the quantum nature of reality into our minds. A living symbol of what it gave birth to, contemplating the double-slit experiment both *reflects* and *effects*—potentially actualizing, i.e., making conscious—the quantum nature of reality. A symbol both reflects and is a portal leading to (i.e., effects) the very experience it is reflecting. This experiment has holographically encoded within it the whole of quantum physics. This is to say that this experiment is the revelation of quantum physics in a form we could only—and did—imagine.

CONSCIOUSNESS

Confronted with empirical evidence that there is no objective universe existing separate from the act of observation (making the world, in Wheeler's words, "a never-never land"[63]) created a previously unimaginable reality crisis in the minds of the most brilliant physicists. It made Einstein seriously ponder deep philosophical questions such as: Does the moon only exist when we are looking at it?[64] This sounds similar to the koan-like question: If a tree falls in the forest and there's no one there to hear it, does it make a sound? Quantum theory is the ultimate Zen koan of our times. As in Zen, the solution to the physicist's dilemma is hidden in paradoxes that could not be solved by logical reasoning but have to be experienced directly through an expansion of awareness. Physics and mysticism seem to be becoming the most unlikely of bedfellows.

Physics is encountering consciousness and is in the beginning stages of consciously realizing it. In essence, consciousness has entered into the physics laboratory, and physicists are not quite sure what to make of this turn of events. Who can blame them? The encountering of consciousness in their experiments (what has been called physics' "skeleton in the closet," a quantum elephant in the physics living room) is, simply put, out of their league. Facing up to consciousness's intrusion into their hallowed halls is forcing physics to come to terms with questions of metaphysics, which for most physicists is not what they signed up for. Quantum physics is itself the greatest threat to the underlying metaphysical assumptions of "scientific materialism," a perspective which assumes that there is an independently existing, objective material world that is separate from the observer.

It can easily seem as if the whole consciousness problem—and its requirement for a new way of thinking—has been forced upon physics against its will by some outside agency or human authorities. But nothing could be further from the truth. The appearance of consciousness in the domain of physics is totally natural and "enforced" by nature herself—it is nature revealing one of her most intimate mysteries to the minds of modern physicists, most of whom were simply not prepared for what nature was beginning to reveal to them. To quote Nobel Prize winning physicist Eugene Wigner, "Through the creation of quantum mechanics, the concept of consciousness came to the fore again. It was not possible to formulate the laws of quantum mechanics in a fully consistent way without reference to consciousness."[65] Quantum physics cannot do without the notion of consciousness—the most important component of the phenomenon of life—while life cannot be understood without invoking the revelations of quantum physics. In a mutually reciprocal relationship, the physical dimension needs consciousness to organize and evolve, while consciousness needs the challenge and limitation of physical reality in order to grow and realize itself.

Most physicists think that something as ethereal as consciousness, or what has been referred to as "the unwanted stepchild of physics," has no place in "real" physics. The prevailing mainstream view is that consciousness, or "philosophy," is not supposed to be studied in a physics department. Anything that isn't testable and can't be measured is of no concern to most physicists. If it can't be quantified, it is considered not real, hence the term "exact science." Mental phenomena, to quote psychologist William James, have come to be treated as "mere waste, equivalent to nothing at all."[66] To the overwhelming majority of physicists, the role that consciousness plays in their experiments seems to be against the spirit of science—which in their view is always supposed to be impersonal and objective. After all, physics is supposed to deal with things, not people. Wheeler comments, "Using such and such equipment, making such and such a measurement, I get such and such a number. Who I am has nothing to do with this finding. Or does it?"[67]

And yet as we study the seemingly objective world, we are discovering that nothing objectively exists "from its own side," separate from our consciousness of whatever "it" is. In the words of Bohr, "The word consciousness, applied to ourselves as well as to others, is indispensable when describing the human situation."[68] Like an uninvited, unwelcome guest at dinner, consciousness refuses to go away.

The standard, mainstream scientific perspective is that of the atheist who embraces the world perceived by the five senses and their extensions through scientific instruments as all that can be legitimately said to exist. In his article about the possible unification of physics and spirituality, Brian Josephson, a Nobel Prize-winning physicist, writes:

> The alternative to this atheistic position is that there exists an aspect of reality—that we may for convenience call transcendental—which embraces the subject matter of religion (or as some may prefer to term it, the spiritual aspect of life) and which is not at present encompassed by science. The question then arises whether some future science may be able to cope with this aspect of reality, or whether it will remain forever beyond the scope of science. The general aim of science being to gain as full and accurate a picture of reality as possible, one would expect logically that scientists in general would take a keen interest in such questions, just as they do in topics such as those of the fundamental constitution of matter, or of the mechanisms of life. In practice however, such questions have been almost entirely split off from scientific consciousness.[69]

It is clear that any deep or complete picture of reality must address the fundamental basis or nature of the invisible nonlocal connectedness of all that arises into perceivable form. Why scientists in general do not "take a keen interest" in questions related to the spiritual or transcendental aspect of reality is an important question that goes directly to the psychological roots of the form that modern science takes. This tendency to ignore or to display an aversion to such questions is a reflection of a pervasive psychospiritual one-sidedness in contemporary humanity which is especially acute in the scientific community in general and the physics community in particular. This one-sidedness translates into an unnecessary and irrational distortion of the practice of science itself, in which matters pertaining to consciousness or spirituality are unthinkingly rejected or considered to be taboo or outside the proper domain of science. This is particularly ironic given that quantum physics, "the crown jewel of the physical sciences," is irrefutably pointing to the fundamental and inescapable role that consciousness plays in the constitution of matter and the ongoing moment-to-moment creation of the physical world as a whole.

• CHAPTER TWO •

OBJECTIVE REALITY HANGOVER

The very structure of human thought changes in the course of historical development. Science, which, in Wheeler's words "is an intensely human activity,"[70] has a great effect on human beliefs and upon the very process of thinking itself. "Science," to quote Albert Einstein and Leopold Infeld in *The Evolution of Physics*, "is a creation of the human mind,"[71] which is to say that science reflects the very mind that created it. The development of science over the last few centuries has changed humanity's thinking, which is to say that it matters what physicists think. Heisenberg writes, "It must be admitted, though, that during the last hundred years there have been such radical changes of thought pattern in the history at least of our own science, physics, that it is perfectly legitimate to speak of one or even several revolutions."[72] The transition from one age to the next can be triggered by a seemingly minute change in a single idea. Oftentimes it is the abandonment of commonly held and cherished "truths" that has propelled science forward.

Some of the most important science-generated beliefs that pervade our world and make up what is thought to be "common sense" are outdated and mistaken ideas that arose in science during the seventeenth, eighteenth, and nineteenth centuries. One such antiquated belief, which is practically a superstition, is the unquestioned assumption of an external, independent, objective universe with its concurrent shallow, limited, and impoverished conception of how humankind—seen as a material object—fits into such an apparently objective world. Having fallen prey to this unreflected-upon belief, we then try to fit this picture of objectivity that exists nowhere except in our heads into the "real" world. Ironically, from the scientific point of view, it is irrational and against the very spirit of science to cling to such a false and obsolete idea of the world we live in. To paraphrase a well-worn adage, it's not what we don't know that gets us into trouble, it's what we think we know with certainty that turns out to be false.

Relating to an afterimage as if it still exists, many physicists are doing quantum physics—and successfully solving their equations—and yet deep in their unconscious are still subtly entranced in a classical mindset that sees the world as independently existing. To quote physicist F. David Peat, "A revolution had occurred in physics, but at a deeper level the same order prevailed. The new wine of quantum theory had merely been put in the old bottles of Cartesian order."[73] In trying to grasp a radically new approach to the world using concepts forged by their familiar "old" mode of thinking, physicists are employing the very modes of thought that the revelations of quantum physics are trying to subvert. Trying to form-fit a newly emerging phenomenon into an already existing framework is, in Feynman's words, due to "the limited imagination of physicists."[74]

Clinging to the idea of an objectively existing world—what Heisenberg refers to as a "limiting concept"[75]—is like holding on to the mistaken belief in a flat earth, all evidence to the contrary. Heisenberg writes, "The hope that new experiments will lead us back to objective events in space and time is about as well founded as the hope of discovering the end of the world in the unexplored regions of the Antarctic."[76] Like the old "flat-earthers," "objective-worlders" are holding onto to an inculcated unconscious belief, reinforced by several centuries of habit, that has now ineluctably been shown to be incorrect, a false make-believe figment of the human imagination. Wheeler openly wonders whether, in an interesting choice of words, we are "sleepwalking"[77] if we think that we aren't influencing the results of our experiments.

Not just the physics community, but the vast majority of our species is suffering from a similar "holding on" to what Stapp refers to as "a known-to-be-false"[78] idea of the world by thinking it inherently exists separate from ourselves. When what we take to be real is recognized to be an illusion, it will initially be a disorienting experience. While in classical physics we pretended that we were *outside* the universe looking in, quantum physics has revealed that we are *inside* the very universe we are trying to understand. We have simultaneously come *into* and *out of* the universe at the same time.

In a true inversion of reality from the agreed-upon consensus reality point of view, the difference between those who are committed to an independently existing, objective reality and those who are not are roughly correlated with the distinction between those who are considered sane and those who are deemed insane. And yet holding on to Stapp's "known-to-be-false" version of the world as existing objectively is nothing other than a form of insanity. It is as though the inmates are running the asylum.

The objective world model, which still has such a pervasive hold on many of us, is a construct, literally a projection of a particular stage of human psychospiritual development. The quantum revolution has revealed that the classical worldview was a reflection of a particular psychological perspective, something that existed entirely within the minds of a certain strain of European humanity. It was then projected outwards on the world and became reified and rigorously mathematically formalized into an orthodox creed and subsequently held the mind of modern humanity in a prison of its own making. Through this process humanity had become entrained by the self-reinforcing fetters of the classical worldview, spellbound by its compelling vision (or rather "hallucination") of the world. Providing a way out of this self-imposed prison, quantum physics heralds the advent of an altogether new stage of human psychospiritual evolution.

Physicists who are still entranced by the notion of an objective universe, with its concurrent exclusion of the observer, are simply unwittingly re-creating the greatest failure of classical physics—its inability to find a place to accommodate us, its creators. Human beings are not likely to thrive or endure in a society ruled by a conception of ourselves that denies the very creative essence of our being. A quick look at the condition of our world shows the devastating impact of this pervasive denial intrinsic to the Western, scientific, materialist worldview, both upon the quality of life for much of humanity, as well as upon the biosphere as a whole. A flaw in our physics has insinuated itself into our day-to-day lives, massively distorting both our psychology—individually and en masse—and its outer collective reflection, our social institutions. The false and disabling worldview that results from this cardinal error in physics can potentially alienate us from our intrinsic creative power as well as from ourselves. Our understanding of the world we live in determines the ethics we live by. Living a life based on a worldview that is an illusion can easily lead to living the wrong life. In re-visioning our idea of the world we live in, we change our perception of the possibilities available in our world, thus opening up previously unimagined pathways of creative and effective action.

"Objective reality" is an unexamined implicit assumption, merely an idea in our mind. It's similar to thinking that a cloud exists on its own, apart from the air; the idea of a cloud is an abstraction. Making a similar point in one of his personal journals, Wheeler writes, "All this talk about an external world—that's what's theory!"[79] Schrödinger was of the opinion that, based on all of the existing evidence, the actuality of the objective world remains a hypothesis. What most of us call objective reality is simply an interpretation of data, the meaning of which is agreed upon by the majority. This can also be called a "consensus reality," or more accurately, a "consensus trance."

It should be noted that a large number of people agreeing on the way things are is not in itself proof that their version of reality is correct. Perhaps their unconsciouses were all similarly activated in such a way that they fell into a collectively shared hallucination that they believe to be real. To quote philosopher Bertrand Russell, "The fact that an opinion is widely held is no evidence whatever that it is not utterly absurd; indeed in view of the silliness of the majority of mankind, a widespread belief is more likely to be foolish than sensible."[80] An inherently existing, objective world, something that has its "own nature" separate from something else, is a form existing only in the imagination. Upsetting the applecart of consensus reality, quantum physics points out that objective reality does not actually exist.

Wheeler was among the physicists who was realizing that instead of talking about the world out there, it was more accurate to talk about *the image* of the world out there. This is a subtle but important distinction, as it is emphasizing the subjective nature of the experience taking place within our own minds, instead of referencing an allegedly objective world outside of ourselves. The apparent world "out there" has its roots in a field of sentience that is inextricably interwoven with the physical world while at the same time being shaped by the world of innumerable observers. Jung asks, "How in the world do people know that the only reality is the physical atom, when this cannot even be proved to exist at all except by means of the psyche? If there is anything that can be described as primary, it must surely be the psyche and not the atom, which, like everything else in our experience, is presented to us directly only as a psychic model or image."[81]

John von Neumann writes, "Indeed experience only makes statements of this type: an observer had made a certain (subjective) observation; and never any like this: a physical quantity has a certain value."[82] What von Neumann is pointing at is not only true in the microworld of quantum physics, but is also true of our everyday experience. Whatever perceptions we are having are just that—perceptions—and as they arise as subjective experience in our minds, they have varying degrees of correlation to whatever is happening in the seemingly outside world. There is not a demonstrable difference between our subjective perception of reality and a stable and durable collective hallucination. Our perceptions, be they of the micro or macro world, oftentimes tell us more about ourselves than the world outside of us, similar to how descriptions of a dream are revelatory of the dreamer.

The notion of an independent, objective reality that exists separate from an observer is a very deep-seated assumption, a habit of mind, which like one of Kant's categories of perception resides at a core level of the human psyche. This assumed viewpoint practically becomes hardwired into the brain, causing us to filter our perceptions so as to reflect back our fundamental assumptions. Regardless of the overwhelming evidence to the contrary, there still exists an underlying unconscious mode of language and type of thinking embedded in physics which conceives of the world as having a type of objective existence that it simply doesn't have. What Philip K. Dick calls "our materialist-atomist blindness" prevents us from seeing the field-like quality of reality that is inseparable from our consciousness.

Thinking that there's an objective reality is a residue of the old materialistic perspective that lingers as an ingrained way of viewing reality, as if many physicists—and the majority of our species—are suffering from an "objective-reality hangover." The word "hangover" implies that we were previously intoxicated and were therefore seeing the world in a distorted, unhealthy, and inaccurate way. "Quantum mechanics," to quote Wheeler, "demolishes the view that the universe exists out there."[83] One of the things that distinguishes Wheeler from many other physicists is his sober refusal to try to save pre-quantum viewpoints, particularly, to quote physicist Anton Zeilinger, "the obviously wrong notion of a reality independent of us."[84]

For many people the idea that there is no independent reality is "unthinkable," an idea so off their map of reality that they can't even imagine it. The projection of an inherently existing world outside of ourselves is a deeply ingrained, seemingly innate and habitual mode of perception. Old intellectual habits die hard. As the adage goes, "One funeral at a time." It can be difficult to let go of familiar, comfortable, and "tranquilizing" ideas about the way the world works. Once we realize, as quantum physics reveals to us, that there is no separation between cosmos and consciousness, the appearance of this separation is recognized to be merely the perspective from which consciousness/cosmos can view its own projection. Just as within a dream (which is a projection of the mind for the purpose of potentially recognizing and thereby integrating its unconscious aspects), we could impute that the reason for this projection is in the service of a deeper integration.

Heisenberg emphasizes, "The idea of an objective real world whose smallest parts exist objectively in the same sense as stones or trees exist, independently of whether or not we observe them . . . is impossible."[85] Thinking any part—big or small—of the universe exists as a "thing-in-itself," a Kantian "*Ding an sich*," is a mistaken conception, an abstract extrapolation with no counterpart in nature. The idea that the universe is made up of objects whose existence is independent of human consciousness is, to quote philosopher and physicist Bernard d'Espagnat, "in conflict with quantum mechanics and with facts established by experiment."[86] In Buddhism, thinking that the external objects that populate our experience exist independently of our consciousness is termed a "wrong view," which is based in confusion and ignorance.

According to our subjective experience the world certainly seems real enough, apparently contradicting what quantum physics is telling us about the world's lack of inherent, objective reality. In the overwhelming majority of cases, the world behaves "as if" it has an independent reality, which furthers our visceral belief in objective reality in what becomes a self-perpetuating and mind-created feedback loop. In other words, because of the quantum, dreamlike (i.e., consciousness-based) nature of reality, once we view the universe "as if" it independently, objectively exists, it will manifest in a way which simply confirms our viewpoint. Nature seems to respond in accordance with the theory and beliefs by which it is approached. The choices we make about what we observe make a difference in what we find. Wheeler wanted to replace the idea of an objectively existing world, or as he puts it, a "hardware located out there,"[87] with a "meaning software" located who knows where.

EINSTEIN'S DEMON

Albert Einstein was deeply disturbed by quantum physics' implication that there is no independently existing, objective universe. He was not able to let go of his strong belief that an external, objective world independent of the perceiving subject existed. To quote Einstein, "The belief in an external world independent of the perceiving subject is the basis of all natural science."[88] Wheeler writes of Einstein, "Nothing made him more unhappy than the thought that the observer-participator has anything to do with the establishment of what one is accustomed to call reality."[89] According to Abraham Pais, Einstein's biographer, the eminent physicist spoke about relativity with detachment, but the topic of quantum theory brought out his passion. Pais wrote, "The quantum was his demon."[90]

Einstein once said that quantum theory reminded him of "the system of delusions of an exceedingly intelligent paranoic, concocted of incoherent elements of thought."[91] Along similar lines, in 1911, referring to the inmates of an insane asylum near his office in Prague, Einstein is quoted as saying, "Those are the crazy people who are not working on quantum theory."[92] Einstein was troubled by quantum theory's implication of the apparent role that the observer played in creating reality, feeling that it seemed incompatible with any reasonable idea of reality. In response, Bohr famously reflected back to Einstein—who he considered his leading spiritual sparring partner[93]—that his "concept of reality is too limited."[94] We should question what it is in the way we think about the world that causes quantum behavior to be so troubling. Our being troubled is a result of the disparity between the way reality actually manifests itself and our ideas of what reality should be. Theoretical physicist N. David Mermin writes, "Experiments have now shown that what bothered Einstein is not a debatable point but the observed behavior of the real world."[95]

Wheeler confesses that he is not troubled at all by what quantum theory is revealing; on the contrary, he feels that it is "a perfectly marvelous feature of nature,"[96] and that "it is just the way the world works."[97] In his autobiography, Wheeler expresses that when he first learned as a student about quantum theory it made him "giddy," putting a gleam in his eye, and at the time of the writing of his life story when he was in his eighties, "it still does."[98]

From his autobiography, we learn that when Wheeler becomes thoroughly relaxed, his "scientific thinking is likely to be almost subliminal."[99] He refers to "the dreamer in me" with reference to his approach to how he thinks about nature. In talking about his initial insights into the quantum realm when he was younger, he writes, "It has the 'purity' of my early dreaming."[100] It gets my attention that Wheeler is connecting his insights into quantum physics with the art of dreaming. Wheeler writes, "I have never been too busy to dream. Dreaming . . . provides necessary sustenance for my brain, as nourishing as any calculation."[101] From a physicist such as Wheeler, this is high praise indeed.

Wheeler readily admits, "I have not been able to stop puzzling over the riddle of existence."[102] He has spent his life trying to find the fundamental laws upon which everything is built, what he calls "deep happy mysteries."[103] Being open and not entrenched in a fixed viewpoint, he confesses that he wakes up every morning and in addition to appreciating the miracle that is life, he is always ready to change his views completely. He writes in his autobiography, "There are many modes of thinking about the world around us and our place in it. I like to consider all the angles from which we might gain perspective on our amazing universe and the nature of existence."[104]

Wheeler's all-angles/omniperspectival approach to exploring the nature of the universe is an expression of his realization that the particular reality that humans experience is always and inescapably perspective based. Our experience is limited and conditioned by whatever perspective we happen to be viewing reality from at any given moment. Wheeler's openness to "try on" and learn from as many perspectives as possible is in the service of gaining the most comprehensive understanding of the vast and multifaceted richness of that mystery of mysteries that we call reality.

Wheeler confesses in his autobiography to have a distaste for what he calls the "herd instinct"[105] that every so often grips physicists as they collectively subscribe to the currently accepted view of reality. Wheeler writes, "When I see a herd running one way, I like to march another way."[106] From his personal writings, it is clear that Wheeler related to physics as a sort of treasure, in his words, "sitting hidden in the back of my mind," which, when it emerged into consciousness, helped him find his calling in life. May we all, in our own way, be so lucky. Or blessed.

MIND VIRUS

To quote Heisenberg, "The Cartesian partition has penetrated deeply into the human mind during the three centuries following Descartes, and it will take a long time for it to be replaced by a really different attitude toward the problem of reality."[107] The Cartesian worldview, in which the world was split into mind and body, into subject and object, is a self-propagating idea—a way of viewing the world—that has "penetrated" the human mind. Not simply an impotent and passive idea, seeing the world as if it exists separate from us actively draws the landscape of the world to manifest itself to our mind as if it is truly other than ourselves, which then "proves" to us the rightness of our unexamined viewpoint in an endlessly self-reinforcing and self-generated feedback loop whose ultimate source is our own mind. The accumulated, agreed upon, and accepted ideas about the nature of our world are unthinkingly passed down through the generations, analogous to the genetic transmission of traits from generation to generation. It should be pointed out, however, that some of these ideas, for example the idea of objective reality, are simply mistaken.

To learn we must be free to err, for correcting our mistakes is how we evolve. But if our mistakes go undetected and uncorrected, we deviate from our nature with potentially destructive consequences that threaten not only our own survival, but our very planet as well. Interestingly, Feynman was of the opinion that the process of unconsciously transmitting and passing on ideas through the generations "had a disease in it."[108] Quantum physics can teach us how certain unexamined assumptions, such as an objectively existing universe, actually "infects" our thinking and language in subtly hidden ways that hinder the realization of our full creative potential.

The idea of an objective world can be likened to a memetic thought virus which has managed to commandeer the facilities of a very powerful host—the overwhelming majority of people, the scientific community, academia, book publishers, distribution services, mainstream media—endlessly and invisibly replicating itself through the ideosphere. When such a fertile idea gains enough traction and momentum, it becomes "self-evident," supplying all the needed evidence to relentlessly prove and ratify the rightness of its perspective. Its way of viewing the world can become firmly entrenched in our minds as the invisible and unquestioned lens through which we instinctively give meaning to our experience. The idea can develop a seeming autonomy and life of its own, driving its own propagation, in other words becoming "self-replicating." The idea of an objective reality links with the idea of a subjective center of psychological operations, the reference point of ego, in such a way that these two ideas reinforce each other, reciprocally assisting each other's survival in the meme pool.

My book *Dispelling Wetiko* explores this virus of the mind. "Wetiko" is a Native American term, but every wisdom tradition has its own creative name for symbolizing this self-replicating memetic thought virus that is transmitted across the generations. By whatever name we call it, this mind virus—whose origin is to be found within the human psyche—is a semantic disorder that functions by deviating the very process by which we attribute meaning to our experience. It is an ideological virus whose means of replication and propagation are through altering the syntax of our ideas in ways that become self-perpetuating. This "bug in the system" alters and subverts the core axioms and interpretive schemas through which the psyche maps its self-created meaning onto its inner and outer experience.

Wetiko works through the projective tendencies of the mind in such a way that, to the extent we are unconscious of it, we unknowingly become instruments through which it acts itself out in the world while simultaneously hiding itself from being seen. The wetiko virus induces a form of psychic blindness in which those afflicted believe they can see.[109] The thought forms, beliefs, and perceptions that this mind-created virus presents to us as "objectively existing" act as an intrinsic, built-in control system defining the limits of what we—both individually and collectively as a species—imagine to be within the realm of possibility.

Wetiko, a collective psychosis, can be likened to an "anti-information" virus. Not only does it block the reception of information, but it substitutes false information for the real thing. It has the power to induce, both individually and en masse, what Philip K. Dick calls a "negative hallucination," in which instead of us seeing what is not there we cannot see what is there.

Quantum physics exposes a flaw in our very process of thinking; it is as if our species is suffering from a thought disorder. Dick writes, "There is some kind of ubiquitous thinking dysfunction which goes unnoticed especially by the persons themselves, and this is the horrifying part of it: somehow the self-monitoring circuit in the person is fooled by the very dysfunction it is supposed to monitor."[110] When we have fallen under the spell of the wetiko virus, we aren't aware of our affliction. From our point of view we don't have a problem, and if there is a problem, the cause is always seen as residing in someone else. Dick continues, "The criminal virus controls by occluding (putting us in a sort of half sleep) so that we do not see the living quality of the world. . . . The occlusion is self-perpetuating; it makes us unaware of it."[111] Wetiko occludes us in such a way that we can't even tell that we are occluded, a situation that Dick refers to as "the most ominous kind of occlusion." Being self-perpetuating, this occlusion in our consciousness will not go away of its own accord; it acts as a feedback loop that perpetually self-generates itself until the spell is broken. The majority of our species does seem to be in a state of "half sleep"; no wonder the recent popularity of zombies in pop culture.

David Bohm contemplates what is "preventing mankind from working together for the common good, and indeed, for survival," and concludes that one of the key factors is "a kind of thought that treats things as inherently divided, disconnected and 'broken up' into yet smaller constituent parts. Each part is considered to be essentially independent and self-existent."[112] It sounds like Bohm is talking about the type of thinking that has unconsciously "in-formed" (given form to) the field of physics for centuries. In this mode of thinking that creates seeming fragmentation between things that are not actually separate nor ultimately separable, Bohm, one of the most original, radical, and important thinkers of the second half of the twentieth century, is pointing at wetiko. He writes, "It's similar to a virus—somehow this is a disease of thought, of knowledge, of information, spreading all over the world. The more computers, radio, and television we have, the faster it spreads. So the kind of thought that's going on all around us begins to take over in every one of us, without our even noticing it. It's spreading like a virus and each one of us is nourishing that virus."[113]

This thought virus, or in Feynman's words "disease," is highly contagious, spreading through the channel of our shared unconsciousness. Its vectors of infection and propagation do not travel like a physical pathogen, however. This fluidly moving, nomadically wandering bug continually regenerates itself through the transmission of meaning-filled and mind-imprisoning informational codes ("info-toxins"). This virus of the mind feeds off of and into each of our unconscious blind spots, which is how it nonlocally propagates itself throughout the field. In this disease of the mind, which is a genuine psychosis, there is a code or logic that affects/infects awareness in a way analogous to how the DNA in a virus passes into and infects a cell.

This psychic disease can be likened to an invisible, immaterial lifeform that has infiltrated our world through the medium of information. Unless recognized, this information virus continually reticulates and arborizes itself, spreading rhizome-like through all the various channels of information media throughout our world, which have become its instruments of propagating itself. This information virus continually "in-forms" our mind, endlessly massaging our brain into the desired shape so as to further replicate itself.

Feynman comments, "Then a way of avoiding the disease was discovered. This is to doubt that what is being passed from the past is in fact true, and to try to find out *ab initio* [from the beginning], again from experience, what the situation is, rather than trusting the experience of the past in the form in which it was passed down."[114] In other words, instead of unthinkingly accepting other people's versions about the nature of our situation—not taking anybody's word for anything—we can think for ourselves. The cure for the disease entails each of us becoming an empiricist and simply inquiring directly into the nature of our present-moment experience. What a radical idea!

SCIENTIFIC REVOLUTION OR MADNESS?

The Scientific Revolution was a deepening of our powers of reason, a flowering of human creativity, and a breakthrough for humanity, helping us to explore our world in ever more profound and ingenious ways. From another point of view that also contains an important truth, the Scientific Revolution, which is now commonly associated with Isaac Newton's ("Newtonian") physics, was also the onset of a particular form of madness.[115] It started as a new worldview that was revolutionary in its power; yet it contained a subtle error that solidified into a widespread delusion which has over time gone pandemic and profoundly enabled the collective psychosis that our species finds itself in.

An essential feature of this madness is the severing between the subject and object, the observer and the observed, as if the scientific imagination thought that in its intellectual examination of the world it wasn't part of, participating in, or thereby affecting that which it was investigating. This was done in pursuit of the ideal of objectivity, which was gradually elevated to the level of an absolute truth about the nature of reality.

Simultaneously, another erroneous assumption was taking hold of the scientific mind, a world model that envisioned this objective world behaving like a giant mechanism, eventually led to seeing the world literally as a machine composed of separate and externally interacting parts, with ourselves as smaller machines within the larger machine of the cosmos. This approach worked remarkably well when it came to dealing with the macroscopic world, enabling unprecedented levels of control to be exerted over the physical world, but in this process of obtaining mastery over the physical plane an unseen cost was being incurred by the human spirit. In viewing life mechanistically, the very thing that makes life the sacred miracle that it is, consciousness itself, is increasingly being ignored, marginalized, and even thought of as a mere illusion. The materialistic worldview is potentially threatening to destroy the most precious essence of what makes human beings human in the first place. To quote Nobel Prize-winning neurophysiologist and philosopher Sir John Eccles, "It must be recognized that monist-materialism leads to a rejection or devaluation of all that matters in life."[116]

This mechanistic, deterministic, and reductionist attitude unfortunately became identified with science itself, thus introducing a number of tacit taboos and limiting assumptions into the otherwise open-ended process of exploration known as the scientific method. The modern scientific attitude which sees the world as objectively existing somehow outside of and separate from itself, "scientific materialism," is actually a deluded view expressing an epistemological blind spot in the very center of the predominating scientific vision of the world, a pervasive blindness of which modern society seems mostly unaware. Materialism—the doctrine that *to be* is to be material—is a "concept of the universe" that Heisenberg refers to as "naïve."[117] Theoretical physicist Amit Goswami writes, "Materialism is like an epidemic disease that has to be healed. And quantum science can be part of the healing."[118]

A strictly materialist understanding of human beings and our place in the cosmos is not part of the solution, but rather, part of the problem; it is feeding into the seemingly never-ending trap that we, as a species, find ourselves in. The materialist worldview, which is based on the idea that things (including ourselves) are freestanding, existing independently on their own and separately from each other, is simply untrue. Denis Postle, author of *Fabric of the Universe*, comments, "This is the lie that we inhabit, which devours us, consuming our energy, poisoning our bodies. . . . If we can come to realize what this lie is made of and how it operates we can begin to be free of it."[119]

Scientist Bernardo Kastrup writes, "Materialism suffuses the core of our being by a kind of involuntary osmosis. Like a virus, it spreads unnoticed until it's too late and the infection has already taken a firm hold."[120] The materialist worldview is itself a living expression of how the aforementioned wetiko virus ("the lie") operates with impunity through our conceptual blind spots. Being the very lens through which we view, interpret, and place meaning on our world and our experiences, the materialist perspective becomes invisible and therefore goes unseen and undetected. When we begin to see how the wetiko virus works through our unconscious blind spots, however, we are on the path to liberating ourselves from its clutches.

Seeing the world as separate from ourselves has become the prevailing and institutionalized worldview of the "academy," a viewpoint that takes the heart, soul, and magic out of the world, reducing it to a dead, inanimate, insensate domain. In this viewpoint, the universe is mechanized and the heavens secularized. Scientific materialism disenchants the world while simultaneously bewitching and casting a materialistic spell over its inhabitants. To quote religious studies scholar Huston Smith, "The greatest problem the human spirit faces in our time is having to live in the procrustean, scientistic worldview that dominates our culture."[121] Increasingly enthralled by science's ever-growing achievements and technological wizardry, few have questioned whether these very advances might at the same time be leading humanity astray from essential aspects of the true nature of our being, slowly dehumanizing our species in the process.

Quantum physics is thus playing a critical role in helping to dissolve one of the greatest impediments to the evolution of human freedom—scientific materialism, the nonscientific dogma that cloaks itself in the laurels of science, using the name of science to justify itself. The advent of a true science for humanity might not correspond to the image many people have of scientific, technological achievement, but rather will undoubtedly have to do with the fall of the illusions that have accompanied science over the last few centuries. Jung comments, "Let us hope that the time is not far off when this antiquated relic of ingrained and thoughtless materialism will be eradicated from the minds of our scientists."[122]

Cultural philosopher Jean Gebser has described how human consciousness, rather than being static, has gone through a number of mutations in our history. Gebser believed that at the current time we are undergoing another such mutation, one of whose flag bearers is quantum physics. In quantum physics, consciousness has entered the field of physics, the realization of which takes place within consciousness itself. This very realization transforms—literally mutates—the very consciousness that realizes this. When human consciousness passes through a mutation, the effects are comparable to entering into another form of reality, as if humanity is becoming a new species.

Before we can accomplish the transition into this new, more quantum-based stage of consciousness, however, the old structures of consciousness invariably enter into a crisis wherein they need to break down so as to make space for the new structure to emerge. This happens when the previous state of consciousness reaches what Gebser calls its "deficient mode," which is when what was initially an asset becomes a handicap. This is an apt way of describing classical physics. It has been very much to our advantage but due to, for example, its bifurcation of subject and object, exclusion of consciousness from being part of reality, and its view of humanity as being separate from the world, it is holding us back from the natural process of evolution.

The newly emerging, in Gebser's words, "integral structure of consciousness" is a radical break from the linear, mechanistic, and deterministic paradigm of the classical scientific worldview. Instead of seeing the world through the lens of causality, linear time, and Aristotelian two-valued logic of "either/or," this emergent new structure of consciousness—totally resonant with the worldview of quantum physics—sees the world through an atemporal, acausal, synchronistic, and holistic lens characterized by the four-valued logic of "both/and." Gebser felt that this new structure of consciousness could potentially reconstruct the fragmentation of humanity in the service of integrating us with the greater whole, what he called the "restoration of the unharmed, original state." It seems more than mere coincidence that Gebser evoked an underlying wholeness at the same time that quantum physics arose as one of its myriad expressions. Gebser was of the opinion that this new mutation of consciousness is a manifestation of latent possibilities that have been present in us all along.

Gebser conceived of this new form of consciousness as not simply a new point of view different than other points of view, but rather, as being "aperspectival," in that it transcends any one seemingly *correct* point of view, reminiscent of Wheeler's aforementioned "omniperspectival" approach towards the universe. Not fixed in any single privileged perspective, seeing aperspectivally is to realize the relativity of all perspectives. The shift to an aperspectival way of viewing the world, bringing to mind the very insight that quantum physics is pointing at, entails the subject becoming part of the seemingly objective scene, as the subject becomes drawn into and begins to merge with the world of objects.[123]

This newly emerging factor that restructures how we see things in the world is mirrored in the inner world of the psyche. Psychologically speaking, an aperspectival approach is analogous to the circumambulation of the *self*—the center of the psyche—a process that ultimately illumines the self, which can be conceived of as the ground of our being. The process of *individuation* or becoming a whole, undivided individual, consists of contemplatively circling around the self's various manifestations, a process which over time sheds light on, makes visible, and potentially realizes the very self which is the underlying source and inspiration for this process. A similar circumambulation is mirrored in quantum physics, as the very nature of the quantum both requires and inspires an aperspectival approach.

It should get our attention that Gebser felt that the integral-aperspectival consciousness that is arising in the Western mind was the very thing that could save our species from complete destruction.[124] This impending destruction that Gebser saw looming over humanity is largely an inevitable result of humanity continuing to operate out of a worldview that makes successfully navigating through the evolutionary challenges that are currently bearing down on us nearly impossible.

Gebser rightfully points out that, once again in alignment with the quantum view that we live in a participatory universe, we are active *participants* in this transition. This isn't a process where we can safely stay on the sidelines as passive witnesses. We are invited—even implored—to join the action. If we don't participate *consciously*, however, Gebser felt that we would *unconsciously* create a catastrophe. The challenge is whether we will participate in the new mutation consciously or simply be unconsciously carried away by its effects, which from the spiritual point of view is akin to being blown by the winds of our own karma.

Like a quantum physicist who is realizing we live in a participatory universe, Jung, who was a friend and colleague of Gebser, writes, "We are not only the passive witnesses of our age, and its sufferers, but also its makers."[125] We are not victims, but are active co-dreamers of this shared waking dream of ours—the question is, do we consciously realize this or not?

• CHAPTER THREE •

REALITY

To quote Einstein, reality—which elsewhere he says "is merely an illusion, albeit a very persistent one"[126]—"is the real business of physics."[127] At the end of the day science is empirical and its theories must be grounded somehow "in reality." But where, and furthermore, what exactly is that reality? Wolfgang Pauli comments, "The important and extremely difficult task of our time is to work on the construction of a new idea of reality."[128] Physicist Hans-Peter Dürr, a longtime coworker of Heisenberg's, comments, "Reality reveals itself primarily as nothing but potentiality."[129] But where is the ontological ground upon which our impression of a "really existing" universe—our idea of reality—rests? Speaking about reality, quantum theory brings the question to the fore: Are we discovering reality, or creating it? And if we are at least in part creating what we call reality, what are we creating it out of?

Wheeler writes, "What we call reality consists of a few iron posts of observation between which we fill in by an elaborate papier-mache construction of imagination and theory."[130] In other words, what we refer to as reality is a "construction" made out of, by, and in our imagination. In the words of John Lennon, "Reality leaves a lot to the imagination."[131] In essence, physics is an imaginative vision of how the world might be put together. We connect the dots between a "few iron posts of observation" so as to create a seemingly coherent picture of our world, the meaning of which is derived from our minds and takes as its basis whatever prevailing theory or set of assumptions we presently think of as being true.

To quote Philip K. Dick, “Regard this as a scientific hypothesis: what we call ‘reality’ is in fact an objectification of our prior thought formations . . . projected onto a pseudo-world.”[132] The idea that reality is composed of projected thought forms that become objectified is a precise description of the dreamlike nature of reality.[133] It is also the deeper message of the wisdom text *The Tibetan Book of the Dead,* which in its essence points out that whatever we are experiencing is not separate from our own consciousness and that recognizing this is a critically important realization that changes everything in terms of the nature, quality, and depth of our experience of ourselves and all of existence.

In the twenty-first century science is the arbiter of what is real. This should give us pause. If we are not careful we can easily fall into a category error, imagining that science’s version of reality is reality itself. “Reality” is just a word in our language. Our sense of reality depends on the structure of our consciousness as well as the ways that we engage and direct our consciousness. To quote Jung, “One always talks of ‘reality’ as though it were the only one. Reality is simply what works in a human soul.”[134]

What we call reality is simply a theory, a mental map, an internalized mental model which at bottom is simply a way of looking at the world. It is a perspective rather than a form of absolutely true knowledge of how the world “really” is. The word *theory* is derived from the Greek *theoria,* which has to do with contemplation, speculation, and sight. A theory is simply the way we are seeing and trying to apprehend the world at any given moment. It is important not to conflate reality with our theories, not to confuse the map with the territory. Our best models and theories are no more than aids to our imagination, by no means are they complete reflections of the nature of reality. Our theories not only “fit” what we know about the world but at the same time they point to what we don’t know and ideally are able to reveal it to us. They can thus become supports for the ongoing process of continually refining our intuition. Theories become “aids to reflection,” as poet and philosopher Samuel Taylor Coleridge would have called them, not final codifications of truth.

There is a fine line between imagination, reality, and illusion. Wheeler writes, “Recent decades have taught us that physics is a magic window. It shows us the illusion that lies behind reality—and the reality that lies behind illusion. . . . Today we demand of physics some understanding of existence itself."[135] In his writings, Wheeler quotes Bohr, "*Deduce the quantum* from an understanding of *existence.*"[136] There appears to be a fissure in what we thought was reality, and quantum physics is the thread that is increasingly protruding through the crack that, if we have the courage to follow where it is leading us, can potentially unravel our ideas about everything. Wheeler uses the image of a rope tied in a knot in such a way that if it is pulled the knot will fall apart into nothingness.

In an interview, Wheeler said, "I continue to say that the quantum is the crack in the armor that covers the secret of existence."[137] This is big stuff indeed, particularly from the physics of the very small. Wheeler's words bring to mind the lyrics from Leonard Cohen's song "Anthem," "There is a crack in everything, that's how the light gets in." There is a sense throughout Wheeler's writings that he recognizes that quantum physics is revealing something previously undreamed of.

Wheeler writes that "it is undeniable that each of us, as observer, is also *one* of the participators in bringing 'reality' into being."[138] Through our acts of observation we are shaping reality while simultaneously being shaped by it. To quote Wheeler, "The observations of all participators, past, present and future, join together to define what we call 'reality.'"[139]

Is there a subtle form of teleology embedded in the observer's role in the quantum world? In other words, are we being shown something, are we being led to a new way of seeing our world which will change not only ourselves, but the way our world manifests to us? It is as if the physical world comes into being so that consciousness can become aware of its own nature.

This is reminiscent of the French Jesuit theologian Pierre Teilhard de Chardin's idea that the universe is directed towards a goal that he called the "Omega Point." He conceived of this as an evolutionary impulse within humanity, as well as within the universe itself, that was ascending towards consciousness, resulting in an individuated consciousness directly (re)cognizing its own nature. At de Chardin's Omega Point there is a direct, nonconceptual comprehension of the ground of "being" by the fundamental cognizant aspect of the ground of "being itself," with human beings as the instruments through which this realization occurs. At the Omega Point our true nature recognizes, comprehends, and illumines itself. From the atemporal point of view, we are already at the Omega Point, and what is happening in our world is the footprint of this realization projected backwards in time. This is to say that the events playing out in linear time are the very vehicle through which the Omega Point realizes itself through us—provided, of course, that we recognize that this is the case.

Nobel laureate Ilya Prigogine echoes Wheeler's perspective when he says, "Whatever we call reality, it is revealed to us only through an active construction in which we participate."[140] Whatever reality is, one thing is beyond doubt—it is not classical but quantum mechanical. Einstein envisioned our universe as a great eternal riddle. Are the insights of quantum physics providing the clue that will help us to solve this cosmic riddle and lead us to a previously undreamed of treasure just waiting to be discovered? Wheeler comments, "Except it be that observership brings the universe into being what other way is there to understand that clue?"[141] It is as if the universe itself is conspiring with us to help us awaken to its (and our) nature, and quantum physics is the theoretical and experimental "instrument" for this deeper insight to reveal itself. According to a remark attributed to the ever-optimistic Wheeler, "Somewhere something incredible is waiting to happen."[142]

In an interview, Wheeler points out, "One of the conditions, I think, for advance in this field, as in any field, is believing that advance is possible."[143] If, on the other hand, we become entrenched in a pessimistic viewpoint, the dreamlike universe will simply supply all of the needed evidence to prove our viewpoint in a self-created and self-fulfilling prophecy. In this case, we are unwittingly investing in and helping to create the very negative events that we are using as evidence to prove the rightness of our pessimistic outlook. We are then using our own genius for cocreating reality in a perverse way that not only doesn't serve us, but is potentially destroying us.

If we are asleep to the dreamlike nature of the universe, it is easy to fall into what Bohm calls "self-deception" (Buddha called this "avidya," i.e., "not seeing," "ignorance") through which, by misunderstanding the nature of ourselves, the universe, and our place in it, we unwittingly and unnecessarily enslave ourselves. By failing to comprehend the nature of our situation we lacerate and cripple both the universe and ourselves, erroneously misidentifying with a false, separate self that is seemingly disconnected from the whole.

If we view the physics community as an individual and quantum physics as its dream, it is as if physicists have "dreamed up" quantum theory in all its glory as a compensation for our intellectual one-sidedness, as a way of showing us our blind spot and reflecting back to us our unfounded unconscious assumptions. As crazy as it might sound, quantum physics, with all of its seeming absurdity, is revealing a deeper order of nature that transcends the bias of the predominant scientific worldview. It is a medicine for the overly materialistic madness we've succumbed to.

Wheeler refers to the old mechanistic viewpoint of the universe as a machine that goes its own inexorable and deterministic way as a "cracked paradigm."[144] It is as if the mind informing the field of the old physics has become "cracked," or lost touch with reality. Our old worldview, in Philip K. Dick's words, "is shabby and cracking apart and fading away."[145] The pre-quantum view of the world being analogous to a huge machine is not correct; the underlying structure of matter is not mechanical. To quote Bohm, "This means that the term 'quantum mechanics' is very much a misnomer. It should, perhaps, be called 'quantum nonmechanics.'"[146] What we construe to be reality is the medium through which the deeper mysterious dimension of the quantum registers and makes itself available to us. Something beyond what we know of as "reality" impinges upon what we think of as reality—what Dick refers to as "a perturbation in the reality field."

Leaving consciousness completely out of our map of the universe is an egregious error that is as absurd as it is insane and tragic. It is the product of a mind clearly in a deep state of unconsciousness and delusion, suffering from a severe dissociative disorder. Once consciousness is factored into the equation of our quantum universe it is the classical picture of a mechanical universe that starts looking bizarre. Something far saner—which from our becoming conditioned to the "cracked paradigm" might look insane—is emerging through the cracks in the mechanistic, materialistic worldview. Quantum physics points at and is an expression of our universe being more like a conscious organism than a robotic machine.

Wheeler felt that when we finally discover what's at the bottom of what's really going on in our universe, we won't find any sort of "glittering central mechanism" at all. Commenting on what he imagines we might find, Wheeler says, "Not machinery but magic may be the better description of the treasure that is waiting."[147] Some of the founders of quantum theory realized they had tapped into what could be accurately described as a quasi-magical process. To quote Wheeler, "There is some magic in this universe of ours."[148] Wheeler dedicates his autobiography not only to his teachers, students, and colleagues, but to "the still unknown person(s) who will further illuminate the magic of this strange and beautiful world of ours by discovering How come the quantum?"[149] In his writings, he makes the point that the magic inherent in the universe by itself isn't sufficient, but rather, what is needed is "magic plus the prepared mind."[150]

It's as if the magic inherent in our universe wanted to reveal itself through the vehicle of hard science and *abracadabra* the result is quantum physics. Wheeler, wondering about the universe in his inimitable style, asks, "Is IT all just a Magic Show?"[151] Seen as a symbol crystallizing out of the dreamlike nature of reality, quantum physics is revealing to us that we don't live in the mechanistic, Cartesian world of classical physics but rather inhabit an enchanted world that is linked to our mind's creative imagination.

IDEAS

Great physics can't be deduced strictly from experimental data; a certain aesthetic sense of the universe and its beauty is required. Beauty is an inspiration for and generator of ideas. This brings to mind Dostoevsky's saying in *The Brothers Karamazov*, "The world will be saved by beauty." Physicists have to develop a certain "taste" for the beautiful. In his book *Across the Frontiers*, Heisenberg cites the Latin motto, "Beauty is the splendor of truth," which expresses the idea that we first recognize truth by its splendor, by the way it shines forth. Heisenberg cites a definition of beauty that stems from Plotinus, "Beauty is the translucence, through the material phenomenon, of the eternal splendor of the 'one.'"[152] It is truly beautiful when the deeper unified field reveals itself through the physical world. Speaking about "beauty," Heisenberg writes, "Let us declare that in exact science, no less than in the arts, it is the most important source of illumination and clarity."[153]

As Pauli told us, all understanding is a protracted affair, inaugurated by processes in the unconscious long before the content of consciousness can be rationally formulated. Speaking of the joy of understanding, Pauli writes that "becoming acquainted with new knowledge, seems therefore to rest upon a correspondence, a coming into congruence of preexistent internal images of the human psyche with external objects and their behavior."[154] It is as if in recognizing this correlation between the external event and the internal image, we remember something we once knew long ago.

Heisenberg writes, "The apprehension of Ideas by the human mind is more an artistic intuiting, a half-conscious intimation. . . . The central Idea is that of the Beautiful and the Good, in which the divine becomes visible and at sight of which the wings of the soul begin to grow."[155] Plato in the *Phaedrus* speaks of the soul remembering something it had unconsciously possessed all along. Plato expresses the idea that the soul becomes awestruck and shudders at the sight of the beautiful, feeling something evoked deep within it that, although catalyzed by the outer senses, has its origins deep in the subterranean realms of the unconscious.

Many of Wheeler's speculations were what he called "an idea for an idea";[156] he was of the opinion that burgeoning researchers need "freedom to experiment with ideas."[157] The conceptual tension that arises between conflicting ideas can potentially become the source of creative insight. Wheeler comments, "Progress in science owes more to the clash of ideas than the steady accumulation of facts."[158] In his writings, Wheeler cites Bohr, who was of the opinion that there is no hope in making any progress in one's subject unless one is confronted with a difficulty or "paradox" (what Jung considers to be one of our most valuable spiritual possessions). If one can find a second difficulty, then one is in luck, as out of this creative tension one can then play off one difficulty with the other and begin to move ahead. To quote Wheeler, "Sad the week without a paradox, a difficulty, an apparent contradiction! For how can one then make progress?"[159]

Jung writes that "only the paradox comes anywhere near to comprehending the fullness of life. Non-ambiguity and non-contradiction are one-sided and thus unsuited to express the incomprehensible."[160] Oftentimes, apparent contradictions have their source in the inadequacy of language itself. When we encounter a paradox, it is a signal that our accepted assumptions and concepts about things are breaking down, indicating that new ideas and new ways of looking at the world are being called for. The paradox does not tell; it points. It is a sign, not the thing pointed to. It is important to not avoid paradoxes and contradictions in order to maintain conceptual consistency; shrinking away from paradoxes comes at the expense of completeness. Rather, if we are able to embrace the paradoxes and hold the tension of seemingly opposed and contradictory perspectives, we are preparing the space for the next step to creatively reveal itself.[161] Currently we are at a transition point, as two contradictory worldviews, the classical and the quantum, are encountering each other. This is not just in physics, but also deep within the human psyche, as it is ultimately the psyche from which all our physics is derived.

It is the leaving behind of commonly agreed-upon truths that have been "outgrown" and shown to be wrong that helps to propel science and human civilization forward. It is as if we have to come up with a new idea regarding how to deal with "reality." Wheeler felt that at the bottom of our universe was not a mathematical equation, but rather a simple idea. In Wheeler's opinion, "the most revolutionary discovery in science is yet to come! And come, not by questioning the quantum, but by uncovering that utterly simple idea that demands the quantum."[162] It is Wheeler's opinion that until we arrive at this basic idea underlying the quantum, we have not understood the essence of the quantum principle. Wheeler felt that everything important is, at bottom, utterly simple. He felt that if we really understood the central point which quantum theory is revealing, "one ought to be able to state it in one clear, simple sentence. Until we see the quantum principle with this simplicity we can well believe that we do not know the first thing about the universe, about ourselves, and about our place in the universe."[163]

For physicists, the description in plain language will be a criterion of the degree of understanding that has been reached. Einstein remarked that our physical theories "ought to lend themselves to so simple a description that even a child could understand them."[164] In a similar vein, Lederman says, "If the basic idea is too complicated to fit on a T-shirt, it's probably wrong."[165] Heisenberg often cited the Latin motto, "The simple is the seal of the true." Wheeler once proposed a challenge—describe quantum mechanics in five words or fewer. Physicist Sean Carroll's response: "Don't look: waves. Look: particles."[166]

What is this "utterly simple idea" that Wheeler is positing that demands a quantum world? Could it have to do with the dreamlike nature of reality, a perspective which embraces the role of consciousness in creating our world? How would we essentialize this idea to fit on a T-shirt, a wristband, a bumper sticker, or a tweet?

Jung writes, "He alone is a philosopher who can transmute a vision born of nature into an abstract idea, thereby translating it into a universally valid language. . . . It is a primordial idea that grows up quite as naturally in the philosopher and is simply part of the common property of mankind, in which, in principle, everyone has a share. The golden apples drop from the same tree, whether they be gathered by a locksmith's apprentice or by a Schopenhauer."[167] The figure that Jung is referring to as a "philosopher" (which can also be called an artist, poet, or a scientist) is someone who is able to translate their unique vision into a communicable idea. This ability is not just available to the credentialed or privileged few, but can be accessed by anyone with sufficient receptivity and attunement. The quantum is a primordial idea that is literally born out of nature herself into the minds of philosophers . . . and physicists . . . and maybe you and me.

Wheeler felt that there are some ideas out there that are waiting to be discovered. It is as though some ideas are in the air, pervading the underlying field of the collective unconscious, just waiting to be tuned into and received. Some ideas suggest themselves, particularly when their time has come. In Wheeler's words, "Ideas can't be locked away in vaults."[168] Thank goodness for this. The time is approaching, and maybe it's already here, for certain formerly radical ideas to be entertained. Revolutionary ideas have the potential to catalyze revolutions, after all.

Philip K. Dick, in writing about "ideas," considers it "quite important" that "in a certain literal sense ideas are alive." He asks the question, "What does this mean, to say that an idea or a thought is literally alive? And that it seizes on men here and there and makes use of them to actualize itself into the stream of human history?"[169] In these questions, Dick is turning our conception of things on their head. Do we think ideas, or do they think us?

As is evidenced throughout history, human beings can become possessed by ideas, which can be either a good or a bad thing, depending on the idea. Jung writes, "Greater than all physical dangers are the tremendous effects of delusional ideas."[170] The notion of an objectively existing universe qualifies as such a delusional idea with its concomitant danger.

Speaking about a novel idea, Dick continues, "He [the person who "had" the idea] did not invent it or even find it; in a very real sense it found him. And—and this is a little frightening to contemplate—he has not invented it, but on the contrary, it invented him. It is as if the idea created him for its purposes."[171] So when the time has come for an idea to emerge into our world, the idea—which according to Dick's conception has a life of its own—enlists a receptive mind to be its purveyor. This line of thinking makes Dick wonder whether the cosmos is one vast entity that thinks. The cosmos, Dick comments, "may in fact do nothing *but* think."[172] Maybe the fact that we are thinking creatures is a reflection that the cosmos itself thinks.

Etymologically the word "idea" has to do with a way of seeing, a perspective through which we view the world. As Feynman says, "There is no authority who decides what is a good idea."[173] One mark of a good idea in theoretical physics is the fun, excitement, and new pathways of creative thought that it generates. If the idea opens up new perspectives on things, enabling novel work to extend in different, previously unimagined directions, then it is probably a good idea. Wheeler writes, "New ideas must correspond to old ones, must include them, but must transcend them!"[174]

Unfortunately, we seem to have fallen into a kind of anti-intellectualism, an intellectual dark age in which many people are frightened of new ideas. In societies with a totalitarian bent (which, many people argue, the US is fast approaching), new ideas can even be criminalized, a process that can become internalized within our minds. In the novel *1984* George Orwell uses the power of the fictive imagination to describe a society in which thought itself was controlled. In Orwell's world children were taught to use a simplified form of English called Newspeak in order to assure that they could never express ideas that were dangerous to the "prevailing order" of society. Orwell's prescient work of imagination could be seen to be descriptive of the very world we live in. From the point of view of the powers that be, the very idea of the quantum, due to its empowering and liberating nature, is precisely such a "dangerous" idea. We as a species are desperately in need, however, of a saving idea. The idea of the quantum perfectly fits the bill.

Speaking about a new idea (what he refers to as a "conceptual dislocation"), Dick writes, "it must invade his mind and wake it up to the possibility of something he had not up to then thought of." In what turns out to be a true collaboration between author and reader, Dick continues that a new idea "sets off a chain-reaction of ramification-ideas in the mind of the reader; it so-to-speak unlocks the reader's mind so that that mind, like the author's, begins to create."[175] A new idea (which oftentimes is said to "fall into our head"; the question arises—from where?) can wake us up to what is possible and, like a key, can unlock the latent creative spirit, a treasure that has been hidden deep within the recesses of our mind, a spirit that has been thirsting to be set free. Maybe the quantum is such an idea entering our mind-stream at this very moment.[176]

Dick writes, "An odd aspect of these rare, extraordinary ideas that puzzles me is their mystifying cloak of—shall I say—the obvious."[177] In other words, some of the most mind-blowing ideas are so obvious that they are staring us in the face, and this is why no one notices them. A judicious questioning of the obvious may well be a mark of genius. Wheeler writes, "Surely someday, we can believe, we will grasp the central idea of it all as so simple, so beautiful, so compelling that we will all say to each other, 'Oh, how could it have been otherwise! How could we all have been so blind so long?'"[178]

Our idea of ourselves—who we think we are—is a primary driving force in human affairs, as who we imagine we are and how we think that we fit into the greater scope of the universe powers the major currents of world history. In pre-quantum, classical physics human beings were conceived of as isolated, impotent beings in a mindless, mechanical universe. The revelations in quantum physics are pointing out that—through our consciousness—we are all integral participants in nature's ongoing process of creation. Instead of being cogs in a giant machine that operates like clockwork, we are creative mental hubs in an ever-evolving and infinitely interconnected network of ideas. Wheeler has described the greatest physicists—and I'm sure he would include everyone in this—as "instruments in an orchestra of ideas,"[179] the "orchestra of ideas" being a reference to the universe.

Classical physics' shallow conception of humanity is one of the main causes of today's growing economic, ecological, social, and moral problems, which obstruct the full flowering of our creative potential. Oftentimes a shift in a single idea can precipitate a transition into a new epoch. Could it be that the most important impending development in science will be ideological (in the realm of ideas) rather than technological, involving a profound re-visioning of science's conception of who we are and our place in the universe? Wheeler felt that finding "a deeper conceptual foundation [ideological] from which we can *derive* quantum theory . . . rather than the experimental [technological] side"[180] was where the greatest hope for progress in quantum physics lay. What quantum physics has unleashed in the realm of technology is the palest reflection of what it can potentially unleash within the human psyche.

IMAGINATION

Max Planck, considered the founding father of quantum theory, writes, "New ideas are not generated by deduction, but by an artistically creative imagination."[181] Whereas Cape Kennedy is the launching pad for sending rockets to outer space, the unfettered creative imagination is the launching pad for the greatest revelations in physics. Not merely breaking the earth's atmosphere, imagination—in Wheeler's words, having "an imaginative eye"—can take us beyond space and time. Freeman Dyson comments, "The Glory of Science is to imagine more than we can prove."[182] Imagine an imagination gone hog wild without having to prove anything to anyone—this is the starting point for the best of what science has to offer.

Regarding the laws of physics that govern the behavior of the quantum world, Einstein is of the opinion that there is "no logical path to these laws," and that "only intuition, resting on sympathetic understanding of experience, can reach them."[183] Speaking about the behavior of quantum entities, Feynman says, "We have to learn about them in a sort of abstract or imaginative fashion and not by connection with our direct human experience."[184] On the one hand, there is nothing like the behavior of the subatomic world in our world and yet, at the same time, the quantum realm is mirroring something deep within us, reflecting back to us something within ourselves. The quantum realm is extremely close to us, closer than close in fact, both unfamiliar and intimately familiar at the same time.

Quantum physics could not have been predicted beforehand or be derived from the previously existing classical physics. Wheeler counsels, "Don't try to derive quantum theory. Treat it as supplied free of charge from on high."[185] In this comment Wheeler is pointing at the revelatory aspect of quantum theory. Great scientific discoveries often come, to quote the renowned physicist Baron Carl Friedrich von Weizsäcker, "as an inspiration or a special gift of grace which comes to the researcher when and as it pleases, like the answer from 'another authority,'" whose origin, he continues, is not from the ego, but from "a more comprehensive self."[186] This higher authority uses the creative imagination as its instrument of revealing itself.

Wheeler personally felt that it was the vividness of his imagination that enabled him to entertain the outlandish notions he was regularly known for, ideas that other more traditional physicists have great reluctance in even considering. Likewise, when asked why he "does" quantum mechanics, Feynman replied, "Because it is a great adventure in imagination."[187] One of the things that makes science so difficult is that it takes a lot of imagination. In physics, Feynman comments, "Our imagination is stretched to the utmost, not as in fiction, to imagine things which are not really there, but just to comprehend those things which are there."[188] In other words, creative imagination is itself an instrument to access reality.[189] Jung writes, "I am indeed convinced that creative imagination is the only primordial phenomenon accessible to us, the real Ground of the psyche, the only immediate reality."[190] In any case, there is a deep and intimate correlation between what we call reality and the reality-creating function of the creative imagination.

Our personal imagination is coextensive with and an iteration of the divine creative imagination. But if it is still entrained by the prevailing classical worldview, it imagines a solid, objectively existing universe that we then imagine has nothing to do with our imagination. This is to say that via our creative imagination we can conjure up an experience of a very convincing physical reality that appears to be utterly independent of our imagination. Ironically, the very imagination whose reality-shaping creative power we are denying happens to be the source of our experience; it is as if the creative genius of our imagination is, through its boundless power to create our experience of reality, denying its own existence. We are then wielding our sacred creative imagination to deny a fundamental aspect of our being, and then conjuring up evidence to support our case.

Wheeler writes that "the pursuit of science is more than the pursuit of understanding. It is driven by the creative urge, the urge to construct a vision, a map, a picture of the world that gives the world a little more beauty and coherence than it had before. Somewhere in the child this urge is born."[191] Hopefully our innate creative urge—the natural impulse within us to imagine and/or create a more beautiful world—is not eradicated by our family, our educational system, or our society before we become capable of bringing our new ideas and visions to fruition. Wheeler elaborates, "Anyone who expects to create, be it as a scientist or artist, scholar or writer, needs self-confidence, even bravado. How else can one dare to imagine understanding what no one else has understood, discovering what no one else has discovered? Where does this confidence come from? Fortunately, every young person is blessed with some of it. It is part of human character."[192]

To quote philosopher John Dewey, "Every great advance in science has issued from a new audacity of imagination."[193] The basis of all of the mind-blowing technology in the world today arose from people who had the courage and daring to imagine; they had what Einstein refers to as "a courageous scientific imagination." Let us bravely follow in their footsteps.

THE LAWS OF PHYSICS

The purpose of physics has always been seen as a search for the fundamental laws of the universe. The idea of the laws of physics, which underpin the entire scientific enterprise, began as a way of formalizing patterns, regularities, and relationships in nature that connect physical events. Reflecting their relationship to the apparently objectively existing universe, many physicists began thinking of the laws as being "real" and existing independently, "out there in the world," separate from their own minds, which resulted in the laws themselves—as compared to the events they describe—becoming promoted to the status of reality. Before the advent of quantum physics, the laws of physics were conceived of as objectively existing in a transcendental realm, lording over lowly matter. It was as if, within physicists' minds, the laws had taken on and assumed a life of their own, a life that paradoxically, in their mental conception, had nothing to do with their minds.

Quantum physics has raised the question: Is the ever-evolving universe like a work of art in progress, making up its laws as it goes along? Are the laws of physics an emergent property of the cosmos, which itself is emergent? Commenting on what quantum physics tells us about the laws of physics, Wheeler famously opined, "There is no law except the law that there is no law."[194] The "immutable" laws of physics are, according to Wheeler, "anything but." This is to say that the laws of physics are malleable, mutating in tune with the universe they support—in Wheeler's words, "of higgledy-piggledy origin"[195]—in the same way that living organisms mutate. To quote Wheeler, "Every law can be transcended."[196]

How can we believe that the laws of physics are eternal if the universe itself is not going to be around forever? To quote Wheeler, "Law cannot stand engraved on a tablet of stone for all eternity . . . all is mutable."[197] Most people have been conditioned to think of the laws of physics as being written in stone, inscribed somewhere in our universe by the universe itself like the Ten Commandments, but Wheeler is pointing out that this is not the case. When asked by physicist Paul Davies what he considered his most important achievement, Wheeler answered, "Mutability!" He meant that nothing is absolute, nothing is so fundamental that it cannot change under certain circumstances, and this includes the very laws of the universe. "Mutability," Wheeler writes, "is a law of nature."[198] This is to say that anything—and everything—is possible.

The laws and the physical universe they describe can only exist together, reciprocally co-arising and in-forming each other. It's meaningless to talk about the laws of physics when the material reality in which these laws are enacted hadn't even come into existence yet. The laws of physics don't exist in some transcendent Platonic realm outside the universe, but cooperatively come into being coextensive with the birth of the universe. The idea that the laws that inform the functioning of reality spring into manifestation out of nothingness fully formed is a nonsensical, preposterous idea. Wheeler comments, "The laws of physics were not installed in advance by a Swiss watchmaker, nor can they endure from everlasting to everlasting. They must have come into being. They could not always have been accurate. They are derivative and superficial, not primary and revelatory."[199] Since we live in a quantum world its laws share in its quantum nature, which is to suggest that they can never be completely known, and that they—like our universe itself—are a continually unfolding work in progress.

The "flexi-laws" advocated by Wheeler evolve and focus in on precisely the forms needed to give rise to the living organisms that eventually observe them. Is observership the ultimate underpinning of the laws of physics, and therefore of the laws of space and time themselves? Quantum theory implies that observer-participants create both the physical laws and the appearance of the material world in which the laws apply. In our questioning about the nature of the universe and its laws, to quote Wheeler, "Could it be that the quantum is trying to tell us the answer?"[200]

In Einstein's opinion, "Everyone who is seriously involved in the pursuit of science becomes convinced that a spirit is manifest in the laws of the Universe."[201] Is science just a modern, secular way of pursuing spirit? In trying to find some deeper structure that underlies the laws of physics, quantum physics is reflecting back to us that it is a mistake to think that as we penetrate to the universe's deeper levels it will terminate at some *nth* level, or that it goes on ad infinitum. Rather, as quantum physics reveals, our inquiry necessarily leads back full circle to the observer with which it began.

In our inquiry we find ourselves in a strange loop—what professor Douglas Hofstadter refers to as a "tangled hierarchy"—in which by moving either upwards or downwards through the levels of some hierarchical system, we unexpectedly find ourselves right back where we started. In quantum physics, the conventional one-way, top-down hierarchy dissolves, revealing instead a holarchical order where every part of the system is the beginning and/or the end, the alpha and/or omega. It is in the ethereal act of observership—and therefore consciousness—where the link that closes the circuit of interdependence between us and our world is to be found.

• CHAPTER FOUR •

THE OBSERVER

The universe and the observer exist as a pair, existing in a timeless embrace. The central and all-encompassing role of the observer in quantum mechanics, what Wheeler refers to as the "magic ingredient,"[202] is the most important clue we have regarding the construction of the universe. Wheeler asks, "Is the architecture of existence such that only through 'observership' does the universe have a way to come into being?"[203] According to Wheeler, the universe is a self-referential "strange loop"[204] in which physics gives rise to observers, who then give rise to information, which in turn gives rise to physics. The universe gives rise to meaning-establishing observer-participants who, in developing the ideas of quantum mechanics, grant a meaningful existence to the universe. To quote Wheeler, "We could not even imagine a universe that did not somewhere and for some stretch of time contain observers because the very building materials of the universe are these acts of observer-participancy. You wouldn't have the stuff out of which to build the universe otherwise."[205] In other words, a universe which doesn't contain observers is unimaginable.

According to quantum theory, observers play no minor part, but rather an indispensably creative role in the genesis of the universe while at the same time being a product of the very universe that they are helping to create. The construction of the universe is such that the observer is as essential to the creation of the universe as the universe is essential to the creation of the observer. The question presents itself: Who or what made the observations necessary to create the observers? In other words, if observers create reality, where do the observers come from?

The question naturally arises—what constitutes an observer? This is one of the central and most burning philosophical questions in quantum physics. In addition to humans, what about a cat, a mouse, a cockroach, an amoeba, or a piece of mica? How does their observation differ from human observership? Are there observers that help shape the universe that do not have physical bodies? The spiritual traditions of the world are replete with accounts of just such kinds of observers, after all. If so, is there a cooperative collaboration between these disembodied observers and the more fully embodied ones in which all the parties involved are cocreating how reality unfolds? Where does consciousness first enter in the elaborate hierarchy of terrestrial life? Where does the capacity to collapse a wave function derive from? Does it come from the presence of consciousness or from some other condition? It's as if observership and its ability to translate unmanifest possibilities into definite actualities is a pervasive feature which is widely distributed throughout the web of life. In any one observation, the entire universe is in some way implicated and participating. From this perspective all life-forms are dreaming together, collectively collapsing the universal wave function of this universe to manifest the way it is moment by moment.

It becomes extraordinarily difficult to state sharply and clearly where the community of observer-participants begins and where it ends; the boundary between the two is continually shifting. Bohr used the famous parable of a blind man with a cane to illuminate the epistemology of the measurement problem, i.e., the difficulty of demarcating the measuring instrument from the object to be measured. Trying to find his bearings on a sidewalk with the aid of the cane, if the man holds the cane tightly, he feels *with* it, feeling as if the tip of the cane were an extension of the self. Becoming part of the subject, the cane becomes an extended arm. For the blind man, the world begins at the cane's tip. But if he holds the cane loosely, then he is more likely to feel the cane as an object outside of himself. In other words, the cane can be either subject or object, depending upon how the man relates to it. The cane can be related to in two ways, but never in both ways at the same time.

Something similar applies to our ideas, concepts, and view of the world. When we view the universe through them, they are part of us as subject; when we examine and think about them, they are objects. Furthermore, imagine extending the man's sense of touch beyond the tip of the cane onto the sidewalk. Where does the blind man end and the world begin? Similarly with a quantum entity such as an atom, each time we attempt to observe it, we become linked to it so that we can no longer say which is us and which is the atom. We can no longer say where we end and where the atom begins, as we and the atom that we are observing become, through our act of observing, part of an indivisible, mutually co-arising system. In this single system it makes no sense, and in fact introduces inaccuracies and misleading distortions, to continue to describe our interaction with the atom as an encounter of two separate systems.

Schrödinger felt that the abrupt change that takes place as a result of measurement was the most interesting point of the entire theory. He writes, "In general, a variable has no definite value before I measure it; then measuring it does *not* mean ascertaining the value that it *has*. But then what does it mean?"[206] Schrödinger is pointing to how the formalism of quantum theory is not only unable to describe what happens during the process of measurement, but it is unable to give an account of what we even mean by measurement. Wheeler felt that measurement, which is an act of choice among possible outcomes, was "the true essence of quantum mechanics."[207] The measurement problem, however, of which the act of observation is key, is no ordinary problem.

The idea of observer-participancy implies that the universe has built into it from the very beginning the potentiality for giving birth to and housing observers. Without observers there is no existence; in Wheeler's words, "There would be nothing rather than something."[208] Philosopher Martin Heidegger considered the question "Why is there something rather than nothing?" to be *the* most fundamental and important question for philosophy. Noted physicist Stephen Hawking has even openly wondered why the universe goes to all the bother of existence at all. Hawking's and Heidegger's questions are another way of asking what is the meaning of life. It sounds like the edge between physics and philosophy is becoming harder to distinguish. In any case, the universe creates the conditions and paves the way for the emergence of the very observers that bestow upon it a certain reality, completing the transaction that allows the stars to shine, so to speak. In a world without a built-in purpose, quantum theory "promotes" the observer to the definer and experiencer of reality as well as its generator of meaning, which is essentially a creator of distinctions, a primordially creative role.

However we view it, we can't get around the fact that we are actively participating in creating our experience of the universe.[209] Wheeler says, "We are inescapably involved in bringing about that which appears to be happening."[210] Not only are we involved in bringing about what seems to be happening, we are intimately involved in creating our experience of ourselves as well. To quote from the wonderful book *Quantum Enigma: Physics Encounters Consciousness* by physicists Bruce Rosenblum and Fred Kuttner, "If our observation creates *everything*, including ourselves, we are dealing with a concept that is logically self-referential—and mind-boggling."[211] As Hofstadter would say, we are one strange loop.

Are physics' insights into the participatory character of the universe, with all of its yet to be realized implications, just the beginning of a profound shift that is truly evolutionary? In his journal Wheeler poetically describes the "quantum principle as tiny tip of giant iceberg as umbilicus of the world." What we currently comprehend of the revelation that is quantum physics—and the corresponding technology that we have made as a result of our insights—is but the "tiny tip of giant iceberg," with the overwhelming majority of the iceberg underwater (in the unconscious), yet to be uncovered. To quote Freeman Dyson, "Fortunately, the ideas of Wheeler give us a basis for believing that the world of physics may be truly inexhaustible . . . there will always be new worlds to explore, new effects of observer-participation reacting back upon the laws of physics, new connections for the Einsteins and John Wheelers of the future to speculate upon."[212]

ART

In modern times, all of the other "hard" sciences, from chemistry to astronomy to biology, are anchored to physics in such a way that physics has been called the "king" of the sciences. Interestingly, Leonardo da Vinci claimed, "Art is the Queen of all sciences." Being a form of insight, physics is a form of art; both disciplines are investigations into the nature of reality. Today physics and art are revealing their interconnectedness such that we are participating in some sort of royal alchemical marriage—a conjoining of the opposites—involving the symbolic coming together of king and queen. Art teaches us a new way to *see* the world; physics formulates a new way to *think* about the world. Maybe "seeing" and "thinking" aren't that different.

Due to the creative nature of scientific theories, Bohm refers to these theories as "art forms." Though quantum physics is a manifestation of the hardest of hard science, the boundary between quantum theory being science or a uniquely creative art form becomes blurred. Are the most creative quantum theorists scientists, artists, dreamers, or an admixture of all three? Bohm was of the opinion that "this division of art and science was temporary. It didn't exist in the past, and there's no reason why it should go on in the future."[213] As an art form, quantum physics is reflecting back to us the part of ourselves that is a creator of our experience. Einstein once remarked, "The greatest scientists are artists as well."[214]

Heisenberg writes that "the history of the development of physics appears not unlike the history of other intellectual fields, for instance the history of art; for even in these other fields the concern is ultimately with no other goal than that of illuminating the world, even if it be the world within us."[215] Oftentimes breakthroughs in both art and physics happen simultaneously, as if artists and physicists were tapping into the same nonlocal realization but expressing and symbolizing their insights differently through their seemingly disparate mediums. For example, some of the most stunning examples of "groundbreaking" art in Western history were made at the turn of the twentieth century, which was the same time that the ground broke open (leaving no ground to stand on) in the field of physics via the theories of relativity and quantum physics. Abstract painting appeared on the scene at the same time that a deeper dimension of abstract thinking emerged in the field of physics. Are these merely coincidences? Or were artists and physicists plugging into something similar, but from different perspectives?

In his book *Art and Physics: Parallel Visions in Space, Time, and Light* Leonard Shlain presents the thesis that revolutionary art often anticipates visionary physics. Artists can be picking up, pointing at, preparing for, and helping to bring into actualization changes that are already taking place in the collective unconscious of humanity. The role of the creative individual is to reveal what is lying dormant in the unconscious. Oftentimes artists give voice to what others only dream. It is common for art historians, however, to emphasize past styles that have influenced the work of artists. Not as often is it pointed out that their work might be presciently anticipating the future of other disciplines such as physics, as though artists were living oracles divining what is yet to come—the canaries in the coal mine of humanity, so to speak. To quote Heisenberg, "In the middle of the nineteenth century, when the growth of mechanical engineering began, there were few indeed—mostly artists, painters and poets—who could sense the first invasion of extremely dangerous, demonic forces into human life."[216] It was the creative and artistically inclined individuals who were intuiting that the genie was beginning to be let out of the bottle, and could be used either for the benefit or detriment of humanity.

Both art and physics attempt to speak about matters that do not yet have words. This is why their languages are so poorly understood by people outside their fields. Radical innovations in art embody preverbal stages of new ideas—and hence, new ways of looking at the world—that could eventually transform a civilization. Whether as an infant, adult, or collectively (as a planetary culture at large) a new way to think about reality begins with the contemplation and assimilation of new images (interestingly, the word "imagine" literally means "to make an image"). Over time these new images can help us see through our unconscious, unquestioned assumptions and begin to develop a new language that is descriptive of our new vision. This new language is art. Shlain opens his book on art and physics by quoting artist and social critic James Baldwin, "The purpose of art is to lay bare the questions that have been hidden by the answers."[217] I imagine John Wheeler would have approved.

LOGIC

Quantum physics simultaneously stretches, boggles, blows, and melts our minds, especially those aspects of our minds that have become conditioned by the classical linear worldview. As we take in and digest what quantum physics is showing us about the universe and our place in it, it psycho-energetically alters, expands, and refreshes our mind. To contemplate—or to coin a new phrase, "quantumplate"—the world of the quantum is like going to the psychospiritual gym in that it changes our "shape." Becoming "quantum-physicized," we learn to think directly and naturally in quantum mechanical language and logic. The concepts of quantum physics are in themselves not difficult; it is our classically-conditioned minds that make them so.

The starting premise through which we view our world automatically deploys a certain type of logic consistent with its viewpoint. If we think that phenomena exist independently as separate parts, having their own intrinsic nature, we then develop and live within a logic based upon that perspective. On the other hand, if we recognize the quantum nature of reality, we develop the aforementioned "dream logic"—a logic in which multiple contradictory things can be true simultaneously—to reflect back the dreamlike quality of our situation. Wheeler went so far as to speculate that neither space nor time nor matter but rather logic is "the nuts and bolts, if you will, out of which the world is made."[218]

Quantum physics is riddled with paradoxes to its core. Thinking "quantum-logically," we must be able to hold paradox in a new way; we can cultivate our ability to appreciate—rather than solve so as to get rid of—"fruitful ambiguities," which opens up all kinds of new possibilities. The apparent paradoxical nature of quantum reality cannot be resolved within the framework of the standard Aristotelian, two-valued logic which is basic to Western analytical thought. Western civilization has been hypnotized by this limited form of "yes/no" logic where things are either true or false, exist or don't exist. Aristotelian logic deals with certainties, thereby subliminally programming us to invent fictitious certainties in a world that is riddled with uncertainty.

Many of the seeming paradoxes of quantum physics are themselves a direct function or artifact of the intrinsic limitations built into the nature of a mutually exclusive, binary, two-valued logic. Having a definite utility, two-valued logic works by contrast, giving attributes to things and making distinctions, thereby limiting them. Something is "this" only by defining it as not "that." Our very language itself, in categorizing things and ideas, conditions us into a dualistic, two-valued, logical way of thinking. The axiomatic set through which we view the world and its logic conditions and shapes our minds and thus affects the state of consciousness we inhabit. To get insight into the non-ordinary reality of the quantum world, we have to introduce a higher form of logic in order to wrap our minds around what we are dealing with. Interestingly, logic has been described as the science of thinking correctly.

Instead of needing one or the other viewpoint to be true, in a radically new form of logic[219]—called "four-valued logic"[220] in Buddhism (also known as "paralogic")—we can hold seemingly contradictory statements together as both being true simultaneously. This higher form of logic is characterized not by the two-valued logic of either/or, but by the four-valued logic of both/and, where things can be true and false at the same time.[221] Two-valued logic is based on the law of the excluded middle in which things are either (1) true or (2) false. By contrast, four-valued logic includes the middle and the ends surrounding it, so that things are (1) true, (2) false, (3) both true and false, and/or (4) neither true nor false.[222]

Which of these four choices seems truer than the others at any given moment depends on what perspective we choose to view things from. (This is analogous to how the results of an experiment depends upon the choice that a physicist makes regarding the experimental arrangement.) Our personal realities are perspective-based through and through. This can be a limiting and blinding mental straitjacket or a path to greater degrees of freedom, depending, as always, upon our perspective. This is to say that the perspective which takes perspectives into account is potentially on the road to liberation from being bound by any one perspective.

Four-valued logic is the logic of interdependence, unlimited wholeness, and the unity of all things. Quantum reality requires that our either-or thinking be replaced by a more nuanced, layered, and fruitful integration of surface and depth, inside and outside, the part and the whole, the root and its branch. The alternatives offered by four-valued logic represent all the possible standpoints from which every problem can be viewed; instead of there being only two extremes (yes or no), we have an infinite spectrum of choices between the extremes. This logic gives new insight into how what may appear to be contradictions at one level can be part of a deeper consistency and completeness from a higher, more inclusive level.

Truly subversive, four-valued logic undermines our ability to hold on to any fixed position whatsoever. By rejecting any one view as well as the ultimate truth of all views, four-valued logic is in essence rejecting the competence of standard Aristotelian reason to comprehend the fundamental nature of reality. We are unable to conceptually understand four-valued logic, however, with a mind that has been conditioned to think with two-valued logic. Four-valued logic helps us to begin to get a sense of what we are dealing with in our encounter with the quantum realm.

Four-valued logic points to and introduces us to a direct experience of reality beyond the straitjacketing conditioning of the bifurcating, duality-creating perspective of two-valued logic. Overcoming the arbitrary confines of the rational mind, four-valued logic deconstructs the conditioned mind into its natural state of seeing holistically. It literally changes the awareness of the mind to allow for a new and expanded understanding of reality, allowing the mind to transcend its own habitual grasping of reality. Seeing the world through four-valued logic gives us greater degrees of freedom of choice. Expressing this higher form of logic, Pauli comments, "It is my personal opinion that in the science of the future reality will neither be 'psychic' nor 'physical' but somehow both and somehow neither."[223]

Wheeler stressed that as we develop more of a capacity to consciously hold paradox, new insights will often emerge. Each new generation will take into itself, learn, and integrate quantum physics' worldview with its quantum dream logic more easily than their teachers learned it. As time goes on, each generation grows more detached from ingrained pre-quantum images of the world, lessening the resistance to be broken down to the seemingly radical worldview of quantum theory. Each person's initiation into this new way of thinking and seeing the world nonlocally affects the whole, transforming the collective unconscious of humanity itself. Feeling more at home with quantum ideas, its theory will be accepted by students from the beginning as a simple and natural way of thinking, because we shall all have grown used to it. This is to say that the newly emerging worldview of quantum physics grows on us with time.

Richard Conn Henry, professor of astronomy and physics at John Hopkins University, writes, "The world is quantum mechanical: we must learn to perceive it as such."[224] In other words, we must learn how to correctly perceive the nature of the world we live in. Henry continues, "One benefit of switching humanity to a correct perception of the world is the resulting joy of discovering the mental nature of the Universe. We have no idea what this mental nature implies, but—the great thing is—it is true."[225] The "mental nature" of the universe is another way of describing the dreamlike nature of reality.

One of the benefits of seeing the quantum nature of the universe that can't be argued with is the feeling of joy, which according to Buddhism, is one of the foundations and expressions of waking up to our true nature. Henry's conclusion: "The universe is immaterial—mental and spiritual. Live, and enjoy."[226] This brings to mind what physicist Victor Weisskopf tells his students when they get depressed by the state of the world: "There are two things that make my life worth living: Mozart and quantum mechanics."[227] Recognizing the quantum mechanical nature of the world is truly uplifting, a natural antidepressant.

As semanticist Alfred Korzybski noted in the 1930s, if all people learned to think in the non-Aristotelian manner of quantum theory, the world would change so radically that most of what we call stupidity and even a great deal of what we consider insanity might disappear, and the intractable problems of war, poverty, and injustice would suddenly seem a great deal closer to solution. Once we catch up with and integrate what science has discovered about our place in the universe, quantum physics will become the lens through which we view our experience, as its simplicity and obviousness will seem utterly natural. We will wonder how we could have been so blind for so long. Or so I imagine. . . .

• CHAPTER FIVE •

SELF-EXCITED CIRCUIT

Wheeler's vision of the universe, to use a metaphor from electronics, is that it is like a self-excited circuit.[228] To say the universe is *self*-excited is to say it is not *other*-excited, which is to say that rather than depending upon an external agent like a god or deity, the universe is self-creating and self-referential. Thus it is able to refer to and reflect and act upon itself and, as a result, endlessly re-create itself anew.[229] This is similar to how in Buddhism our true nature is described as self-excitatory intelligence, likened to the self-originating heart of the sun, or the unceasing main channel of a great river. We are, according to Wheeler, the universe looking at—and simultaneously creating—itself.

We live, in Wheeler's words, in "a self-observing universe,"[230] where we are the instruments through which the universe becomes aware of its creative nature. The question then becomes: How do we hold up a mirror to ourselves when we ourselves are the mirror? For we are simultaneously the mirror, the light it reflects, and the eyes that see the reflection. Everything is part of one unified quantum system with no separation to be found anywhere.

The universe is self-excited in the sense that the observation comes from within the universe, not from without. In Wheeler's words, "The universe is a grand synthesis, putting itself together all the time as a whole."[231] In a sense the parent to itself, the universe is the cause and effect of its own existence. Seen as a self-excited and self-actualizing circuit, the physical universe bootstraps itself into existence, laws and all. In his personal journal Wheeler writes, "We form what forms us."

For him, reality was a kind of Mobius strip, like Escher's hands drawing themselves. This brings to mind the ancient philosophical definition of God as "the uncaused cause."[232]

As a self-excited circuit, the universe gives rise to observers who, in completing the circuit, potentially give meaningful reality to the universe. Cosmologically speaking, conventional science assumes a linear, logical sequence: cosmos gives rise to life, which gives rise to mind. Including both a linear and circular sequencing of events in his vision of the universe, Wheeler closed this chain into a loop: cosmos gives rise to life, which gives rise to mind, which in turn gives rise to the cosmos. As observers, we look back to both observe and create the very universe that bore us. Wheeler says, "The universe is to be compared to a circuit self-excited in this sense, that the universe gives birth to consciousness, and consciousness gives meaning to the universe."[233]

In our act of observing the universe, we are actualizing the universe. Since we are part of the universe, this makes the universe, as well as ourselves, self-actualizing. This is similar to the nature of the psyche. The life of the psyche arises out of organic life, while at the same time the biologically based human psyche transcends its origins through its own self-creation—a process which takes place at an immaterial level of consciousness. The psyche has the unique quality of creating itself through its own activity. Is this a mirror of the quantum nature of the universe?

The emergence of consciousness in the universe is as epic and epochal an event in cosmic history as the first big blast of its materialization in the supposed big bang. The self-excitation is caused by the innate fundamental tendency for consciousness's self-referential self-perception—i.e., for illuminating itself in ever new ways with the light of consciousness—built into the very ground of being. In this process of radiant self-cognition, the universe is able to turn back upon itself so as to explore its nature via its various life-forms as it endlessly creates and re-creates itself through innumerable acts of observer-participation. The universe generates an interactive feedback loop of cosmic intelligence within itself that becomes the internal guidance system and source of its own continually unfolding genesis. Contrary to the mechanistic worldview of classical physics, the universe as a self-excited circuit implies a participatory universe that endlessly creates itself through innumerable acts of participatory self-perception.[234] To quote Wheeler, "Directly opposite to the concept of universe as machine built on law is the vision of *a world self-synthesized.* On this view, the notes struck out on a piano by the observer participants of all times and all places, bits though they are in and by themselves, constitute the great wide world of space and time and things."[235]

In such a self-referential cosmology, the nature of which is a self-generating feedback loop of pure creativity, we are dreaming up the universe while at the same time the universe is reciprocally dreaming us up, as the seemingly subjective and objective realities interblend and cocreate each other. The world and consciousness are intermingled in such a way that they mutually co-arise in a deeper unified sphere of being. It is impossible to say which initially caused the other, as their relationship has no beginning in time. Their relationship is reciprocal—now one side and then the other acts as a cause. This realization of the acausal and synchronic co-arising of the world and consciousness "orients" us towards the universe, as well as ourselves, in a whole new way that opens up vast realms and domains of possibilities that were simply not available to us while operating from a more linear and deterministic worldview. Quantum physics' new perspective can potentially liberate our mind from the mental straitjacket of the classical mechanistic worldview, revealing new creative orders of freedom that we can step into at any moment—even right now!

Uncountable small acts of observer-participancy have over eons built up the tangible appearance of the material world. As observers, there is no getting around the fact that each of us are participants in bringing reality into being. In his personal journal Wheeler writes, "I can't make something out of nothing, and you can't, but altogether we can." The universe is a collectively shared dream that is too seemingly dense and solidified for any one person's change in perspective to transform, but when a critical mass of people get into alignment and consciously put together what I call our "sacred power of dreaming" (our innate power to dream the universe into materialization), we can, literally, change the (waking) dream we are having.

As agents of cosmic evolution, we are being invited to contribute to the growing edge of the universe's creative unfoldment into uncharted territory. This is truly evolution in action, as we discover that we can actively participate in our own evolution, and in fact are being called to do so. We become (or maybe we always have been but just didn't know it) a channel for the universe to autopoietically re-create itself in a novel and evolutionary way. Or maybe I'm just dreaming.

TIME

"Self-excitatory," to quote Wheeler, "the universe is a grand synthesis, putting itself together all the time as a whole. Its history is not a history as we usually conceive history. It is not one thing happening after another after another. It is a totality in which what happens 'now' gives reality to what happens 'then,' perhaps even determines what happened then."[236] In the quantum realm, the "time ordering of events" is without meaning, a mistaken way of speaking. Wheeler writes, "There is no such thing as spacetime in the real world of quantum physics. . . . With time gone the very idea of 'before' and 'after' also lose their meaning."[237] There is no linear time in the original quantum formalism. There is no explanation of how one moment becomes the next, because there is only one moment in quantum physics—the moment of observation.

To quote Einstein, "People like us, who believe in physics, know that the distinction between past, present, and future is only a stubbornly persistent illusion."[238] It is an unexamined assumption that time can be defined entirely in terms of the behavior of clocks. The notion of time as a linear sequence of moments is a deeply ingrained assumption symptomatic to our classically-conditioned way of thinking; it is easy to fall under the false imagination that a timeline of moments from the past to the future actually exists and that the present moment is just a single point on this long timeline.[239] We are always at the moment of observation—which is the singular present moment, the only moment that has ever existed—where all the action of creation takes place. There are no other lived moments except in our imagination, which is the very place that the notion of time is conceived. Summing up the current worldview emerging from the new physics, Wheeler simply says, "Time today is in trouble."[240]

The notion of time emerging from quantum physics has a distinct teleological feel to it. In Wheeler's terms, it is "teleology without teleology." As a self-excited universe, time gets turned on its head—it's not just the past causing the present, but the future causing the present too. This enlarges the idea of linear causality in that it implies that the effect is not only produced by a cause, but that the effect can also precede the cause. It is as if enfolded within the present moment are not only traces of the past that are informing it but also the potentiality of the future that is influencing it as well.

No moment exists on its own. The present can be envisioned as a seed in which the past and future creatively intermingle. Out of the present sprouts the results of the past while concurrently being drawn into the possibilities inherent in the future, which could be said to be *causing* the present moment as much as the past does. The present moment is like a magical doorway in which the past and future come together in a radial matrix via their mutual in-forming of the present. In talking about the past and the future, it is important not to reify them, since this makes it easy to forget that they are just constructs made up by the mind. It is an error to think of the past or future as distinct from the present moment, as the idea of past or future is never experienced separate from the present moment. Stepping and relaxing into the present moment (where we all are right now) and releasing our habitual and chronic contraction against it, spontaneously introduces us to our true creative nature.

This idea, called "retrocausality," implies a symmetrical treatment of time in which both past and future events can play a role in informing the present moment to happen the way it does. Such a perspective collapses our notion of sequential time as always flowing in one direction, that is, from the past to the future, as it allows causal movement in two directions simultaneously. The present moment—the point where our power to shape reality is to be found—is the place where the "handshake" completing this transaction between the past and the future happens.[241] Wheeler comments, "There is no more remarkable feature of this quantum world than the strange coupling it brings about between future and past."[242]

To quote Einstein, "Time is not at all what it seems. It does not flow in only one direction, and the future exists simultaneously with the past."[243] From the point of view that takes linear time as unassailably true, Einstein's words make no sense. Quantum physics requires us to step into a radically new way of thinking about time. Wheeler writes, "Heaven did not hand down the word 'time.' Man invented it."[244]

Time is a human-made construct built so as to order our experience. To quote physicist N. David Mermin, "Clocks do not measure some preexisting thing called 'time'; our concept of time is simply a convenient way to abstract the common behavior of all those objects we call 'clocks.'"[245]

Time in itself does not exist in a fundamental, absolute sense as an entity or external parameter that can be measured in a way that physical things can be measured in the third dimension, but is an abstraction constructed by the mind; time is a mental construct. Time in and of itself cannot be found in the three-dimensional material world; rather, it is to be found within the domain of our minds. Time is a qualitative aspect of our consciousness that we pretend to quantify and measure with clocks. Our erroneous linear conception of time infects both our thinking and our language, which is to say it deranges our mind, not to mention our global civilization as a whole (as is evidenced all around us). As Wheeler writes, "TIME is anything but simple."[246] He asks, "Is everything—including time—built from nothingness by acts of observer-participancy?"[247]

In a situation where the psyche and the quantum realm seem to be mirroring each other, Jung says, "We cannot apply our notion of time to the unconscious. Our consciousness can conceive of things only in temporal succession, our time is, therefore, essentially linked to the chronological sequence. In the unconscious this is different, because there everything lies together, so to speak . . . we are left standing in the shadow cast by a future, of which we still know nothing, but which is already somehow anticipated by the unconscious."[248] It is as if all possible universes exist in potential in the unconscious of humanity, and which one will actualize itself into form depends on how we engage with the unconscious, in other words, how we dream it.

Compare Jung's idea that in the unconscious "everything lies together," to physicist Louis de Broglie's words: "Each observer, as his time passes, discovers, so to speak, new slices of space-time which appear to him as successive aspects of the material world, though in reality the ensemble of events constituting space-time exist prior to his knowledge of them."[249] The underlying space-time continuum of the plenum, or the collective unconscious of humanity (depending upon if we are talking as a physicist or psychologist), is like a cauldron, an overflowingly full maternal womb out of which all events in our world—be they big or small, inner or outer—exist in a state of potential, preparing themselves to enter manifestation as living experiences. The arena in which these potential events are actualized is the interfacing of our third-dimensional universe of space and time with our minds.

Contemplating the whole fabric of space-time, Wheeler, noting that we are not limited to moving in space in only one direction, wonders, "Why do we seem to move through time as if it were a one-way street? Why does time have an arrow? Why do we remember the past but not the future?"[250] As Lewis Carroll writes, "It's a poor sort of memory that only works backwards!" Sometimes Wheeler felt that an idea seemed "so crazy that it just might be right." Contemplating the direction of time, he wrote, "One direction might be quite likely, the other direction incredibly, ridiculously unlikely. Yet unlikely is not quite the same as impossible."[251] An infinitesimally small or "nonzero" probability is radically different than something that is impossible; we should be very careful what we assign to the trash bin of the impossible. In his typical "far-out" way, Wheeler wonders whether built into the fabric of our minds is the capacity "in which the old past is remembered less well than the immediate future."[252]

Regarding the so-called past, Wheeler writes, "Ah, but 'what has already happened' is not so easy to say."[253] Talking about one of the most startling features of a radical thought experiment based on a cosmic version of the famous double-slit experiment that he dreamed up called the "delayed-choice experiment" (which has since been empirically verified), the act of observation, Wheeler says, "reaches back into the past in apparent opposition to the normal order of time."[254] Wheeler's thought experiment used the imagination creatively in order to tease out a little more information from nature. He discovered that "a choice made in the here-and-now has irretrievable consequences for what one has the right to say about what has *already* happened in the very earliest days of the universe, long before there was any life on Earth."[255] This is to say that acts of observer-participancy in this moment give tangible "reality" to the universe not only now but back at its beginning. In an interview, Wheeler expressed this idea very simply, "We are participators in bringing into being not only the near and here but the far away and long ago."[256] This is not far-out science fiction but hard science that is actually stranger than fiction.

To quote cosmologist and astrophysicist Martin Rees, "In the beginning there were only probabilities. The universe could only come into existence if someone observed it—it does not matter that the observers turned up several billion years later. The universe exists because we are aware of it."[257] Though what the delayed-choice experiment is revealing to us seems utterly paradoxical, this is only because we are still unconsciously caught in a pre-quantum viewpoint. There is no paradox in the delayed-choice experiment if we let go of the idea that there is a fixed and independent physical world that exists even when we are not observing it.

Wheeler comments, "It is wrong to think of that past as 'already existing' in all detail. The 'past' is theory. The past has no existence except as it is recorded in the present. By deciding what questions our quantum registering equipment shall put in the present we have an undeniable choice in what we have the right to say about the past."[258] In other words, the questions we ask today, in Wheeler's words, form "an inseparable part of a phenomenon that in earlier thinking one would have said had 'already happened.'"[259] The idea that some event has "already happened" is a remnant of our classically-conditioned minds. From the point of view of quantum physics, the seemingly past event that we think of as already having happened has not really happened until it is observed, whenever that occurs. The act of observation that helps to actualize the past only takes place in the present, which is to say that the present moment in a very real sense encompasses and is inseparable from the past. Wheeler points out that the question "What was the reality of, for example, the *photon*[260] before it is observed?" is an "utterly meaningless" question. Our observation in the here and now has an undeniable impact in bringing about that which appears to have happened. The past is never finished once and for all.

Classical physics describes the present as having a particular past. Quantum physics, on the other hand, because of its probabilistic nature, enlarges the arena of human history such that the past is an amalgam of all possible pasts compatible with the version of the present moment we are currently experiencing. The quantum universe is polyhistorical—the past involves a wide range of possible pasts all coexisting in a state of unmanifest potential. Wheeler writes that "there is no unique history that one can ascribe to the universe. Instead there is a certain probability of this, that, and the other history of the universe."[261] The act of observation collapses the wave function[262] (a mathematical construct that describes all of the system's possible states) in such a way so as to evoke a particular universe in the present moment while simultaneously reaching backwards in time to create a history appropriate with our present moment experience. To quote physicists Stephen Hawking and Leonard Mlodinow, coauthors of *The Grand Design*, "We create history by our observations, rather than history creating us."[263]

There is no way to say unambiguously what the past was really like until we know its future in the form of the present; as in a work of art, each part of the universe acquires its full meaning only in its relation to the whole. Hawking and Mlodinow write, "Quantum physics tells us that no matter how thorough our observation of the present, the (unobserved) past, like the future, is indefinite and exists only as a spectrum of possibilities. . . . The universe, according to quantum physics, has no single past, or history. . . . The fact that the past takes no definite form means that observations you make on a system in the present affect its past."[264] In any case, in quantum physics it certainly appears "as if" an observation made in the present moment reaches back and influences the past. Wheeler writes, "Our very act of measurement not only revealed the nature of the photon's history on its way to us, but in some sense *determined* that history. The past history of the universe has no more validity than is assigned by the measurements we make—now!"[265] Through our observations in this moment, Wheeler writes, "we decide what the photon *shall have done* after it has *already* done it."[266]

The connection between the observer and the observed not only cannot be separated in space, but has no distinction in time as well. This perspective turns our conception of linear time and causality on its head. To quote author Graham Smetham, "The entire universe appears to be a kind of collective delayed-choice experiment in which inhabiting sentient beings somehow determine the manifested nature of the universe even backwards in time!"[267] This introduces a self-referential circularity in which the laws of quantum physics can allow for their own self-modification backwards in time. The implication is that as observers we are participants in the genesis of the universe, a process that Wheeler calls "genesis by observership."[268] The moment of the world's creation lies in the present, in the eternal now, with us somehow playing a "starring" role. Wheeler writes, "We don't understand genesis and we never will until we rise to an outlook that transcends time."[269]

Classical physics was of the opinion that if it had sufficient information, it could predict how the universe would unfold in the future. After the discoveries of quantum physics, Bohr, doing his best Woody Allen impersonation, allegedly said, "It is difficult making predictions, especially about the future."[270] With the advent of quantum physics, however, the whole of science had suffered a dimly perceived yet drastic, fundamental, and sobering change. In quantum physics, to quote Wheeler, "prediction comes to the end of the road."[271] Speaking of science, physicist Banesh Hoffmann, an associate of Einstein, writes, "Its proudest boast, its most cherished illusion had been taken away from it. It had suddenly grown old and wise. It had at last realized it never had possessed the ability to predict the detailed future."[272] To predict the future, we must fully know the present, and the present is not fully knowable. For not only, as Heisenberg pointed out, is there always an element of uncertainty in trying to ascertain complete knowledge of any system, but in trying to know whatever present universe we inhabit we inevitably alter it. This results in liberating us from the deterministic straitjacket—the linchpin of classical physics—that seemed to have been constraining us. According to quantum theory, the best way to predict the future is to invent it ourselves. Welcome to the world of quantum physics!

COSMOGENESIS

Without an observer, the universe wouldn't exist (after all, who would it exist *for*?), let alone evolve, which is to say it might as well be dead. Quantum theory reflects back to us, to again quote Wheeler, "that the universe would be nothing without observership as surely as a motor would be dead without electricity."[273] In the act of observation the physical reality of the world becomes actualized. In this self-generating, self-referential feedback loop, it is the physical world that generates observers who are responsible for bestowing seemingly tangible reality to its existence. To quote Stapp, humans weren't brought into the universe to passively observe it but to "contribute to the actual unfolding of the actual."

From an atom's point of view, it would be an impoverished universe without physicists since physicists are the atom's way of knowing about itself. This is similar to how light, out of its thirst to be seen, creates an eye in order to simultaneously illumine and be illumined. The eye, being solar (light-based) in nature, owes its existence to light. Two lights brighten our world—the light provided by the sun, and the light of the human eye as it realizes the sun's radiance and responds to it. Only through their interaction do we see; lacking either, we are blind. It is not that one moment the photon of light leaves the sun and after a certain amount of time enters the eye; rather eye-consciousness and light are intimately, irreducibly linked in such a way that they lose their separate distinction within an overall quantum process. The creative and immaterial aspect of light has called forth and precipitated out of itself, in fully materialized form, an organ similar to itself so as to reveal itself, be seen, and known. Similarly, physicists—macroemanations of the subatomic world—are the reflecting mirror for atoms to experience and become aware of their own nature. It's as if atoms wished to know about themselves and dreamed up physicists (which from the atoms' point of view are vast, self-reflective ensembles of themselves) to accomplish this.

The observer-participant is both a result of an evolutionary process and, in some sense, the cause of its own emergence. Wheeler wonders, "Is observership the 'electricity' that powers genesis?"[274] In other words, mind-boggling as it is to contemplate, are we, as "observer-participants" playing a role in the genesis of the cosmos in this very moment? According to Wheeler, "It is incontrovertible that the observer is participator in genesis . . . it is difficult to see any other line that lends itself to exploration. What other way of genesis is there?"[275] Wheeler is reflecting that we play a role in the creation of the universe that has been normally reserved for the "gods." Wheeler delighted in thinking about the conceptual frontiers of cosmology, what he referred to as the "flaming ramparts of the world." His far-out speculations might be more easily dismissed if it weren't for his unassailable credentials and his countless important technical contributions to the field of physics. He has earned the right to have his (oftentimes mind-stretching) opinions be taken seriously.

In contemplating these practically theological questions, Philip K. Dick writes, "We are all but cells in a colossal mad brain that both makes and perceives reality . . . there is some relationship between the creating of reality and perceiving of it . . . the percipient is cosmogenitor, or, conversely, the cosmogenitor wound up as an unwilling percipient of its own creation."[276] To the extent we are unaware that the reality we are perceiving is the very reality we are creating (think of an ordinary, nonlucid dream) we are deluded—actually deluding ourselves. We are then but individual cells in a "colossal mad brain" whereby we are contributing, through our delusion and lack of awareness of what we are doing,[277] to the greater collective psychosis that has taken over the body politic of our species. Psychosis is defined as a condition in which there is a profound loss of contact with external reality. How else other than a newly recognized form of psychosis would we describe a situation in which someone believes there is an actually existing external reality when in fact there is no such thing? It should be pointed out that the nature of delusions is that the delusional always believe they are nondelusional, while believing that the nondelusional are the ones who are deluded.

I can only imagine what it must have been like for the founders of quantum physics to stumble upon the quantum realm. They must have felt like explorers from a faraway land coming across something completely unknown and mysterious. Wheeler uses the example of someone seeing an automobile for the first time. Conjecturing on what it is like to encounter this mysterious phenomenon, Wheeler writes that thoughts arise such as, "It is obviously meant for use, and an important use, but what use?"[278] In his example, the automobile is the quantum: One opens the door, cranks the window up and down, flashes the lights on and off, perhaps even turns over the starter, all the while without knowing what it's really for. Similarly, we use the quantum in a transistor to control machinery, in a molecule to design an anesthetic, in a superconductor to make a magnet. All are great advances that we are using to our advantage, but are we missing the main idea? Wheeler asks, "Could it be that all the time we have been missing the central point, the use of the quantum phenomenon in the construction of the universe itself? We have turned over the starter. We haven't got the engine going."[279]

This is similar to if Copernicus used his equations to calculate planetary positions but missed the main point, i.e., that the Earth goes around the Sun. This brings to mind Einstein's assertion that what is most basic in physics is not the mathematics but rather the underlying concepts. In quantum physics words cannot replace mathematics, and mathematics cannot replace words; both are needed for a full understanding. Until we understand the central idea that is offered to us by the quantum, to quote Wheeler, "we have not met the challenge that is right there."[280] Could it be that quantum physics' revelation is so profoundly simple and obvious that in its light the previously held ideas of science would seem as laughable as the doctrine that the earth is flat?

In talking about the quantum, Wheeler evokes the creation of the Hawaiian Islands, which were built on the fertility of volcanic ash and decomposed lava, literally formed by "the moving fire beneath the surface." Each stage of our understanding of the quantum he likens to a further eruption. He writes, "So the quantum, fiery creative force of modern physics, has burst forth in eruption after eruption and for all we know the next may be greatest of all."[281] Is the eruption after eruption into physics of the quantum the doorway into deepening our understanding of the very architecture and engineering of the creation of the universe itself?

Wheeler refers to quantum phenomena as untouchable, indivisible "elementary acts of creation"[282] that reach into the present from billions of years in the past, while at the same time reaching into the distant past from the present moment, thereby realizing the particular universe we find ourselves inhabiting. Wheeler views these elementary acts of creation as the building material of all that is.

Out of all the signs that testify to quantum phenomena as being the elementary acts of creation, Wheeler considers their *untouchability* to be their most striking feature. We have come no closer than before to penetrating to the untouchable interior of the quantum phenomenon. This can be equated to how a finger can't touch itself, an eyeball (which is what is doing the seeing) can't spin around and see itself as an object (without a mirror), or a flashlight can't shine on itself—all of these examples only lead to an infinite regress. The experience of touch is related to the idea of substantiality. Physicist, mathematician, and astronomer Sir James Jeans writes, "For substantiality is a purely mental concept measuring the direct effect of objects on our sense of touch."[283]

This brings to mind Buddhism's notion of "emptiness," which is the intrinsic nature of all of the seemingly substantial forms of our universe, but yet is similarly untouchable. This is analogous to how a mirage of water in the desert doesn't make the sand wet. Similarly, the empty nature of the quantum realm is beyond—transcendent to—the merely physical dimension of the universe, even though it is what "makes up" the physical. Quantum physics reveals that the seemingly solid and substantial physical world is actually composed of something that is not physical at all, but rather mental.

Wheeler openly wonders, "Are billions upon billions of acts of observer-participancy the foundation of everything?"[284] The very fact that we can ask, in Wheeler's words, "such a strange question"[285] shows how far we are from understanding the deeper foundations of the quantum and its ultimate implications. Wheeler is openly wondering whether "billions upon billions of acts of observer-participancy" by innumerable beings over countless eons is the very quantum process which has created our world, literally dreaming our world into materialization in this very moment. Wheeler ponders whether the very term "big bang" is merely a shorthand way to describe the cumulative effects of these billions upon billions of acts of observer-participancy.[286] In his textbook on Quantum Theory, Wheeler asks, "Beyond particles, beyond fields of force, beyond geometry, beyond space and time themselves, is the ultimate constituent, the still more ethereal act of observer-participancy?"[287] Are we, through our ongoing acts of observing the universe, the ultimate constituent which makes the whole universe?

Regarding how the universe came into being, Wheeler asks, "Is the mechanism that came into play one which all the time shows itself?"[288] Is enfolded within our seemingly simple present-moment experience the primordial creative act which reflects the genesis of the entire cosmos? Does the mystery of the world's ongoing creation lie in the present moment, in the eternal now? Is the key to the mystery of existence hiding in plain sight, encoded within the simplicity and naturalness of our ordinary everyday awareness, within and as our own mind?

Wheeler continues, "For a process of creation that can and does operate anywhere, that reveals itself and yet hides itself, what could one have dreamed up out of pure imagination more magic—and fitting—than this?"[289] What more "fitting" physics could we have, in Wheeler's words, "dreamed up" out of pure imagination to reflect back to us the "magic" of our dreamlike world? A process which itself is an expression of the dreamlike nature, we have "dreamed up" quantum physics to reflect the dreamlike nature of the universe back to us. In trying to understand nature, as if by magic, physics is helping us rediscover our nature.

Reflecting upon the importance of asking the right question, Wheeler writes, "We ask today, 'How did the universe come into being?' realizing full well that how properly to ask the question is also a part of the question."[290] We live in a universe that is capable not only of harboring life, but of cultivating life which is intelligent enough to wonder and ask about its origins. Why do we think that the universe isn't imbued with conscious intelligence when it gives birth to conscious intelligences? The fact that consciousness is an emergent property of the universe implies, given the underlying wholeness of the universe, that consciousness is one of the intrinsic ingredients of the universe. It is as though the universe is curious about itself, and through its intrinsic sentience, is capable of reflecting upon itself. As if living in a dreaming universe, in this process of self-reflection, the universe, just like ourselves, has the capability of becoming a lucid dreamer. Through us, the universe questions itself and tries out various answers on itself in an effort parallel to our own to decipher its own being. In the process of observing and reflecting upon our universe we are actually changing the universe's idea of itself.

In his textbook on quantum theory, Wheeler tells the following story:

> An old legend describes a dialogue between Abraham and Jehovah. Jehovah chides Abraham, "You would not even exist if it were not for me!" "Yes, Lord, that I know," Abraham replies, "but also You would not be known if it were not for me."
>
> In our time the participants in the dialogue have changed. They are the universe and man.[291]

In Wheeler's vision, humanity is the cipher through which the universe—and in the old legend, God—becomes aware of itself. To quote scientist Carl Sagan, "We are a way for the universe to know itself."[292] In an imaginary dialogue, Wheeler replies to the universe, "Yes, oh universe, without you I would not have been able to come into being. . . . [And yet] you could never even exist without elementary acts of registration [i.e., observation] such as mine."[293] Compare this with the words of theologian Meister Eckhart, "Man cannot live without God, but God cannot live without man either. Without man, God wouldn't know he existed."[294] It should get our attention that one of the greatest scientists of the twentieth century is saying things that sound just like one of the most eminent mystic visionaries from centuries past. In expressing the point of view that we play a key, participatory role regarding the coming into being of the universe, Wheeler is in the good company of wisdom-holders the world over.

In becoming conscious, we become the instruments through which whatever we call "it"—God, the universe, the creator, the universal mind, etc.—becomes aware of itself. Speaking about the divine service that humanity can render, Jung writes in his autobiography it is so "that light may emerge from the darkness, that the Creator may become conscious of His creation, and man conscious of himself."[295] Every advance, every conceptual achievement of humanity has always been connected to an expansion of self-awareness. This expansion of self-awareness is necessarily an expansion of the universe's awareness of itself. Wheeler comments, "And then at last an inspiration: a feeling that we who felt ourselves so small amidst it all are, in the end, the carriers of the central jewel, the flashing purpose that lights up the whole dark universe."[296] His words bring to mind Jung's similar reflection that "man is indispensable for the completion of creation; that, in fact, he himself is the second creator of the world."[297]

• CHAPTER SIX •

STRANGER THAN FICTION

To quote Wheeler, "If I had to produce a slogan for the search I see ahead of us, it would read like this: that we will first understand how simple the universe is when we realize how strange it is."[298] It has been said that the universe is not only stranger than we imagine, it is stranger than we can imagine. Wheeler felt that if you haven't found something strange during the day, it hasn't been much of a day. Physics has taught us that nature has a way of being a little stranger than we think it ought to be. Wheeler writes in his autobiography, "The strangeness of the quantum world, from which Einstein incessantly sought escape and from which Bohr saw no escape, is real."[299] Elsewhere, Wheeler ominously adds, "whether we like it or not." Quantum entities are incomparable, having no physical analogy to anything else; atoms behave like atoms and nothing else.

The quantum realm—the world of the really small[300] (a world, in Bohr's words, "hitherto closed to the eyes of man")—is composed of objects that are unlike any other objects we have ever imagined. Quantum entities, to quote Feynman, "behave in their own inimitable way. . . . They behave in a way that is like nothing you have seen before."[301] Subatomic objects don't exist as "things"[302] but rather, as events, as happenings, as dynamic, ever-changing, interactive psychophysical processes. In the elaborate mathematical structure of quantum theory separate objects have, in fact, no representation. Only constantly changing events have reality in the quantum realm, and these events are intimately and inextricably interconnected.

What these elementary entities *are*[303] and what they *do* are inseparably intertwined. The mystery of what they are can only be inferred from what they do.[304] The aspects of nature represented by quantum theory are converted from elements of "being" to elements of "doing," which basically replaces the world of material substances with a world populated by actions, events, and processes; ceaseless dynamic activity is the very essence of a quantum entity's being. It should be noted that people are not "things" in the same way that subatomic particles are not things.

Not located in time or space but in an abstract realm, the elementary quantum phenomenon, to quote Wheeler, "is the strangest thing in this strange world."[305] The strangeness of these subatomic entities is highlighted by our inability to even conceive of them separate from our participating in their genesis. It turns out that the elementary particles out of which matter is supposedly composed are not really "elementary" at all, but rather of a secondary, derivative nature. Rather than providing the concrete "stuff" from which the world is made, these elementary particles are essentially abstract constructions based upon and arising out of acts of observation. This is to say that our acts of observation are more elementary than the elementary particles that are evoked in the process of observation.

One of the major challenges of modern physics is to figure out how the seemingly contradictory theories of relativity and quantum mechanics can both coexist in the same universe. Wheeler writes, "In short, there is no hope in comprehending the 'big picture' unless one takes account of both relativity and quantum mechanics."[306] In addition to their insistence on looking at the world as an undivided whole, another similarity in both theories is their emphasis that perspective—our point of view and frame of reference—matters. Our viewpoint not only determines how we see things, but how things appear. This common ground between these two hard to reconcile theories is often not acknowledged. In science as well as art, our choice of frame is of fundamental importance.[307] To quote Wheeler, "No picture is a picture that does not have a frame."[308] The new physics requires the dissolution of the rigid, narrow, and limited/limiting frame of concepts through which we unconsciously were entrained to view our world. Like a lens through which the phenomenal world is filtered and interpreted, the pre-quantum frame of reference through which we tried to apprehend the essential nature of reality was unnecessarily too restrictive. Speaking about these two theories, Wheeler says, "Which will be ultimately the deeper principle time only will tell. I'm prepared to believe that the quantum principle is the deeper one."[309]

Our worldview expresses the prevailing spirit of the times we live in; the general prevalence of a worldview that still sees things mechanistically leads to and is an expression of fragmentation. How we view the world, what David Bohm calls our "self-worldview," includes our vision of ourselves relative to the world, thereby having a tremendous effect on our sense of ourselves. Even people who think they don't have a self-worldview have them tacitly, below the level of their conscious awareness. Thinking they don't have a self-worldview is itself a self-worldview. Our self-worldview has a profound effect on the individual and on society as a whole, not only physically, but also psychologically, spiritually, and ethically. It subliminally operates as a set of givens, of core axioms that comprise the focal settings through which we view our world. This mental lens filters out of our perceptions, and thereby out of our awareness, all features of the world that are not part of our self-worldview, potentially creating impoverishing blind spots in the process.

Einstein once admitted, "I thought a hundred times as much about the quantum problems as I have about general relativity theory."[310] He confessed that fifty long years of "conscious brooding" had not brought him any closer to understanding the quantum. The advent of quantum mechanics represented a far greater break with the past than had been the case with the coming of special relativity or of general relativity. The theory of relativity does not stretch human understanding or credulity in the way that quantum theory does. As compared with Einstein's theory of relativity, which seems less and less strange the more deeply we think about, the more deeply we think about quantum physics, the stranger it seems. However the universe's mind-bending strangeness is part of its charm. In science oftentimes the greatest insights are won from nature's strangest features. And yet at a certain point the universe and quantum physics' strangeness will seem utterly natural—or so I imagine. Wheeler is fond of quoting Gertrude Stein's view of modern art, "It looks strange and it looks strange and it looks very strange, and then suddenly it doesn't look strange at all and you can't understand what made it look strange in the first place."[311]

The quantum realm lacks phenomenality. To use an architectural metaphor, quantum physics has discovered that there are no fundamental "building blocks" of reality. In a quote attributed to Bohr: "There is no quantum world. There is only an abstract quantum description. It is wrong to think that the task of physics is to find out how nature is. Physics concerns what we can say about nature."[312] Quantum entities aren't real in the way we usually think of "real"—they have no independent, intrinsic existence, they don't exist "on their own" (in Buddhist terminology, they have no "svabhava," i.e., "own nature"), and they cannot be said to exist when not being observed. Heisenberg famously said, "The concept of the objective reality of the elementary particles has thus evaporated."[313] There is nothing that science is able to say about the intrinsic nature of the atom—its fundamental nature is truly inconceivable and unspeakable. To quote a sacred Vedic scripture from ancient India, "It is that which is inconceivable but by which all things are conceived."[314]

Having no well-defined boundaries, elementary particles exist in a state of open-ended potentiality, "inhabiting" (if we can even talk about location for a nonexistent object) at the same time every possible universe they could potentially manifest in. To quote Heisenberg, "But the atoms or elementary particles themselves are not real, they form a world of potentialities or possibilities rather than one of things or facts."[315] Elementary particles don't "exist" in the commonsense meaning of the word, but if physicists treat them as if they do, then they manifest as though they do and physicists get the right results in their equations. Everyone is happy, as long as no one asks what it all means.

Elementary subatomic particles are simply a construct, a convenient way of talking about what is nothing but a set of mathematical relations concerning different observations. For the modern physicist, Pauli writes, "matter" has become "an abstract *invisible reality.*"[316] Because an atom does not have an independent, preexisting reality, it is meaningless to ask, for example, what an atom really is. Atoms are only concepts physicists use to describe the behavior of their measuring instruments and the outcomes of their experiments. An idea such as an atom emerges from the interaction between the observer and the observed, mediated through the particular measuring devices used to make any specific observation. The behavior of atomic particles is something made up after the fact to "explain" and make sense of observations.

Just because we have a label for something doesn't mean that what we are labeling is real. In actuality an elementary particle is an idea in a physicist's mind that serves the purpose of giving them a way to talk about nature. The properties of microscopic objects are inferred from the behavior of the physicist's measuring apparatus, and are then treated as if they are real physical things. It is easy to mistake the model for reality and think of subatomic particles as being actually real.

In quantum physics the wave function is not a wave of material things, but rather a probability wave, a wave of possibilities. The wave that is being described is, in a sense, not of this world. The wave function is not real; it is a purely formalistic device or abstract idea. Both the wave function and the atom are essentially ideas, and outside of these ideas, both the wave function and the atom are not there. The wave function does not have any physical correlate—it is not itself an observable thing. And yet, in typical quantum style, it combines both objective and subjective elements. Speaking of the wave function, according to Heisenberg, "It introduced something standing in the middle between the idea of an event and the actual event, a strange kind of physical reality just in the middle between possibility and reality."[317] The wave function is essentially passive, in that it cannot stimulate action from within itself. It requires an agent to make a choice among its probabilities for the three-dimensional world to be formed. This is where we come in.

Quantum theory implies that immaterial factors that have more of the nature of images and ideas are the blueprint for our universe, actually in-forming and shaping the evolution of the universe as a whole. Similar to when we are inspired by images in our mind, physical reality appears to be driven by deeper primordial images and ideas. To quote Stapp, "We live in an *idea-like* world, not a matter-like world."[318] This world we live in is like an idea or thought giving itself form, or like a dream that seems unmistakably real while we are in it. The primal stuff of the quantum realm is dreamlike in character, like ideas rather than matter. Stapp continues, "The actual events in quantum theory are likewise idea-like."[319] Wheeler goes so far as to liken the universe itself to an idea. This brings to mind the opening lines of philosopher Arthur Schopenhauer's *The World as Will and Idea*, "The world is my idea."

The concept of matter has undergone a great number of changes in the history of human thinking. From the results of its experiments, quantum physics has replaced the metaphor of the universe being composed of building blocks to a universe that is composed of an interrelated network of ideas and observations. Arnold Mindell writes, "The most elementary substance of the physical world is dreamlike."[320] In the quantum world, there is no "place" for matter, in the same way that in the classical world there is no "place" for mind. Classical physics' theory of a world of matter is converted by quantum physics into a theory of the relationship between matter and mind. Unveiling a great mystery, quantum physics is pointing out that the ultimate nature of the universe is more mind-like than matter-like. The "matter" of this world seems more akin to the phenomena of dreams rather than that of a solid, independent reality.

As quantum physics has lifted the veil to our understanding the connection between mind and matter (and hence of consciousness), it can't help but deepen our insight into the nature and operations of our own being at the same time. "Sciences made by man," as Pauli writes in a letter to Jung, "will always contain *statements about man*."[321] We can't give a consistent description of the universe without also describing ourselves. We are inside of and part of the universe we are contemplating. What happens inside of us is not only happening inside the universe, but—as if the universe has gotten turned inside out—outside in the universe too. Recognizing the synchronistic correlation between the inner and outer—a chief feature of the phenomena of spiritual awakening—is to enter the realm of the quantum where there is ultimately no separation between things at all. Could it be that quantum physics, in leading scientists to rigorously and rationally understand the permeable and ultimately artificial boundary between the outer physical world and the inner mental subjective world of ideas and imagination, is part of a larger process of the spiritual awakening of our species?

DEPENDENT CO-ARISING

To quote the Dalai Lama, "The world is made up of a network of complex interactions. We cannot speak of the reality of a discrete entity outside the context of its range of interrelations with its environment and other phenomena."[322] Things forge their identity in relation to other things; nothing stands alone. Quantum entities exist relationally with other interdependent quantum objects that themselves don't exist as separate things but rather in relation to other interrelated quantum objects ad infinitum. This is to say that there is no independent, objectively existing quantum object that has a reality in and of itself; there is solely the quantum field. "The field," as Einstein famously said, "is the only reality."[323]

The quantum field is the repository of this infinitely interdependent interconnectedness that characterizes and in-forms all conditional manifestation. For example, think of being in a plane high enough above the ocean so that when you look down you aren't able to see the individual waves, but rather you only are able to see the white crests on the waves of an agitated ocean. Though these crests might appear to be isolated from each other—each a seeming independent entity—they are all conjoined in and expressions of one giant movement of the whole. The whole in this case is comprised not only of the entire ocean, but the atmosphere, the wind, the earth, and when we get right down to it, the state of the whole universe in that very moment (all of which, when seen together, make up one indivisible quantum system). Each of the seemingly separate movements of these crests is correlated to each other in a web of seemingly infinite interdependence. In this example, the ocean is the stand-in for the underlying field that connects, contains, and in-forms all of its seemingly separate manifestations.

Quantum physics is pointing out that the basis of the visible world doesn't rest on a material foundation, but rather on a realm of immaterial forms, as though our world is afloat on an invisible ocean. Whereas in classical physics the fields were interpreted as real, physically existing entities spread throughout space, quantum fields are immaterial information waves of probability, in other words, waves of pure possibility.

Subatomic particles have no existence or meaning in isolation, but only in relationship with everything else. At its most fundamental, elemental level matter is an irreducible network of interactions; at bottom it is indivisible. To quote Heisenberg, "By getting to smaller and smaller units, we do not come to fundamental units, or indivisible units, but we *do* come to a point where division has no meaning."[324] Nature, at the atomic level, is not composed of fundamental building blocks, instead it is a dynamic network of relations with no separate parts in this interconnected web. Whatever we call a "part" is merely a pattern that has some temporary stability and therefore has caught our attention. There is only the whole. Thing-ness has dissolved into a state of "no-thingness," a web of mutual interactivity with no fixed reference point to be found anywhere. This lack of "thingness" is equivalent to discovering that we are a verb (a dynamic process), not a noun (a static thing).

The Buddhist sage Nagarjuna[325] expresses this same state of "(no)-things" when he says, "Things derive their being and nature by mutual dependence and are nothing in themselves."[326] In Buddhism, this state of "no-thingness" or emptiness is characterized by a process known as "dependent co-arising" (also called interdependent co-origination),[327] which is considered to be the very condition of and process by which empirical reality is constituted. In dependent co-arising, we are dreaming up the universe, but in a circular, nonlinear, and acausal process that exists outside of time, the universe is dreaming us up to dream it up, ad infinitum and vice versa. In other words, every part of the universe evokes and is concurrently evoked by every other part in a seamless expression of undivided wholeness. This simultaneity of cause and effect, of "creating" the seemingly outer world while at the same time being "created" by it, is a manifestation of the fundamental correspondence and ultimate indivisibility of the inner and the outer. The synchronistic simultaneity of dependent co-arising is an expression of the inseparability between the universe and ourselves. To quote Nagarjuna, "Something that is not dependently arisen, such a thing does not exist."[328]

Dependent co-arising is not something created by Buddha, rather it is something that is continually being rediscovered. Dependent co-arising is not a belief or a theory to which one assents, but an insight that one is invited to experience and encouraged to win through focused and disciplined inquiry. This view is not a final affirmation about reality, as it doesn't seek to define a reality external to the observer, instead it is a way of seeing that focuses on how our experience of the world and ourselves arises. Oftentimes referred to as the "great noble truth," it is considered the true heart of Buddhism and the key to understanding the Buddha's realization. When asked by what authority he spoke, the Buddha would always cite the law of dependent co-arising; not any supernatural entity ruling our world, but the dynamics at work within our world as well as our minds. Similar to quantum physics, dependent co-arising is considered to be a milestone in human thought; its ever-deepening realization is one of the greatest and furthest reaching conceptual breakthroughs and cognitive revolutions in the history of our species.

Dependent co-arising is a precise articulation of the very same deeper, primary, and fundamental process that underlies, informs, and gives shape to material existence that quantum physics is bringing to light. This is to say that *Buddha* (which literally means "one who has awakened to the dreamlike nature") discovered the underlying principles of quantum physics over two thousand years before its formal scientific discovery. He did so by looking within his own mind.

Contrary to *linear* cybernetics, in which feedback loops take place through space and time, creating a delay or time lag in the processing and coordination of information, dependent co-arising is characterized by a process of *synchronistic* cybernetics in which the feedback loops are circular, instantaneous, and timeless. Everything is simultaneously *inter-causing* (causing and being caused by) everything else in a cybernetic chain that is fastened back on itself. This is to say that everything that emerges into manifestation is reciprocally co-arising with everything else in a nonlinear and nonlocally coordinated way in the one singular and eternal "now" moment.

That quantum entities dependently co-arise, existing not in isolation from each other but in infinitely interconnected co-relation, is a reflection of our own inseparability from everything, a key feature of our true nature. In a sense, we are quantum entities that don't exist as separate objects, but rather are interdependently interrelated with each other as well as the whole universe. We reciprocally and interdependently co-arise together. The emptiness of the quantum world is an out-picturing, a seemingly externalized reflection of our own inner condition of openness, infinite interconnectedness, and freedom.

From the deeper perspective emerging from quantum physics, when we look at photons, for example, we are not looking at separate, independent, individual entities, but rather interconnected members of a greater whole (a beam of light). In the organization of this greater whole, the seemingly separate individualities of the photons are merged, not merely in the superficial sense in which an individual is lost in a crowd, but rather as a raindrop is immersed in the ocean and becomes one with it. We can translate what happens in the quantum realm to the everyday world of human beings. Though space and time seem to be inhabited by distinct individuals, to quote Sir James Jeans:

> When we pass beyond space and time, from the world of phenomena towards reality, individuality is replaced by community. . . . As it is with light . . . so it is with life; the phenomena may be individuals carrying on separate existences in space and time, while in the deeper reality beyond space and time we may all be members of one body.[329]

VOODOO FORCES

When two quantum entities interact, they become intermingled in such a way as to remain forever linked together, joined together at the hip, so to speak.[330] Exhibiting a form of contagious magic, each seems to telepathically "know" what the other is doing. Once connected, their wave functions become phase entangled with each other, such that there are no longer two independent wave functions but one which encompasses both quantum entities forevermore. After their interaction each one leaves part of themselves with the other, leaving behind traces on each other. At that point they are no longer separate in the way that they once seemed to be. In fact, even when separated by vast amounts of space and time they behave in concert, as if they were one entity.

Their quantum telepathy is due to the fact that they are not, and have never been, separate. Seeing quantum entities as separate is a delusion, an artifact of our limited, classically-conditioned perspective. In actuality there is just the underlying, unified, information-filled quantum field giving rise to transitory patterns which, from one perspective, may appear to be distinct entities, but which ultimately are expressions of a singular and indivisible field.

Speaking of their enduring nonlocal entanglement—what Schrödinger calls the "central mystery of quantum physics"—Schrödinger writes, "I would not call that one but rather *the* characteristic trait of quantum mechanics, the one that enforces its entire departure from classical lines of thought."[331] Moreover, quantum entities do not exist in isolation, but are always coupled with an environment (the measuring apparatus, the mind of the physicist, as well as the rest of the universe). The system which is treated by the methods of quantum mechanics is in fact always and necessarily part of a much bigger system. The act of "measurement"—in Wheeler's words, "the act of turning potentiality into actuality"[332]—is not a private affair but a public event in which the whole universe participates.

What if the quantum system under investigation is the whole universe, in which case there is nothing outside of itself to interact with? In Wheeler's words, there is "no circumferential highway"[333] which would allow us to observe our universe from the outside. In other words, there is no external position from which to contemplate the universe. We are *of* the universe. There is no inside, no outside. This is what is meant by *oneness,* a oneness in which the whole universe, including ourselves, is contained. If we think of the whole universe not in relative terms (what would it be relative to?) but in absolute terms, there is no place outside of it (how do we get outside the absolute?) where we could take a stance to describe it, for if we could get outside of it, it wouldn't be absolute. One of the key lessons of quantum theory is that we are part of the nature we seek to understand.

No matter how much we scrutinize something seemingly outside of ourselves (such as physical objects made of matter), we are ultimately the subject of our own investigation. Quantum physics shows us that there is no inside versus outside when it comes to the ultimate reality of ourselves and the things of the world. The entire material world is essentially made of the same mysterious essence as we are. Everywhere we look, within and without, we only find our own self-nature assuming a vast diversity of differing forms, functions, and aspects. In inquiring into the world of the quantum, we are ultimately confronted with our own true face.

If, as quantum physics tells us, the whole universe is quantum to its core, this suggests that the universe is inseparably phase entangled with itself as, ultimately, there is no part of itself that the universe is not nonlocally connected with. This entanglement brings us face-to-face with the fact that what seems far off in space and time might be closer than our heartbeat. The quantum field is instantaneously interconnected with the totality of itself at each and every moment. This holistic perspective helps us understand David Bohm's words: "I'm saying that everything is the observer and everything is the observed."[334]

Each and every non-separate part (and particle) of the universe is synchronized, orchestrated, and coordinated with the states and movements of every other part (and particle) in such a way that there is no lag time in information transmitting itself across space. Space, and the distance that it implies, does not pose any obstacle to the correlation of quantum states for the simple reason that the universe is actually one singular, indivisible system. The seemingly separate parts are a false appearance that occurs only within our minds as a result of our "infirm" viewpoint. Interestingly the word "infirm" simultaneously connotes both sickness, "infirmity," as well as being "in" a "firm"—or rigid, fixed—perspective. Quantum physics can serve as a medicine that helps us to correct the distortion in our materialist myopia. Its medicinal effect is particularly unique and has its own mode of healing, which can be called, in true quantum style, "participatory medicine." In order to receive its benefits we have to engage and participate with it through our efforts to understand what it is revealing to us.

Nonlocality is an expression of the indivisible wholeness of the universe. It represents an observable and measurable indication of the deeply singular and unitary nature of the cosmos despite its apparent physical diversity. In a quantum universe such as ours, the universe is a unity, one big entangled state composed of its interdependent constituent parts.[335] Just like the whole contains its parts, each part contains and is not separate from the whole. Thinking of these parts as separate has nothing to do with the reality of things, but is purely a fabrication of the mind that does not correspond to the actuality of the world. A separate entity is encountered only in thought, and is never found in reality.

The quantum world has never heard of or experienced space or time—as if in some strange way a quantum entity thinks of itself as always being in one place at the same time, a place that is not localizable because it is, in a sense, everywhere at once. For example, a photon of light "traveling" from Point A to Point B does so, from its point of view, in no time—meaning that, in some sense, the two points aren't actually separate. The separation is only apparent as an artifact of the way we've been conditioned to view things. From the photon's perspective, it is not moving at all but simply standing still. At the speed of light, the beginning and the end of the light ray are separated by neither time nor space. All of space shrinks to a single point—from the photon's perspective, distance has no "place" in its existence—and all of time collapses to an instant, a single instant in which, paradoxically, everything that has ever happened and ever will happen takes place.[336] In the realm of light, therefore, there is no before or after, there is only *now.* Interestingly, while light does not experience space or time in its own reference frame, it is the measurement of light that defines space and time in ours.

In our reference frame, we experience ourselves as moving very slowly compared to the speed of light and yet the matter composing our bodies (which provides the locus for our conventional human frame of reference) is itself composed of nothing but crystallized light (which is in its vibratory essence always moving at light speed), so light and light speed are inescapable fundamentals in the structure and construction of the universe. Matter, and the appearance of a sub-light speed frame of reference that accompanies it within space-time, are secondary derivatives from the fundamental universal frame of reference of light itself. Another interesting feature of light is that it can never be caught up with and "overtaken"; like the horizon—no matter how fast we go, it is always the same distance away. Similarly, we can never catch in our grasp the elusive quantum.

The elementary particles that make up the quantum realm appear to be involved in a continual, never-ending exchange of information. Transcending the conventional, third-dimensional rules of space and time, nonlocal interaction is characterized by instant informational exchange, where one part of the universe, in no time whatsoever (outside of time), appears to interact, affect, and communicate with another part of the universe in an immediate, unmitigated, and unmediated way. Imagine, in baseball terminology, a throw from deep centerfield to home plate—only the outfielder is on the other side of the universe, and the ball takes zero seconds to arrive. To quote Huston Smith, "If both mind and matter are nonlocal, we are on our way to regaining what was lost in Newton's time—a complete, whole world in which we can live complete, whole lives, in the awareness that we are more interrelated than we had thought."[337]

Nonlocality's "action-at-a-distance" is an expression of an underlying and outflowing, information-filled field which connects and inextricably links every part of the universe with every other part in no time. For something to be nonlocal in terms of space is to be omnipresent. Nonlocality, according to Henry Stapp, could be the "most profound discovery in all of science."[338] The recognition that our universe is nonlocal has more potential to transform our conceptions of the "way things are," including who we are, than any previous discovery in the history of science. Being that quantum physics is the most accurate means of describing how atoms turn into molecules, and since molecular relationships are the basis of all chemistry, and chemistry is the basis of all biology, entanglement and nonlocality could well be the secret to life itself.[339]

The nonlocality of the universe is a reflection of a part of our nature that is likewise unlimited. Jung comments, "I simply believe that some part of the human Self or Soul is not subject to the laws of space and time."[340] The nonlocality of our universe is reflected in the realm of psyche. Speaking of the collective unconscious, Jung comments that "it seems to be like an omnipresent continuum, an unextended Everywhere. That is to say, when something happens here at Point A which touches upon or affects the collective unconscious, it has happened everywhere."[341] The implication is that when any of us becomes more awake to the unconscious within ourselves, that quanta of consciousness registers throughout and thereby changes the whole universe.

The seemingly separate parts of the universe are connected in such a way as to nonlocally, over inconceivably vast distances of space and time, influence and provide instantaneous feedback for each other, as if telepathically communicating with each other faster than the speed of light. This is another aspect of quantum reality that greatly troubled Einstein—what he referred to as "spooky action-at-a-distance." The superluminal (greater than the speed of light) interaction seemingly involved in a nonlocal universe is not any form of interaction we are familiar with, as it doesn't involve any expenditure of energy or exchange of information in the conventional manner. And yet, innumerable experiments in physics have all conclusively shown that what Einstein derided as "voodoo forces" do indeed exist—at least as much as we do.

• CHAPTER SEVEN •

WHOLENESS

There is truly nothing like our universe; having no frame of reference outside of itself, there is nothing to compare it with. Our nonlocal universe's spooky action-at-a-distance is an expression of the fundamental, indivisible wholeness of the universe, which is radically different from classical physics' previous conception of the universe as composed of separate parts. Thinking that the universe is made of composite parts is an abstraction from the real state of affairs. At the quantum level, there is the radically new notion of intrinsic unbroken wholeness, a seamless interconnectedness among all of the universe's seemingly separate parts; at the quantum level, the universe is "one" with itself. In a quantum universe, everything is related to everything else. At the moment of observation, the observer and the observed compose a single, unified whole. The quantum universe, as Bohr could not emphasize enough, can be properly conceived of only as an intricately interconnected dynamic whole.

In Buddhism, a beautiful image portrays this interconnectedness called "Indra's Net."[342] In this image the whole universe is represented by an infinitely interpenetrating net of glittering jewels (the gems are set at every knot on the net). Each jewel in the net reflects the light reflected in all of the other jewels. Each of those jewels in turn reflects the light from all of the jewels around them, and this multifaceted mutually reflective process goes on forever into infinite space. Each of the jewels contains and reflects the radiance of the whole universe, which is to say that the whole universe can be seen reflected in each of its parts. On such a net, no jewel is in the center or at the edge, while, depending on our point of reference, the opposite is also true–each and every jewel is at the center and at the edge. Each jewel can be conceived of as being an individual, localized spark of consciousness, which is simultaneously reflecting and being reflected by the whole universe. The polishing of any jewel in the net makes it more reflective, thereby enhancing the overall radiance of the whole net. This is to say that as each of us expands our conscious light of self-reflective awareness, the whole universe shines that much brighter.

To quote David Bohm, the "inseparable quantum interconnectedness of the whole universe is the fundamental reality."[343] An expression of this undivided wholeness—the fundamental reality—is that consciousness is no longer separated from matter but somehow is essential to it. Consciousness is not one thing and matter another thing that it interacts with; on the quantum level consciousness and matter are indistinguishable. As Bohm points out, if we don't see this "it's because we are blinding ourselves to it."[344] Later in his life Bohm would use the analogy of trying to separate the north and south pole of a magnet to illustrate the impossibility of separating consciousness from matter. If we try to separate the north and south poles of a magnet by cutting the magnet in half, we simply generate a new north and south pole for each of the new halves.

Bohm writes, "Thus we could come to the germ of a new notion of unbroken wholeness, in which consciousness is no longer to be fundamentally separated from matter."[345] Just as Einstein's theory of relativity abolished the dichotomy of "space" and "time," and modern psychosomatic medicine dissolved the distinction between "mind" and "body," quantum physics undermines the final remnants of the traditional Cartesian dualism of "matter" and "consciousness." After a health scare that included a supposedly near-death experience, Wheeler realized that the one thing in physics that he felt more responsible for than anything else was how everything fit together. In his last years, he was particularly interested in whether the mind was irrelevant to the structure of the universe, or central to it. From the sound of it, Wheeler's brush with death refocused his attention on the big picture, i.e., on what existence is all about.

As science increasingly sheds light on the deeper intrinsic wholeness of the universe, it runs into its own edge, as science, by definition, can't define wholeness any more than mathematicians can define mathematically an empty set, or cosmologists can define the universe before its origins. Definition requires difference between at least two points of reference. Science can only deal in correlations between the behavior of parts. "Science," to quote Robert Nadeau and Menas Kafatos, authors of *The Conscious Universe*, "can say nothing about the actual character of the undissectable whole from which the parts are emergent phenomena. This whole is literally indescribable in the sense that any description, including those of ordinary language, divides the indivisible."[346]

Creating duality (in our example, between the seeming opposites of matter and consciousness) out of something that is inherently, seamlessly nondual is as impossible as trying to separate heat from a flame, wetness from water, or sweetness from sugar. Our attempts at bifurcating what is intrinsically one with itself is a project that not only is doomed to failure, but is at bottom an insane enterprise that is the generative root for the wetiko mind virus. Bohm had strong faith that a proper understanding of the nature of matter could help bring human consciousness into order.

Our universe is an emergent universe in which the whole is greater than the sum of any of its parts and greater than any of its parts can even imagine. In the same vein, human beings are also emergent phenomena. Playing off the famous saying "Less is more," Wheeler had a fondness for the expression "More is different."[347] A substance made up of a great number of molecules, for example, has properties that no one molecule possesses; its difference is qualitative rather than quantitative. An example is water, which is made up of two gases, hydrogen and oxygen. Neither hydrogen nor oxygen exists in a fluid state by itself, but when combined the "liquidness" that emerges from their aggregation is an emergent property that arises from their mutual interaction, which neither of the separate parts contain in themselves. This is similar to how single letters of the alphabet lack the quality of poetry, but when arranged in a certain way, can have undreamed of effects. At the holistic level of quantum reality, relationship creates identity. To try to understand the whole system by studying its seemingly discrete constituents apart from one another only leads so far. In quantum reality, relationship is truly creative.

Similar to a hologram, where every part contains an image of the whole even when divided, in each part of the quantum system the whole is encoded. Due to the holistic, indivisible, and holographic nature of the quantum realm—in which the whole is implicitly encoded in each of its seemingly separate parts—by focusing on a seemingly partial aspect of reality such as the smallest fragment of the microworld, we gain, in a practically magical way, access to the whole. Schrödinger writes, "This life of yours which you are living is not merely a piece of the entire existence, but is, in a certain sense, the 'whole,' only this whole is not so constituted that it can be surveyed in one single glance."[348] Heidegger offered an analogy to describe the wholeness of truth, comparing truth to a drinking glass: As you turn the glass in order to see one aspect, you necessarily conceal another aspect. You can never see the whole glass although it's all there in whatever aspect you see. It's as if the whole is enfolded within each of its parts, or to say the same thing differently, each part is an expression of the whole. Each of the particular aspects contains a distinct perspective on the whole, while each part contains the whole. William Blake's "Auguries of Innocence" comes to mind:

To see a World in a Grain of Sand
And a Heaven in a Wild Flower
Hold Infinity in the palm of your hand
And Eternity in an hour.

INFORMATION

Wheeler comments, "The rich complexity of the universe as a whole does not in any way preclude an extremely simple element such as a bit of information from being what the universe is made of. When enough simple elements are stirred together, there is no limit to what can result."[349] When enough bits of information come together, who knows? The behavior of the whole ecosystem cannot be described in terms of the language or qualities that only apply to any one of its parts. Moreover, an emergent global property can feed back to influence the individuals who produced it in an interlocking, creativity-generating, self-sustaining, and life-supporting positive feedback loop. As consciousness gains momentum in our world, it becomes a unique form of currency (or medium of exchange) in that it continually augments and reinforces itself rather than becoming depleted with use. Thus individuals and groups can begin to consciously tap into the creative energy that makes up the quantum realm, the zero point energy of creation itself, in a way which changes everything.

An observing consciousness does not "cause" the collapse of the wave function in the way we normally think of one thing linearly, mechanistically causing something else. At the quantum level the "material" world has melted away into an apparently immaterial field of quantum potentiality which is somehow synchronously, cybernetically, and synergistically entangled with the minds of observers.[350] What we call matter is, at the quantum level, not separable from aspects of the observer's mind, as if the quantum entities are somehow embedded in the observing consciousness itself. So-called material reality, which from the quantum point of view is considered to be a series of momentary phenomenal events, does not originate purely from the side of the external world alone, but rather is contingent on a complex causal and ultimately acausal nexus that necessarily includes an extraphysical factor—the observer's mind.

The quantum is simultaneously the source and the result of the interaction between the human mind and the seemingly external world. The quantum is not to be found solely in the external world; neither is it to be found solely within the human mind. Rather the quantum exists in the interface between mind and world. The quantum realm isn't simply made up of events; it in itself is an ongoing event expressing the intrinsic union between the human mind and the universe. Mathematically speaking, there is an equation between the human mind and the seemingly external world, and the quantum field emerges out of the ceaseless creative interaction of these two halves that neither completely possesses or encompasses by themselves. Our mind picks up the image of the cosmos, while simultaneously the cosmos reflects our mind back to us in a back-and-forth reciprocally co-arising synchronistic feedback loop without end. It is as if half of the quantum is inside of us, the other half is outside of us, and the information-rich quantum field reveals itself out of the coincidence of the two.

The collapse of the wave function is mediated by the observing consciousness in a similar way that the witnessing consciousness within a dream affects the unfolding of a dream. The degree of lucidity of the observing consciousness plays a significant role in the way a "night dream" (a dream had while asleep)—or the waking dream called life—develops. There is a profound parallel between the process by which material events are understood to unfold in physics and the process by which consciousness is involved in creating and shaping how a dream manifests. The dynamics of material processes appear to have a deep similarity to the dynamics of dreaming. The correspondence between these two dynamics suggests that they might be one and the same process. Quantum physics provides strong supportive and compelling evidence for the dreamlike nature of the universe.

Once these atomic events are registered in consciousness they are transformed into meaningful "information" (which itself is a meaningless idea without some sentient being who relates to and thereby "knows" and experiences the information). This information somehow nonlocally loops back in no time and in-forms the information patterns comprising the atomic realm in what Wheeler refers to as a "meaning circuit." We can think of a seed as an example of something that contains encoded information. The information within the seed (of a tree, for example) interacts with other elements in the environment (air, water, soil, light), coordinating them to self-organize in such a way so as to eventually produce a tree. Similarly, the information that the physicist's experiments yields interacts with the mind of the physicist to produce a result that can grow beyond the initial seed of information.

Information can be conceived of as sitting at the core of physics, just as it sits at the core of a computer. In a sense, the information itself is more real than any of the vehicles that carry it, similar to how a digital signal is the source of, and hence more primary, than the particular form (music, image, text, etc.) it takes on (its "read-out"). Wheeler wasn't satisfied that in physics there is a disjunction between how matter and information are conceived. To express what a significant role information plays at the very foundation of physics, Wheeler coined a pithy slogan, "It from bit."[351] The "it" refers to all aspects of the physical world—atoms, electrons, physical objects, every field of force, even the space-time continuum itself. The "bit" is a unit of information that relates to and generates the appearance of an "it." Wheeler was of the opinion that every "it" derives its function, its meaning, and its very existence from informational "bits." This is to say that the physical universe is fundamentally informational in nature, and that matter is a derived phenomenon from these bits of information.[352] The laws of physics themselves are informational statements: they tell us something about the way the physical world allegedly operates.

Based on the fact that everything we discover about the universe ultimately boils down to bits of information, Wheeler regards physical existence as "an information-theoretic entity."[353] This is to say that the universe either processes information or is itself living information. Speaking of the informational nature of the universe, Philip K. Dick writes, "It is a titanic biological organism that is evolving; as it does so it 'subsumes its environment into arrangements of information.'"[354] From this information life-form's perspective, reality is not a thing, but a series of ideas, since the info life-form is itself an idea. According to Wheeler's worldview, information is the primary entity from which physical reality is constructed. From this perspective, in its very earliest moments every piece of the universe was processing information so as to register and compute itself, long before sentient beings came onto the scene preparing the way for our arrival, so to speak. Wheeler writes that "we may never understand this strange thing, the quantum, until we understand how information may underlie reality."[355]

Wheeler's idea is based on the vision of the universe as fundamentally being an information-processing system. It should be pointed out that for a system to be able to process information implies intelligence; we, both individually and collectively, are active participants in this process. The universe is like a single gigantic quantum computer. It is no secret that in this computer age that we live in we are in the midst of a never before seen or dreamed about information-processing revolution that is accelerating at an exponential rate. Other examples of information-processing revolutions include art, science, mathematics, psychology, the development of the spoken word, language, and writing.

In essence, the physical state of the universe acts to alter the mental state, which then instantaneously feeds back into and changes the physical universe through a kind of informational feedback system. Once a bit of information is added to what we know about the world, at the same moment in time that bit of information determines the structure, creating the reality of the time and place, of one small localized part of the world. Wheeler speculates, "Information may not be just what we *learn* about the world. It may be what *makes* the world."[356] Bohm likewise felt that information occupies a central place in physics. He felt that information must be placed alongside matter and energy as one of those factors underlying the living processes of the universe.

In a certain sense, Wheeler's universe as an "information-theoretic-entity" is not pointing to something other than or outside of itself; it simply *is* the thing that it describes. This is to say that this informational life-form is not just *describing* something; rather, it is the doorway connecting us to what it is describing. Interestingly, this is a description of a living symbol that doesn't just point at what it is symbolizing, but is itself the portal introducing us to that which it is symbolizing. Symbols are the language of dreams. Quantum physics is a purely symbolic procedure revealing that the universe is a symbol pointing at itself, seeking to be deciphered. This is analogous to how a symbol in a dream can both be an expression of the dream while simultaneously pointing at the dreamlike nature of which it itself is a living manifestation. Speaking about the "symbolic character of the entities of quantum physics," Eddington comments, "There is nothing else to liken it to."[357]

The universe not only processes information, but is itself living information, as if it is a constantly evolving organism imbued with intelligence that endlessly subsumes, incorporates, and structures itself into ever-new arrangements of information—appearing in, as, and through the forms of the material universe—thus revealing and re-presenting itself in novel ways. This information-theoretic entity has no body (soma), but somehow extends itself into and makes use of the physical world to serve as its body in such a way so as to transmit itself into the human mind. The mind, like a mediating instrument, serves as the interface between this information life-form and the physical world; the mind is like a "groove" on a record to which the information life-form is the "music."

In some mysterious way, the ultimate source of the messages from this information life-form is to be found within the human mind itself, as if we ourselves, at some deep level, are the authors of the information coming our way. This brings to mind how the unconscious sculpts its images to the particular state of the dreamer, depending on where we are in our evolution. Perhaps a deeper, already awakened, higher-dimensional part of ourselves that exists outside of linear time is planting messages—psychospiritual alarm clocks—in the fabric of the waking dream so as to help us to snap out of our forgetfulness and remember ourselves. This is to say that the universe (which is not separate from ourselves) is conspiring with us in order to help us to wake up. From this point of view, it is no accident that quantum physics appeared on the scene when it did. Quantum theory, being a symbolic procedure, can itself be likened to a symbol in a dream that is being "dreamed up" by humanity as a way to awaken ourselves to our quantum nature. We are clearly ready to awaken to the quantum universe, or we wouldn't have dreamed up the revelations of quantum physics in the first place.

Information distinguishes itself in the quantum world; it is utterly unique in that no prior information is required for information to exist—information arises out of no information. This is a reflection of the idea, found in both quantum theory and Buddhism, that the forms of this world arise from and are an expression of an underlying and all-pervasive emptiness. Information breaks the infinite chain of regression in which we always seem to need a more fundamental law to explain the current one. "Tomorrow," to quote Wheeler, "we will have learned to understand and express *all* of physics in the language of information."[358] This is to imply that physics itself is potentially becoming a new language, and hence, giving rise to a form of art.

Similar to most artists and scientists, Wheeler's career went through different phases (think of Picasso's Blue Period). First was the "Everything is Particles" period, followed by his "Everything is Fields" phase. Wheeler refers to his third and last period as his "Everything is Information"[359] phase. Interestingly, Jung describes the "archetypes" which underlie and in-form our experience of the world as "informational fields of influence." Potentially altering the recipient, information has the possibility of informing, which is to say it has a "mental" aspect. Putting information at the forefront of physics at the same time implicates the mind as being a primary factor in the new physics, a physics that is emerging not just in the world, but within our minds as well.

MEANING

Information is organized by a second invisible element—meaning. The quest for meaning is what drives cutting edge physics. Information is meaning, it represents and carries meaning in one form or another and is therefore based upon awareness, which is the ground of all meanings. "Physics," to quote Wheeler, "is the child of meaning even as meaning is the child of physics."[360] We should not forget that the meaning that arises from information, at bottom, is made of observation.

Whether we talk about meaning, information, observation, measurement, interpretation, logic, ideas, or language, deep down at the most fundamental level all of it has to do with consciousness. A meaningful reality is always an experienced reality, i.e., a reality arising within a sentient subjective field. Meaning is never found separate from—and is always to be found within—a cognizant field of awareness. Meaning has no meaning unless it is occurring within a mind. Meaning cannot come from outside a person but always emerges from within. Meaning-making is an inside job.

There is no meaning that is complete in and of itself; meaning is always context-dependent. To quote Bohm, "We have to say that all meaning is always in a context. . . . There is always the question of content and context: a given content of meaning always depends on its context.... The one who understands meaning is part of the context."[361] In other words, meaning, which is a part of reality, only takes on its true meaning in relation to the whole as well as in relation to whom the meaning occurs. To quote Schrödinger, "The great spectacle of nature acquires a meaning only in relation with the spirit who contemplates it."[362]

Many physicists still unconsciously objectify meaning in a similar way that they objectify the seemingly external world. Physicists who still think of the world as existing objectively imagine that meaning somehow inheres in the seemingly objective material world, thereby disowning the mind's role in meaning creation.

Those operating out of a materialist mindset, however, do not all agree about the nature of meaning. Some think that meaning objectively resides in the outside world, independent of the mind. Others conceive of the universe as having no inherent meaning other than what the mind projects into it, which is to say that all meaning is considered to be purely subjective. From this latter perspective, it is a major fallacy to project the inner contents of our own minds onto the outer objects of the world, for this taints the outer world's pure objectivity with the whims of the subjective mind's imaginative, and thereby unreal, constructs. From the materialist point of view, this would be to conflate the strictly distinct categories of the inner with the outer, which is seen as a state of delusion. From the psychiatric point of view, to confuse the inner with the outer is to have fallen into magical thinking and suffer from what are called "delusions of self-reference." From this allegedly "sophisticated" materialist viewpoint, this "naïve" and primitive perspective of not differentiating the inner and outer worlds would be seen to be intellectually undeveloped, something to be outgrown. The materialistic viewpoint is blinded, however, to their questionable and unreflected-upon axiomatic viewpoint that presumes that the inner and the outer are categorically distinct realms with no interconnection.

The advent of quantum physics, however, has helped to shed more light on the nature and genesis of meaning. In quantum theory, similar to the materialists who think of meaning as being projections of the mind, meaning is seen as coming from the mind. Rather than "degrading" meaning (thinking that all meaning is "merely subjective"), however, the quantum viewpoint interprets the fact that all meaning arises within the mind as the reason to "promote" the mind to the creative arbiter of meaning. Instead of being an anthropocentric delusion, an example of human hubris and self-aggrandizement of cosmic proportions, recognizing that the mind is the source of meaning is a clear-sighted realization of our place in the cosmos. To quote Bohm, "Meaning is the bridge between the mental and the physical."[363]

In proving that there is no such thing as objective reality, quantum physics has revealed the intrinsic interconnectedness between the seemingly outer world with our inner consciousness, which is to say that the two, the outer and the inner, are a joint cooperative venture. To say this differently: The seemingly outer world possesses an interior aspect that is linked to and continuous with human subjectivity. Instead of thinking that all meaning is merely subjective, or that meaning inheres in the outer objective world, quantum physics is revealing that it is the mind's task to create/discover meaning via its ongoing interaction with the world, a world which is recognized to not be in any way separate from our own consciousness.

Jung writes that "nature has put a high premium precisely upon the development of consciousness. . . . The importance of consciousness is so great that one cannot help suspecting the element of *meaning* to be concealed somewhere within all the monstrous, apparently senseless biological turmoil."[364] Meaning is one of the essential features of consciousness; it is the very currency of our conscious mental lives. Thoughts and ideas are themselves made of meaning. Without meaning our mind would have no frame of reference with which to orient and guide our ongoing actions and choices in relation to the world. Without meaning our subjective mind would be featureless, perhaps only experiencing raw sensory data that would have no significance or utility.

Language is a repository of meaning for the whole of society. Speaking about the meaning of meaning itself, Wheeler says, "'Meaning'—and what else is 'objective reality' if it is not meaning? . . . is the joint property of those who communicate."[365] Being our shared property, it is as if we are the creators and owners of meaning or "meaning generators." This is to say that ultimately the meaning of our world and our experiences originates within our minds and truly belongs to us; we are the ultimate "author"-ity when it comes to meaning. Putting on his philosopher's hat, Niels Bohr says, "The meaning of life is this, that it has no meaning to say that life has no meaning."[366]

Wheeler writes in his journal, "Last night before falling asleep I could not see how anyone could doubt it is the individual who gives meaning to existence—where else except in my mind is the world I seek to explain?" In contemplating not the conventional but the ultimate meaning of things, Wheeler is being led back to his own mind. It should be noted that the aforementioned wetiko virus of the mind—the "disease" that Feynman was pointing at—deviates our mental syntax, the rules of how we form language, thereby distorting our semantics and the meaning we place on our experience of ourselves and the world. Wheeler comes up with what he refers to as a "preposterous idea," wondering whether "meaning itself powers creation. But how? Is that what the quantum is all about?"[367]

It is worth noting that our civilization has been suffering from a failure of meaning. Describing the malaise that modern humanity suffers from, author Colin Wilson coined the phrase "the fallacy of insignificance" to connote the underlying sense of meaninglessness that is so prevalent in our world. In his autobiography, Jung writes, "Meaninglessness inhibits fullness of life and is therefore equivalent to illness. Meaning makes a great many things endurable—perhaps everything."[368] This brings to mind the hard-won insight of Viktor Frankl, himself a survivor of the Nazi concentration camps, who felt that the ability to create meaning was the determining factor for surviving the death camps. Speaking of the relation between meaning and health, Bohm commented, "An incoherent meaning means that we are in some way unhealthy. Our being will not be right. Just simply the aim to be whole inspires the search for a coherent meaning."[369] In other words, individuation, the process of becoming whole, is indistinguishable from the creation/discovery of meaning.

Bohm suggests that meaning is a form of being. In the very act of interpreting the universe, we are creating the universe, which is to say that the perception of a new meaning is a creative act. Through our meanings we change the tone and quality of being. We literally change the world that we inhabit by what we choose it to mean (i.e., what it "sign-ifies") to ourselves. Learning to more consciously exercise our power of giving meaning to our experience enables us to creatively paint our world with meaning that enriches it in ways that are more in alignment with—and thereby feeds—who we are.

Bringing his thoughts on meaning back to quantum physics, Bohm points out, "The electron, in so far as it responds to a meaning in its environment, is observing the environment. It is doing exactly what human beings are doing. . . . It is gathering information about us, about the whole universe. It is gathering-*in* the universe and responding accordingly. Therefore it is observing, if you take that in its literal sense."[370] Is the quantum realm simply mirroring us, providing us with a mirror of our mind?

This brings us back to Wheeler's idea of the importance of asking the right question. The question "What is the meaning of quantum physics?" more and more seems inextricably linked and entangled with the deeper existential question "What is the meaning of life?" This naturally brings up the question: "Who is asking?"

• CHAPTER EIGHT •

A PHYSICS OF POSSIBILITIES

Just when we think we have elementary particles pinned down, they seem to evade our grasp, turning into something else. To quote D. H. Lawrence, it's "as if the atom were an impulsive thing always changing its mind."[371] Until the moment it is observed, a quantum entity exists in a realm of potentiality, in what is called a state of "superposition," where all possible quantum states coexist simultaneously (a state, interestingly enough, which has never been observed). The hallmark of an unobserved quantum entity is to hover in a ghostly ethereal state between the extremes of existence and nonexistence, where it can be said to both exist and not exist at the same time (reminiscent of the aforementioned four-valued logic). This is to say it exists in all possible states (each one a parallel world), not fully occupying any possibility until the moment it is observed.

The moment of observation is when, in Wheeler's words, "an elementary quantum event" takes place. It is the act of observation that forces nature to "make up its mind" and manifest itself in a specific state that we experience, thus becoming a determinate feature of our world. Heisenberg points out that the change in an object from potentiality to actuality always correlates with a change in our knowledge about the object. Being that consciousness is the vehicle through which we are able to attain knowledge, what Heisenberg is saying is that the change of potentiality into actuality necessarily involves a change in consciousness.

Not existing in space-time, the appearance of the quantum realm in space-time at the moment of observation is a quantum event in which an atemporal process manifests itself in what appears to be time. In so doing, the quantum event helps to create the material basis upon which the illusory appearance of linear time is projected. Wheeler expresses the central point of quantum theory in a single, simple sentence when he says, "*No elementary phenomenon is a phenomenon until it is an observed phenomenon.*"[372] The necessity for this demarcation is the most mysterious feature of the quantum, for it holds the clue to the central principle of the construction of everything out of nothing. This tenet changes our traditional view that something has happened before we observe it; as Heisenberg writes, "The term 'happens' is restricted to the observation."[373]

At the moment of being observed, the wave function collapses in no time at all into a particular manifestation, while all of the other potentialities vaporize as if they had never existed.[374] Due to an observation *anywhere*, the quantum entity's wave function instantaneously collapses *everywhere*. Not all of the infinite potentiality, however, is converted to actuality in a finite amount of time. To quote Wheeler, "There are innumerable clouds of probability running around in the universe that have yet to trigger some registered event in the macroscopic world."[375] I love Wheeler's image of "innumerable clouds of probability running around the universe."

From the quantum point of view, everything that might have happened influences what actually does happen. This has to do with Feynman's "sum over histories" idea, which points out that the particular event that actually takes place is the sum of all of the virtual effects of what might have possibly happened. Using his typical baseball analogy, Wheeler comments, "It is as if a baseball pitcher, instead of throwing a single ball toward the batter, could launch simultaneously a thousand balls that travel a thousand different paths through space and time on their way to the batter. Each of these thousand baseballs has a 'history' as it flies from the pitcher's mound to plate. What the batter sees and swings at is the result of all of these histories combined. A mind-bending idea, to be sure, but it's just what happens in the quantum world."[376]

In a quantum universe such as ours, everything ultimately exists in a state of open-ended potential, what Heisenberg calls "transcendent potentia." Quantum theory implies that the whole universe, including ourselves, is re-created and re-creating itself anew every moment based on observership, in other words, how we are dreaming it up.

Once the seemingly infinite and open-ended world of quantum potentiality manifests in whatever particular and actual way it does due to being observed, we find ourselves (relatively speaking) in a very convincingly stable and seemingly classical universe. We subjectively experience the illusion that there is only one fixed and solid classical universe that exists, and everyone else, at least on the surface, appears to be living within the same world that we do. All of the potential, parallel, and unmanifested alternative universes seem to disappear as if they never existed. We are then like the proverbial horse with blinders limiting our vision, partitioning ourselves within and only seeing the one classical world we find ourselves inhabiting, with no awareness of the potential alternative universes existing all around (as well as inside) of us. This is to say that at each and every moment there are a myriad of alternatives to choose from. To quote physicist Roger Penrose, "The behavior of the seemingly objective world that is actually perceived depends on how one's consciousness threads its way through the myriads of quantum-superposed alternatives."[377]

Observation is the very act through which the quantum realm "discloses" itself. In quantum theory the moment of observation is where physics gets personal and the rubber meets the road, which is to say, where abstract theory and empirical data meet and a specific actuality is realized and manifested out of a vast array of possibilities. It is important to note that we are always "at" the moment of observation, which is to say that we're there right now! There is no other moment but the one eternal moment of observation. The tendency to think that the moment of observation is just one single discrete moment in a linear sequence of other moments is due to the long ingrained habit of thinking in terms of linear sequential time (a "linear time hangover").

In our mind's role as observer-participants, we are truly on the cutting edge of the "big bang"—itself a thoroughly quantum event, which billions of years later led to our appearance on the scene, wondering what original impetus led to our existence. We live on the forefront of the moment of creation that is always taking place in the moment, in the here and now. To quote Schrödinger, "For eternally and always there is only *now*, one and the same now; the present is the only thing that has no end."[378] In a similar vein, speaking about the nature of the mind, Schrödinger reminds us that it has a "peculiar timetable, namely mind is always *now*. There is really no before and after for the mind."[379] This brings to mind philosopher Ludwig Wittgenstein's statement, "If we take eternity to mean not infinite temporal duration but timelessness, then eternal life belongs to those who live in the present."[380] The fact that a new beginning appears every time we participate in the act of observation is an expression that the universe—of which we are a part—is always refreshing itself and is thus perpetually creative at its most fundamental level.

Quantum theory is revealing to us the creative nature of our moment-to-moment experience. It should get our highest attention that observing these quantum objects is the very act that brings them into existence. When we observe an atom to be someplace, quantum physics tells us that it is our looking that caused it to be there at the exact moment that we observed it. When a physicist observes an elementary particle—which, from the quantum point of view, "causes" the particle to exist—it is as if the physicist is "dreaming up" the quantum entity in the same way that a dreamer dreams up their own dreamscape. At the same time (if we let our creative imagination run wild) it is as if the elementary particle is reciprocally dreaming, as it dreams up the physicist to observe it and hence, bestow upon it existence. The physicist and the subatomic particle are, in a timeless process that actually takes place in time, mutually dreaming each other up, indivisible aspects of one unified quantum field.

Quantum theory reveals that there is nothing inherently solid or objectively real about the properties of an object that we measure. Before the advent of quantum physics, the hallmark of "reality" was its measurability—for something to attain the status of being real, it had to be able to be measured. Quantum physics has startled the physics community by seeing through and dissolving this long-held and cherished notion. It is revealing that we ourselves are intimately involved in producing the results of our own measurements. Our discovery of a quantum entity in a very real sense "causes" it to be there, which implies that there is no physically real world independent of our observation of it. Before these entities are observed they don't really exist. There is nothing we can say about them; they are "unspeakable."

Wheeler sometimes used a baseball analogy to illustrate the nature of quantum reality. Talking about how they call balls and strikes, some umpires say, "I call them the way I see 'em," which is an expression of the subjective, projective nature of our perception.[381] A second umpire might say, "I call them the way they are," which is an expression of there being an objectively existing reality not dependent on observation, which was Einstein's point of view. Wheeler then quotes a quantum umpire who would say, "They ain't nothing till I call 'em," which is an expression of a quantum baseball game in which nothing exists until it is observed. The properties of quantum objects aren't inherent to the object, but instead emerge from and are created by interactions with their environment as well as their relationship to observers and their inescapably creative acts of observation.

We can use light as an example. It is well known that light displays either wavelike or particle-like qualities depending upon the experimental setup and how it is observed. These two conditions can never appear simultaneously, however. To ask whether light is a wave or a particle is a meaningless question. According to quantum physics, there is no way of knowing what light "really is." To be more accurate, the wavelike or particle-like behavior that we observe in light is not a property of light per se, it is a property of our *interaction* with light. If, as quantum physics attests, there is no independent, external objective reality, then light, be it in its wavelike or particle-like aspect, cannot be said to exist separate from our interaction with it. In other words, we cannot talk about the properties of light without including ourselves in the equation. What we are saying about light is true of everything; what we experience is not external reality, but our interaction with what our minds construe to be an external reality. It can't be repeated often enough: Our observing psyche is an integral part of the process being observed.

The fact that the properties of these quantum objects are a function of our observation and that there is no substance, no separately existing intrinsic quantum object separate from its properties, is an expression of these quantum objects having no independently existing objective reality. They are not real in the way we commonly think of something being real. And yet we ourselves, as well as the experimental instruments physicists are using to measure these "not real" quantum objects, are made up of a conglomeration of the same quantum stuff that itself isn't real in the ordinary sense. The revelations of quantum physics put us into a very unusual situation—everything we call real is composed of things that themselves cannot be regarded as real.

Quantum physics points out that mass is not an inherent property or primary quality of the ultimate building blocks of nature. According to quantum theory, there is in fact no such thing as mass. The appearance of mass is constructed entirely from the energetic interactions between massless, insubstantial, and phantomlike elementary particles, which themselves are the result of interactions between immaterial, information-filled quantum fields of potential. To quote pre-Socratic Greek philosopher Heraclitus, "Latent structure is master of obvious structure." To put a slightly different spin on our contemplation: How can abstract quantum entities generate physical events? It is only possible to answer this question if we recognize a dimension that is part of the cosmos but is not in the dimension of space-time, being somehow beyond it.

Light is made out of photons, which have what Banesh Hoffman calls their "flighty propensity . . . of jumping into and out of existence." "Photons," to quote Hoffman, "were mere will-o'-the-wisps, evanescent and insubstantial . . . they were free to come and go; to come out of nothingness and return to nothingness; to materialize as radiant, lustrous wavicles and melt away again. . . . They could multiply like rabbits. You could never be sure how many you had. You might even start with none at all and suddenly find yourself overwhelmed by them. In an instant . . . there appears a stupendous plenitude of photons, a dazzling flash of light where previously all was darkness."[382]

The massless, intangible, and immaterial photon, which has zero weight, somehow gives rise to the massive weight of the whole material universe. The question naturally arises: How does the universe precipitate out of a field of pure light? For example, massless photons are routinely observed becoming electrons (which have mass) if they interact in a specific way. This remarkable phenomena of mass spontaneously appearing out of massless light and then dematerializing back into light seems to indicate that mass is not as substantial as it seems and may be more of an appearance arising in the theater of space and time than a solid reality.

Simply put, there aren't any nuts and bolts (i.e., fixed attributes) at the quantum level. We can't visualize the quantum world, not because we know too little, but because we know too much. To quote Heisenberg, "If we think we can picture what is going on in the quantum domain, that is one indication that we've got it wrong."[383] If we insist on making an image to represent the quantum realm, it is, to quote Sir James Jeans, "like making a graven image of a spirit."[384] Instead of an icon, our image will then be an idol; the logos will have turned into a logo. Though impossible to visualize, the quantum realm is not impossible to think about. We just have to learn to think in completely new quantum-compatible ways. Though beyond our conventional imagination, it is certainly not beyond nature's imagination (which is where our biological existence, wherein our human imagination is rooted, ultimately springs). Nature has no trouble producing such quantum entities; indeed, such entities are what this whole wide world, including ourselves, is made of.

The universe's appearance is deceptive; the universe appears in one way, but exists in another. Buddhism distinguishes between the absolute or "ultimate" mode of reality—the way things really are—to the relative, "conventional," or "seeming" mode of reality—the way things appear. Behind the apparent solidity of everyday objects lies a world of open-ended potentiality. In order to more fully understand conventional reality, we must take into account other dimensions of a deeper reality. There is a nonphysical higher dimension that is not located in space-time in which our world of space-time is contained—a dimension which we must factor into our equations of what constitutes reality. In other words, nature (and our nature) is not contained solely in space-time.

Wheeler writes, "Every item of the physical world has at bottom—a very deep bottom, in most instances—an immaterial source and explanation."[385] This idea bears repeating: all of the materialized universe arises out of "something" (a "something" which is akin to "nothing") that is immaterial and mind-like in nature. To quote Wheeler, "We go down and down from crystal to molecule, from molecule to atom, from atom to nucleus, from nucleus to particle, and there's still something beyond both geometry and particle. In the end we have to come back to mind."[386] When we get down to it, we get down to mind. Along similar lines, regarding physics' attempt to penetrate to the innermost center of things, Heisenberg writes, "I cannot regret that this center is not material, that it has to do rather with the ideas than with their material images."[387] In other words, the human endeavor to access the deep, fundamental structure of nature, the very heart of matter, has resulted not in finding anything material, but rather has led to the subjective and immaterial realm of ideas.

Physics has penetrated to the very core of material, seemingly objective reality and has found nothing that can be said to ultimately exist beyond or outside of our observation of it. It is as if objective reality has slipped beyond our grasp, beyond concepts, beyond even the concept of existence and nonexistence. To quote Eddington, "We have found a strange footprint on the shores of the unknown. We have devised profound theories, one after another, to account for its origins. At last, we have succeeded in reconstructing the creature that made the footprint. And lo! It is our own."[388] When we peer into the deepest recesses of matter or to the farthest edge of the universe, we see nothing but our own puzzled faces staring back at us.

Exploring the farthest reaches of the outside world brings us right back to our inner selves. Russian philosopher Nikolai Berdyaev writes, "All attempts at external perception of the world, without immersion in the depths of man, have produced only a knowledge of the surface of things. If one proceeds from man outward, one can never reach the meaning of things, since the solution of the riddle of meaning is hidden within man himself."[389] We can never speak about nature without, at the same time, speaking about ourselves. There is a paradox at the heart of reality, as if at the heart of everything—the answer to the riddle of life—is a question, not an answer. Poetically expressing the same realization, Wheeler asks the question, "What is Out There? T'is Ourselves?"[390]

RAINBOWS

The appearance of a rainbow is made up of the interaction of raindrops, sunlight, and our consciousness. By itself a rainbow has no existence in our universe prior to being observed. This is to say that there is no intrinsic, independent, objective rainbow that exists separate from the observing consciousness. Being a function of our observation, if no one were there observing it, there would be no rainbow. There is no physical rainbow existing somewhere out there in space; the rainbow is only to be found within, and not separate from, the very mind that is observing it. The highest Buddhist teachings say, and quantum physics agrees, that our physical universe is similar in this regard to a rainbow.

In the same way that a rainbow can't be said to exist until the moment that it is observed, quantum entities can't be said to exist until the moment of observation. The act of observation is truly creative. The quantum entities that make up what seems to be solid matter (let's use, for example, a tree) are no more like the thing we call a tree than the raindrops are like the thing we call a rainbow. Just as a rainbow is the outcome between the raindrops and our consciousness, the tree is the result of the interaction between the quantum entities that compose it and ourselves. This is to say that in the case of the seemingly solid, three-dimensional world, there is one necessary ingredient: an observing consciousness. This isn't some New Agey gobbledygook, but simply the logical outcome of following what quantum physics is revealing to us about the nature of the universe we live in.

If more than one person is seeing a rainbow, there is an unexamined assumption that this means that the rainbow is "really there," i.e., that there is an objectively existing rainbow that they are all seeing. If a number of observers are seeing what seems to be one and the same rainbow, however, it is not accurate to say they are seeing the same rainbow. A rainbow appears in a different place for each observer. In fact, when any one of us sees a rainbow, each of our eyes sees a slightly different rainbow. Its position is context dependent, rather than being innate; if we move, it moves. In a very real sense, there are as many rainbows as there are observers; each observer is seeing their own private rainbow. There is an infinite superposition of rainbows existing in a state of potentiality, each one inhabiting a virtual world until the moment it is observed.

Similar to how there is not one objective rainbow that exists as an object per se, quantum physics has discovered that there is no invariant way the universe "really is." To quote Philip K. Dick, "For every person there is a different universe which is the result of a mutual participation between him and the macrocosm, a field that is a syzygy between them."[391] There is no single reality that all observers share. Wheeler writes in his journal, "Idea, surely not new, that there is not 'one world' but as many worlds as observers."

Dick continues, "If reality differs from person to person, can we speak of reality singular, or shouldn't we really be talking about plural realities? And if there are plural realities, are some more true (more real) than others? What about the world of a schizophrenic?"[392] Are some of us more "in touch" with reality than others? It is easy to presume that the differences between people's worlds are caused entirely by the subjectivity of the various human viewpoints—in other words, that people are just interpreting the one, objectively existing world differently. Quantum physics suggests that our situation might, however, be one of plural realities superimposed onto one another (akin to the potentiality of the wave function) like so many film transparencies. At any given moment, based on our observation, one of these transparencies takes on substantial form and appears, to our mind, to be the real and therefore only existing reality, while the other potential universes disappear as if they never existed.

The idea that there is an objectively existing world that we all share is a flawed assumption that creates the seemingly unsolvable paradoxes that riddle the quantum physics world. This state of affairs makes me think of how many conflicts in the world, both big and small, are the result of people arguing over the idea that their version of reality is the correct one. But in actual fact there is no "true" reality. The only true reality there is, is that there *is* no true reality. To quote Robert Livingston, one of the founders of the discipline of neuroscience, "Our individual experiences are so different from one another that the world consists of a couple of billion people and a couple of billion worlds."[393] If there are indeed plural realities, problems arise due to breakdowns of communication between the different realities. This puts the various conflicts in our world in a new context. Maybe instead of fighting among ourselves to determine who is in possession of the true reality, we can learn to build bridges to connect the multitude of realities. Quantum physics is such a bridge.

• • •

NOTHINGNESS

Wheeler was of the opinion that everything—the whole universe—emerged out of nothingness. He writes, "Everything came from nothing."[394] In his journal he comments, "Nothing! Nothing! You start with nothing to get everything." Bit by bit, measurement by measurement, observation by observation, the airy nothingness crystallizes into physical form as we collaboratively dream up our world out of the primordial nothingness from which we ourselves arose. In a dialogue with an imaginary colleague, Wheeler writes, "Preposterous we have to agree is the idea that everything is produced out of nothing—as preposterous, but perhaps also as inescapable, as the view that life had its origin in lifeless matter."[395]

Wheeler's nothingness is completely void of structure, law, and plan. In his autobiography, Wheeler writes, "And, just as life arose from nonlife on Earth, something arose from nothing in the universe. That 'nothing' from which something arose should not, however, be confused with the emptiness of a vacuum. It is nothing in a profounder sense. It is nothingness."[396] As Wheeler himself has pointed out, this "nothingness" is filled with enormous energy density,[397] and is actually a "plenum"—an overflowing fullness of pure creativity—effectively disguising itself as a vacuum.[398] This nothingness is a cauldron of activity, bubbling more vigorously the further we go down the quantum rabbit hole to look at smaller bits of space and tinier intervals of time.

Classical physics is based on a concept of space as an empty, inert receptacle or vessel that can be filled with something from the "outside." It is seen simply as a location in which matter exists and things can happen. To quote Wheeler, "No point is more central than this, that empty space is not empty."[399] From the pre-quantum idea of simply being an empty container in which physical phenomena move, the modern conception of the void contains infinite potentiality—being a field of powerfully fluctuating energetic relations—a dynamic process in its own right with enormous importance. To quote the Chinese sage Chang Tsai, "When one knows that the Great Void is full of ch'i, one realizes that there is no such thing as nothingness."[400]

This same idea is poetically expressed in the Buddhist *Heart Sutra,* "The Heart of the Perfection of Transcendent Wisdom," in which it states: "Form is emptiness. Emptiness is form. Form is none other than emptiness. Emptiness is none other than form."[401] These words are pointing at the inseparability of the seemingly solid forms of the world and the emptiness out of which these forms arise; matter cannot be separated from the empty space out of which it arose and in which it exists. Matter is in fact simply a dynamic modification of empty space, a structure mysteriously constructed out of pure structurelessness. Nature's lack of a fixed essence is essential to what it is. Just like in a dream, forms are empty of inherent, independent existence, not having a nature of their own. It is the nature of emptiness to take on form, which is to say that emptiness itself is appearing in the "form of form." In the context of quantum theory, "emptiness" is the field aspect and "form" the particle aspect. Quantum entities are not objects in space and time, rather they are processes through which the void manifests itself and in so doing gives rise to the domain of space and time. In other words, space and time are produced in and through the making of phenomena.

It should be noted that the reason quantum entities can spontaneously change their form is because they are in direct contact with the void, the ground of its (and our) being. The physical world is a momentary crystallization and expression of the underlying emptiness, which is not a vacuum but an overflowing fullness or "plenum." This is similar to the way in which a whirlpool in a stream is not separable from the water that makes up the stream, but rather is an unmediated expression of it. Just as the whirlpool is nothing but a momentary patterning of the flowing water, the so-called particles that make up our physical world are nothing but dynamic patternings in and of the featureless plenum of space. Before the advent of quantum physics (to quote Wheeler), "one thought of particles in space as really important and the space around as relatively unimportant."[402] Modern physics reverses this perspective, as it recognizes that particles are constructed out of seemingly empty space (the plenum). What seems to be mutually exclusive opposites—form and emptiness—necessitate each other. They are in actuality two aspects of the same underlying reality which coexist and are in continual connection and cooperation.

"The plenum," Bohm writes, "is the ground for the existence of everything, including ourselves. The things that appear to our senses are derivative forms and their true meaning can be seen only when we consider the plenum, in which they are generated and sustained, and into which they must ultimately vanish."[403] The plenum is something (a something akin to a "no-thing") with which we are always in touch with, contained within, and of which we are expressions. The emptiness of the plenum constitutes the essential being of all forms, both outside in the world and within our mind. Everything arises out of the all-pervasive common ground of the plenum, which is thereby immanent in each and every form, making each form indirectly immanent in each other. Though the idea of the plenum seems really "far-out," the plenum itself is actually "far in," pervading and composing the core of our very being.

The revelatory insights of quantum physics are leading us towards a grand synthesis in which the essential nature of the physical world of matter and sentient awareness (the essence of our minds) are recognized to be different modes or expressions of one and the same underlying sentient field (the quantum plenum) that pervades and informs all that arises both within and without. This newly emerging visionary understanding of the inseparability of mind and matter is for the first time finding a rigorous and precise articulation in the technical mathematical language of modern physics.

The Buddhist word for "emptiness" is "shunyata." The root of "shunyata" is "sunya," which alternately means zero, nothing, hollow, or void, and which refers to the zero point, the cosmic seed of emptiness that is pregnant, swollen with potentiality. Within the emptiness of the plenum is an internal intelligence, a primordial sentience with innate cognizance, for how else could the cognitive faculties of sentient beings arise?

Modern theoretical physics has taken our gaze off of the visible and directed it to the underlying field. This reminds me of when we learn to "see" as an artist, we typically are taught not to draw the object per se, but rather the space around the object. In a similar figure/ground reversal, shifting our view from content to context, physics is shifting our vision from the foreground (the forms) to the background (the empty space) in which the forms arise, and is realizing that the two—the foreground and background, the forms and the emptiness—are ultimately inseparably one. Form *is* emptiness. Emptiness *is* form.

TWENTY QUESTIONS

Wheeler likens how we create "reality" out of nothing but our interactions to a slightly skewed, surprise version of the party game "twenty questions." In the regular version of the game, someone leaves the room, and everyone decides on a word. The person is allowed to ask a series of yes or no questions until they feel that they have enough information to guess the word. Wheeler tells the story of a party where he was the one sent out of the room, and when he came back and began asking his yes or no questions, his friends were taking longer and longer to answer. The tension was building in the room, until he finally guessed the word to be "cloud," at which point the whole room burst into hysterical laughter. His friends explained to him that they had decided to not decide on a predetermined word, and were playacting "as if" they had decided on a particular word based on nothing but the answers they were giving, the only rule being that every answer had to be consistent with all previous answers. There was no word that existed until the very moment of Wheeler's guess.

Wheeler's questions and interactions with his friends helped conjure up the word in the same way that physicists and their measuring apparatuses' interactions with the subatomic realm actually create the elementary particles they are measuring. To talk about the word "cloud" existing "in the room"—i.e., in the "minds" of Wheeler's friends—before Wheeler's guess is not accurate in the same way that the elementary particle wasn't "in the universe" before the experiment, having no existence prior to being measured. Similarly, in our inquiries into the nature of the universe it is easy to imagine that the final answer already exists, which we will one day uncover, without realizing that the very questions we ask and the actions we take condition and help to create the answers we get back. If Wheeler had asked different questions or the same questions in a different order, he would have ended up with a different word. The idea that the word "cloud" was sitting there, waiting to be discovered, is in Wheeler's words "pure delusion and fantasy."

In discussing the surprise version of the game of twenty questions as illustrative of how physicists participate in producing the results of their experiments, Wheeler painstakingly makes the point that the power he had to bring about the word "cloud" was only partial. Similarly, the experimenter has some substantial influence on what will happen to the electron by the choice of experiments he will perform ("the questions he will put to nature"); but there is always a certain unpredictability about what any given one of his measurements will disclose ("what answers nature will give"). This unpredictability is because the rest of the universe is always inescapably involved in any observation that we make. The world is neither wholly determined nor arbitrary but, like Wheeler's "cloud," an intimate amalgam of chance and choice.

• CHAPTER NINE •

QUANTUM BUDDHA NATURE

Quantum entities don't "have" or "possess" intrinsic properties. Jonathan Allday, author of *Quantum Reality: Theory and Philosophy*, writes, "Our whole manner of speech . . . rather naturally makes us think that there is some stuff or *substance* on which properties can, in a sense, be glued. It encourages us to imagine taking a particle and removing its properties one by one until we are left with a featureless 'thing' devoid of properties, made from the essential material that had the properties in the first place. . . . Now, it seems, experimental science has come along and shown that, at least at the quantum level, the objects we study have no substance to them independent of their properties."[404] When we remove the last property and the quantum entity dissolves into itself, there is nothing left, no thing that "has" the properties. The quantum field is empty of substance, and yet this substanceless substance appears to be what everything that appears to be substantial is made of.

This brings to mind the Buddhist practice of inquiring into the mind so as to discover who the thinker of our thoughts is. In this contemplation, the practitioner discovers through their own empirical experience that no matter how long they look, they cannot find the thinker anywhere at all. Many people ignore the profundity of what not being able to find the thinker is revealing to us; other people simply try harder, with as much success as trying to catch an ever-receding rainbow, as they double up their efforts to grasp the ungraspable. The Buddha would point out that our inability to find the thinker is extremely significant in that it puts us face-to-face with the empty essence of our own mind—we are simply invited to recognize that there is nothing to recognize. Our true and most essential nature has no features whatsoever and is empty of all formal attributes. This is precisely analogous to the invisible and indivisible plenum of space, the all-pervading quantum field of pure potential out of which all forms and structures of the material universe arise.

This is why the Buddha, who was a true empiricist, counseled his followers to not take his word for it, but be skeptical and do the experiment and look within for themselves. In Buddhist practice, experiencing the "unfindability" of the one who thinks our thoughts is precisely the point of the whole enterprise, as our discovering that there is nothing to discover effortlessly releases us into the emptiness of our true nature. From the Buddhist point of view, recognizing that there is no thinker separate from the thought, which is to say that the thought thinks itself and is nothing other than what it is, is itself considered to be liberation.

The empty nature of the psyche seems to mirror the empty nature of the quantum realm. To quote the Dalai Lama:

> If you subject anything—space, time, matter, whatever you like, even the mind itself—to a certain type of close scrutiny, looking for its actual nature independent of other phenomena, it will dissolve under analysis every time you look for it. Then, if you agree that it is empty of inherent existence, and you try to seek out the very nature of that emptiness, the emptiness itself is not to be found.[405]

In other words, emptiness is itself empty of inherent existence—there is nothing there. It is not that in our analysis we "find" emptiness; on the contrary, we don't find anything, which is what is meant by the term "emptiness."

His Holiness comments, "Comprehending the reality of emptiness transforms the mind such that one's previous mistaken views are banished."[406] In other words, realizing the empty nature of phenomena, be they outer or inner, transforms our mind. This realization frees our mind from its ingrained, habitual tendency to ascribe definite, fixed, and objective attributes of either the inner or outer side of our experience as being "real," existing independently "on their own side." When our mind confronts the all-pervading reality of emptiness, it is emancipated from the core delusion that any independent, objective features of the world exist apart from the mind that is experiencing them. In addition to realizing the emptiness of the seemingly outer world, the mind can potentially self-reflect upon itself and realize that its nature is empty of features, attributes, and forms as well, thus completing the circle—realizing the empty nature of both matter and mind.

His Holiness comments, "It's ironic that analysis approached purely from a physicist's point of view, and confined to physical phenomena, seems to reach a point where it may just be opening the door to Buddhist emptiness."[407] To all appearances, it certainly seems like Buddhism and quantum physics are pointing at the same deeper mystery. To quote Wheeler, "One has the feeling that the thinkers of the East knew it all, and if we could only translate their answers into our language we would have the answers to all our questions."[408] Oppenheimer opines that the "discoveries in atomic physics are not in the nature of things wholly unfamiliar, wholly unheard of or new." Mentioning the similarity to Buddhism, Oppenheimer feels that the insights in modern physics are "a refinement of old wisdom."[409] It's as if, with our advanced technology, quantum physics is translating the ancient, intuitive wisdom of our ancestors into a modern scientific idiom. Schrödinger comments, "Our present way of thinking does need to be amended, perhaps by a bit of blood-transfusion from Eastern thought."[410] Make no mistake about it—quantum physics is very much in agreement with the Buddhist notion of emptiness.

We should be careful, however, in citing the insights of quantum physics, thinking that they "prove" the validity of the Eastern mystical traditions, in the same way we should avoid invoking the insights of the Eastern religions as evidence of the rightness of quantum physics. Unfortunately, many New Age guru types have jumped on the quantum bandwagon with various outlandish mystical claims based on their (mis)interpretation of quantum physics. The undeniable similarity between quantum theory and Eastern spirituality should, however, warrant our highest attention. And hopefully it will inspire further research into their potential complementarity and/or sameness or lack thereof.

Misunderstanding potentially arises when what are called "parallels of identity" (referring to the same "object") are confused with "parallels of analogy" (referring to different domains of objects and different levels of reality). Many physicists, in justifying their denial of the connectedness between the new physics and the Eastern spiritual traditions, claim that their seeming similarities are simply parallels of analogy, not identity. On the other hand, commenting on the "striking similarity of the views on reality" in Buddhism and the new physics, philosopher Andrey Terentyev writes:

> I'd like to stress that we are not just considering analogies in different fields of human endeavor; in fact, both Buddhist thinkers and modern physicists, using very different methods, arrived basically at the same description of [the] reality we live in. This is the point where the parallel worlds of Buddhism and Physics unexpectedly touched each other, and the deeper meaning of this is yet to be appreciated by both parties.[411]

I like Terentyev's image of "parallel" worlds (which are supposed to never intersect) "unexpectedly" touching each other. The deeper meaning of their surprising meeting is still in the process of being decoded and unpacked by both Buddhists and physicists alike. In my ever-deepening studies of both disciplines, it certainly seems that in their striking similarities we are not dealing with superficial "analogies," but encountering fundamental truths concerning the very nature of reality, as if they are both describing "identical" realities.

I have noticed that the typical physicist who disparages the similarity between physics and Buddhism always seems to have little to no experience in the Eastern contemplative traditions, while the ones who have actual experience in these spiritual practices are typically the ones who point out the similarity to the quantum gnosis. I am of the opinion that the reader should do their own empirical/experiential research and make up their mind for themselves.

Here's my two cents: The revelations of quantum physics and the Buddhist worldview certainly seem to be pointing at exactly the same reality.

To quote the Dalai Lama, "It can't be a pure coincidence that both science and Buddhism come to more or less the same conclusion on the nature of emptiness when taking physical objects as the focus of analysis."[412] Buddhist psycho-metaphysicians pointed out the interconnectedness between apparently external objects and the mind of observers over two thousand years before the discoveries of quantum physics. His Holiness comments, "I think Buddhist philosophy and Quantum Mechanics can shake hands on their view of the world."[413] When Buddha realized the true nature of his mind, I find myself imagining that he had directly realized the quantum nature of reality. What I call our quantum nature and what is referred to as our Buddha nature are the same nature, at least in my imagination.

ALCHEMY

The empty nature of reality is revealing itself from opposite directions—through physicists looking into the core of the outer material world, and contemplatives putting their attention on the essence of their minds. To say this differently, the inner "empty" nature of the mind is revealing itself not only through introspection, but also via the medium of the outside world through the revelations of quantum physics. This is reminiscent of Jung's insight into the ancient art of alchemy that the transformation alchemists saw in their flasks was actually a reflection of what was happening within their own psyche.

There is something unknown in both the psyche as well as in matter. The concept of matter has undergone a great number of changes in the history of human thinking. Jung writes, "The real nature of matter was unknown to the alchemist. Inasmuch as he tried to explore it he projected the unconscious into the darkness of matter in order to illuminate it. In order to explain the mystery of matter he projected yet another mystery—his own psychic background—into what was to be explained."[414] In other words, humanity tried to explain one mystery by unknowingly using another.

In conflating these mysteries, the alchemists' created confusion, as these two mysteries are in fact one and the same mystery being explored through two different approaches. This confusion led to the dismissal and rejection of alchemy, which coincided with the start of the modern scientific phase of history where mind and matter were completely severed from each other. This separation between mind and matter can be viewed as the *separatio* phase of the deeper archetypal alchemical operation writ large in and *as* history itself. With the advent of quantum physics, however, the possibility of a resolution concerning the longstanding mystery of how mind and matter go together has at last become available.

As Jung reminds us, the unconscious, once it is activated, is projected into matter, which is to say that it approaches people circuitously, from the outside. The fact that the unconscious appears projected into the seemingly outer world is therefore not surprising, as there is no other way by which it might be perceived. It is as if the psyche "fleshes out" its immaterial nature through the embodied forms of our universe. Pauli writes about the psyche, "From an inner center the psyche seems to move outward, in the sense of an extroversion, into the physical world."[415] Compare Pauli's words about the psyche to Stapp's words describing quantum entities, "An elementary particle is not an independently existing unanalyzable entity. It is, in essence, a set of relationships that reach outward to other things."[416] Interestingly, Pauli completes his quote by saying that once the psyche extends itself into the world, "the spirit serenely encompasses this physical world, as it were, with its Ideas."[417]

The psyche, as a higher-dimensional "no-substance" that is not located in the third dimension of space or time, is able to affect our ordinary lives and reveal itself by mysteriously interpenetrating into and synchronistically configuring our three-dimensional world. Jung writes, "'At bottom' the psyche is simply world."[418] Inhabiting the world, we find ourselves inside of the psyche, just like we do when we become lucid in a dream.

Henry Stapp comments, "The fundamental process of nature lies outside space-time but generates events that can be located in space-time."[419] A higher dimension—whether it resides within the psyche or God knows where—is the repository of nature's core processes. The fact that the most fundamental processes of nature are not even located in space-time is a major clue indicating that an unexamined assumption that underlies scientific materialism is flawed and unnecessarily limiting. This underlying and tacit assumption is that the only things that are real are those things that are observable or measurable in space-time; anything outside of space-time is considered to not only not be real, but to not even exist. The limiting and circular logic built into scientific materialism keeps those who subscribe to its viewpoint locked within the box of third-dimensional space and the treadmill of linear sequential time.[420]

The art of alchemy was based on, and was an expression of a psychophysical unity. Alchemists had little or nothing to contribute to the field of chemistry, least of all the secret of gold-making. But only our overly one-sided, rational, and intellectualized age could miss the point so entirely and see in alchemy nothing but an abortive attempt at chemistry. On the contrary, to alchemists, chemistry represented a degradation and a "Fall," because it meant the secularization and commercialization of a sacred science. To understand the true meaning and function of alchemy, we must not judge it based on the potential chemical insights it contains—this would be tantamount to judging poetry based on scientific data that it contains or its historical accuracy.

Jung writes, "Medieval alchemy prepared the way for the greatest intervention in the divine world that man has ever attempted: alchemy was the dawn of the scientific age, when the daemon[421] of the scientific spirit compelled the forces of nature to serve man to an extent that had never been known before. . . . Here we find the true roots, the preparatory processes deep in the psyche, which unleashed the forces at work in the world today."[422] Needless to say, this process of compelling "the forces of nature to serve man" had a shadow side. Modern corporatized science isn't so much interested in nature as it is, but rather what we can do with it, what "practical" use and profit can be derived from it.

Commenting on this situation, Heisenberg writes, "The progress of science was pictured as a crusade of conquest into the material world. Utility was the watchword of the time."[423] The "purposiveness" (according to Heisenberg, one of science's "catchwords") of science can lead to chaos if the purposes are not understood as parts of a greater whole, a higher order of things. In his book *Across the Frontiers*, Heisenberg cites the following quote, "Purposiveness is the death of humanity."

Philip K. Dick refers to this way of looking at the world (what can I get out of it?) as "use-view," which he correlates with the fall of humanity. This "use-view" perspective is expressed in English philosopher Francis Bacon's words concerning nature, which should be "bound into service . . . and made a slave . . . the scientific goals were to torture nature's secrets out of her."[424] The essence of all forms of slavery is the belief that the slave, in this case nature herself, is an object separate from ourselves. When we view the world in terms of use, in seeking to overcome the world as object in order to enslave it, the world responds by enslaving us.

Rather than viewing the world from the "what's in it for me?" perspective, the words of Richard Feynman come to mind: "Our greatest advances come from researchers not aimed at use but just for fun, curiosity, and desire for understanding."[425] This desire for understanding is an expression of our true nature, which is not bent on dominating and controlling nature, but rather on treating the natural world as if it had an inherent value and life of its own. Along similar lines Wheeler writes, "Fortunately the human heart has the power to seize on the hard rock of 'truth for survival' and jewel it over into a pearl, 'truth for truth's sake.'"[426]

In alchemy, the idea was that existing within matter was a spirit awaiting release. Alchemists were, in their imagination, helping to liberate this divine spirit; the benefit they received was only secondary. As Jung emphasizes:

> For the alchemist, the one primarily in need of redemption is not man, but the deity who is lost and sleeping in matter. Only as a secondary consideration does he hope that some benefit may accrue to himself from the transformed substance as the panacea, the *medicina catholica*. . . . His attention is not directed to his own salvation through God's grace, but to the liberation of God from the darkness of matter. By applying himself to this miraculous work he benefits from its salutary effect, but only incidentally.[427]

The alchemists weren't trying to get something for themselves through their endeavors; they were serving something beyond and greater than themselves that they felt was their very reason for being alive. The notion that there existed a spirit imprisoned in matter needing liberation is a symbolic reflection of what quantum physics has revealed—i.e., that the world of physical matter does not exist objectively, separate from consciousness.

Alchemy is a timeless, sacred art, as the alchemical art involves becoming an instrument for the incarnating deity to make itself real in time and space. Once they had found the "gold," alchemists didn't feel that they possessed or owned it, but were rather its stewards and ministers. Alchemists were having the archetypal imagination that the deity had become imprisoned in matter, bewitched by its own genius for reality creation, and needed their help to become liberated. In other words, the sacred, metaphysical art of alchemists was to liberate the creative spirit of the cosmos from the prison of matter. Creative artists of and for the soul, they were touching their own soul while being in service to the soul of humanity. Alchemists—just like modern day quantum physicists—were a channel for the universe to autopoietically re-create itself in a uniquely evolutionary way. Quantum physics has unleashed this spirit; the question is, will this be recognized and will it be used for the betterment or destruction of our species?

What Wheeler calls "the dream of the alchemists" was to turn something of little value into something precious. The "philosopher's stone" or "lapis" was the fruition of their art. "The lapis," Jung writes, is "a psychological symbol expressing something created by man and yet supra-ordinate to him."[428] The alchemists considered the stone to be the universal medicine, the panacea for what ailed humanity, the elixir of life, a rejuvenating and universe-transforming magical potion. It was the "living stone" mentioned in the New Testament,[429] and was referred to as a "stone that hath a spirit," and also called the "Savior of the Macrocosm." Had the imagination of the alchemists simply run wild? In combining in one image a *stone* (a symbol of materiality) and its seeming opposite *spirit,* they had conjured up what Jung would call a "reconciling symbol," a symbol that reconciles and unites the opposites.

In any case, there seems to be a connection between what the alchemists were pointing at and what quantum physics is illumining. This brings up the question—is the quantum the modern day equivalent to the alchemical *lapis philosophorum*? Are quantum physicists, in their attempt at trying to find the secrets of living matter—and with this knowledge proceeding to transmute nature into something that it couldn't do by itself—unknowingly the living representatives of the ancient and sacred alchemical art?

In one sense, the alchemical opus is a work against nature, an *opus contra naturam,* a work against nature which nature itself desires. The alchemists saw the "prima materia" (the primal matter)—the chaos and raw material out of which the "gold" is made—as an "imperfect body" in need of being perfected. The alchemists, at least in their imagination, were helping nature do what she could not do herself. An alchemical maxim states, "What nature left imperfect, the art perfects." Although the impulse to become conscious exists in nature, within the unconscious of humanity, a conscious human ego (an observer) is needed as a transformative vessel to realize and perfect this natural urge. In alchemy, as in quantum physics, the interaction between subject and object reciprocally transforms both parties.

Jung encountered the historical counterpart of the psychology of the unconscious in alchemists. This is to say that the alchemical stages symbolically expressed the developmental phases of individuation (the process of becoming whole) of an individual. He comments, "The alchemical operations were real, only this reality was not physical but psychological. Alchemy represents the projection of a drama both cosmic and spiritual in laboratory terms."[430] As Jung realized, the entire alchemical procedure could just as well represent the individuation process of a single individual. There is something in the unconscious of humanity that is wanting to become conscious, and it found in both the operations of alchemy and the experiments of quantum physics a "hook" that attracted it so that it could express itself in some way.

To put it simply, alchemists had unconsciously stumbled onto and were participating in what centuries later quantum physicists were tapping into—that what we experience as matter is inseparable from our own mind. The alchemist, Jung continues, "experienced his projection as a property of matter, but what he was in reality experiencing was his own unconscious. . . . Such projections repeat themselves whenever man tried to explore an empty darkness and involuntarily fills it with living form."[431] Any prolonged preoccupation with an unknown, mysterious object acts, in Jung's word, as "an almost irresistible bait" for the unconscious to project itself into the unknown nature of the object and to accept its resultant perception as objective. The more enlightened alchemists (and quantum physicists) were beginning to realize that what they were seeing was not "objective," but was rather a mirrored reflection of the nature of their own minds.

For alchemists, matter had a divine aspect. The alchemists, to quote Jung, "saw their soul, not in themselves, but in chemical matter."[432] The "prima materia" is the "famous secret" and the basis of the entire alchemical opus. The prima materia is called *radix ipsius* (root of itself). It is an *increatum* (a transcendent creation), an uncreated, autonomous, self-generating, spirit-like entity which is rooted in itself, is dependent on nothing (as there is nothing outside of it), and has everything that it needs.[433] Without beginning or end, and in need of "no second," it can by definition only be something of a divine nature. The noted alchemist Paracelsus conceived of matter as being an *increatum,* thereby coexistent and coeternal with God. The prima materia is related to the God-image of the alchemists, *Mercurius,* who interestingly enough begets himself. Notice the similarity to Wheeler's idea of the universe as a self-excited circuit. Similarly, Philip K. Dick writes about "reality experienced as a unified self-governing field (it initiates all its own changes acausally in synchronization)."[434]

Not knowing what they were trying to articulate (truly "not knowing what they were talking about"), the unconscious itself was simultaneously living through the alchemists as it revealed itself to them. Is something similar happening in quantum physics? Something in the alchemists' (and the quantum physicists') minds wanted to make itself known and become conscious, and it projected itself into matter to do so. The self, the intrinsic wholeness of the human psyche, wanted to actualize itself, so it created the art of alchemy (and the field of quantum physics) as a medium for and symbolic expression of its realization.

There is an alchemical saying "solve et coagula," *dissolve and coagulate,* which corresponds to the alchemical operations "solutio" and "coagulatio." Alchemists were dissolving and regenerating elements of their experience both over time, as well as in each and every moment, so as to potentially distill, reconstitute, and create something new. Jung, who many consider to be a modern day alchemist, writes, "I take these thought-forms that have become historically fixed, try to melt them down again, and pour them into moulds of immediate experience."[435] Desolidifying, deliteralizing and deconstructing their experience in each moment empowered alchemists to actively participate in their own transformation and evolution. Alchemists saw the essence of their art as both analytic and synthetic, as it involved separation, discrimination, and analysis on the one hand, and synthesis, consolidation, and integration on the other.

As previously mentioned, physics can be considered a form of art. Though their coupling at first seems strange, art and physics might be an example of Bohr's idea of complementarity, where two seeming opposites actually complement and complete each other. On the surface, physicists break down nature into its component parts and *analyze* the relationship of these parts, whereas artists *synthesize* different levels of reality such that the work of art is greater than the sum of its parts. Physics isn't all about analysis at the exclusion of synthesis, however, while art isn't solely about synthesis while having nothing to do with analysis—both disciplines are alchemical in the sense that they partake of a continual back and forth of analysis and synthesis, of dissolving and coagulating their experience.

Theologically speaking, the shape-shifting *Mercurius* could be considered to be a further emanation of the resurrected body, using the creative imagination of the alchemists as its vehicle of incarnation. Similarly, the quantum could be the modern-day iteration of the alchemical "glorified body" (*corpus glorificationis*—the philosopher's stone), an incorruptible subtle body that is simultaneously using the canvas of the physical world, the art of physics, and the mind of humanity as its joint instruments of revelation. Similar to how alchemists "dreamed up" *Mercurius* as their image of divinity, have we, as modern people, "dreamed up" the quantum to reflect back something similar in ourselves? In any case, both alchemists and quantum physicists are potentially having the epochal, revolutionary, and evolutionary realization that we play a crucial, participatory role in how things turn out. As Jung points out, we don't want to be "like the foolish Parsifal, who forgot to ask the vital question because he was not aware of his own participation in the action."[436]

BOUNDARY BETWEEN THE WORLDS

Quantum physicists' excursion into the microscopic realm of atoms is an adventure that could be compared with the great journeys of discovery that astronomers took into the outer limits of celestial space. The exploration of quantum physics into the realm of the really small is a mirror image of those who explore the outer world, but in the opposite direction. And yet, quantum physics has discovered that there is a mysterious connection between the worlds of the really small and the really big. It is a mistake to think that quantum theory is limited to the microworld. To quote physicist N. David Mermin, "To restrict quantum mechanics to be exclusively about piddling laboratory experiments is to betray the great enterprise. A serious formulation will not exclude the big world outside the laboratory."[437]

Quantum theory points out that the "real world" is not classical, but quantum mechanical. Rather than the quantum realm being illusory, quantum physics points out that the appearance of the macroscopic, conventional world can be likened to a holographic optical illusion produced by the interaction of our sense faculties with quantum reality. Quantum theory insists that our everyday world is embedded in quantum reality, that our day-to-day world is quantum through and through, which is to say that the quantum realm is not separate from the world of ordinary objects. The world of the very small is coextensive with the world at large. There are not two separate domains of nature, one macroscopic and one microscopic; our world is quantum at all scales.[438] The quantum permeates all of existence. Wheeler writes, "So what does the quantum have to do with the universe? Perhaps everything, because in any fundamental theory of existence, the large and the small cannot be separated."[439] Quantum theory applies to big things as well as small; we can't get to first base without quantum theory in dealing with such large-scale objects as stars, for example. Speaking of the classical and quantum worlds, Wheeler writes, "Yet these two worlds are linked. They must be linked. Looked at closely enough, the classical world around us is nothing but a collection of quantum systems."[440]

And yet, our everyday world, with its chairs, trees, and people, seems, at least to all appearances, not to be quantum at all, but quite real and solid, very much in alignment with classical physics' version of reality, with its one-at-a-time sequence of definite actualities. When we throw a baseball, for example, it has a continuous trajectory that can be measured. This is very different from probabilistic quantum entities, which are discontinuous, can take multiple routes to get somewhere at the same time, and get to where they're going in no time at all. And yet, quantum theory tells us that baseballs are quantum objects, too. They have a cloud of probability which collapses from uncertainty to certainty, but their quantum fluctuations are so microscopically small compared to their macroscopic size that they are entirely below the threshold of observation. The elementary particle and the baseball differ only in scale, not in principle. To quote physicist Hideo Mabuchi, it is as if "the universe were ruled by atoms' aversion to the public embarrassment of quantum behavior writ large."[441]

In the transition from the random uncertainty of the quantum realm, where particles ceaselessly spring into and out of existence, to the seeming solidity and orderly certainty of our everyday world, the question naturally arises, where is this boundary between the quantum world, where things don't actually exist in a real way but in a state of potentiality, and our everyday world, where things at least appear to exist in a solid-seeming way? Wheeler asks the question, "If the world 'out there' is writhing like a barrel of eels, why do we detect a barrel of concrete when we look?"[442] How do the classical and quantum worlds join together?

To quote Allday, "It is very difficult to see how all the funny business going on at the atomic scale can lead to the regular, reliable world we spend our lives in."[443] The quantum reality of the microworld is inextricably entangled with the classical reality of the macroworld, as the part has no meaning except in relation to the whole. Paradoxically, in quantum physics the macroworld determines, through the act of macroscopic observer-participancy, the microscopic reality that it itself is made of. And if the ordinary-seeming classical realm manifests out of the underlying quantum domain, where did the "weirdness" or "funny business" of the quantum realm go?[444]

The moment of observation appears to be the link between the uncertainty of the quantum world and the apparent certainty of the classical world, for observation is the point at which what might happen crystallizes into what does happen. As Heisenberg writes, "The transition from the 'possible' to the 'actual' takes place during the act of observation."[445] This brings up the question: *How* does the act of observation, of gaining mere information (knowledge or "software"), modify the state of macroscopic things ("hardware")? "Good question," I imagine Wheeler saying.

According to quantum theory, the whole universe is in a quantum state, which is to say that, at least in principle, there ultimately is no boundary between the microscopic/quantum realm and the macroscopic/classical realm. Though some physicists still cling to the idea that these two realms are separate, others consider it delusional to conceive of there being a sharp distinction between the two. Still others are of the opinion that the line separating the classical from the quantum world is to a large degree arbitrary, of our own choosing. Wheeler playfully comments, "I know that in that empty courtyard many a game cannot be a game until a line has been drawn—it does not matter where—to separate one side from the other. . . . Even if neither you nor I know how to define that line."[446] In any case, the line between the classical and quantum worlds is a mobile boundary that is hard to pin down.

In another example of how far from the traditional physicist Wheeler is, in *The Frontiers of Time*, he has an imaginary dialogue (an example of active imagination) with "a colleague in another realm of thought."[447] His imaginary colleague asks Wheeler, "May I question you now about the game itself? How would you describe it if forced to commit yourself?" Wheeler responds, "Let us try to squeeze an answer into three sentences. . . . The universe is a self-excited circuit. As it expands, cools and develops, it gives rise to observer-participancy. Observer-participancy in turn gives what we call 'tangible reality' to the Universe."[448]

There is a linguistic gulf between the world of quantum phenomena and the classical world of everyday experience. The quantum domain and the classical domain require two different ways of speaking. From one point of view, the location of the line demarcating the classical from the quantum reflects the point at which physicists stop using one vocabulary and start using the language of the other. The best way to become confused is to try and talk about the classical world in the vocabulary of quantum theory, or vice versa. In other words, we run into problems when we try to import classical (large-scale) concepts into the subatomic domain—concepts that have no business being there. The necessity of shuttling back and forth between the two forms of language can easily create misunderstandings, since in many cases the same words are employed in both languages. We should not intermingle these two languages, which is to say we need to develop our capacity for more subtle and nuanced modes of thinking. Ludwig Wittgenstein was of the opinion that many of the great "philosophical problems" were really nonproblems, in that they are confusions generated by the misuse of language. This brings up an interesting question: What follows from talking about reality in one way or the other—what do we gain and what price do we pay for adopting one vocabulary and giving up another?

Things that are transforming from one state to another have an interesting logic. When something changes from having a property to not having a property, there is a moment in between when, in some sense, it both has the property and does not have the property. Aristotle was quite interested in this "in-between" stage; for example, when an egg is changing into a chicken, there is a moment when it is both egg and chicken and neither egg nor chicken. In any case, it certainly seems as if the boundary that simultaneously connects and separates two different states, be it egg and chicken or the quantum world and the everyday, classical world, is an extremely interesting place, the exploring of which could bring about great insights.

This interfacing between the classical and quantum world is mirrored in the realm of the psyche. Jung writes, "Somewhere our unconscious becomes material. . . . Somewhere there is a place where the two ends meet and become interlocked. And that is the place where one cannot say whether it is matter, or what one calls 'psyche.'"[449] This is the place where it is impossible to distinguish between matter and psyche, as if the opposites are revealing their inseparability and interconnection with each other to the point where we can't tell them apart.

The boundary between the classical and quantum worlds is mirrored in the psyche by the boundary between the conscious and the unconscious. Wheeler sounds like a depth psychologist when he writes, "The line between the unconscious and the conscious begins to fade."[450] Jung would add that in actuality there is no clear boundary demarcating the conscious from the unconscious—one begins where the other leaves off. The question is: What happens at the boundary in the moment when one turns into the other?

Where is the line between us and the world? Is the need to even draw such a line an old habit of mind that the emerging quantum gnosis is helping us to erase by revealing it to be unnecessary, misleading, and possibly even nonexistent?

• CHAPTER TEN •

UNCERTAINTY

There is always an element of uncertainty in describing quantum entities; they can never be known in their totality. Heisenberg discovered the uncertain nature of nature, which is therefore referred to as "The Heisenberg Uncertainty Principle" (or more simply, "The Uncertainty Principle"). For example, we can never know—experimentally or in principle—a quantum entity's position and momentum (considered to be the two measurable variables which are the cornerstones of classical physics) at the same time. This makes it impossible to pin these quantum objects down; they defeat all our attempts at complete surveillance.

We can decide what we want to measure, but we can't measure all properties of a system at once. In what Wheeler calls "the great lesson of quantum mechanics,"[451] if we choose to measure one thing, we prevent the measurement of something else. It is pointless to speculate upon whether the missing information "exists." It is not a question of building better technology to one day know both of these properties. Instead it is "as if" these quantum entities simply don't possess both qualities at the same moment. So this is not merely a matter of our lack of knowledge; our lack of complete clarity is a clear reflection of the ontological condition of the quantum entity. It turns out that one of the biggest breakthroughs in modern physics is the recognition that whatever we say about the state of a physical system is always limited and approximate.

It is meaningless to consider, for example, an electron to have a precise location and motion at one and the same time. Being that quantum entities don't have any real attributes until they are measured, and being that these different attributes can't be measured at the same time is to say that certain attributes can't exist at the same time. Quantum physics has shown that not only is the full description of these quantum entities unknown, but, because they do not exist prior to being measured, they are ultimately unknowable. Not only is there a limit to our knowledge of these quantum entities when we are looking at them, we have no idea at all what they are doing when we are *not* looking at them.

If we know where these quantum entities are, we pay a price, for then we don't know where they're going. Similarly, if we know where they're going, we don't know where they are. Imagine if measuring our height changed our weight, and then when we measured our new weight, our height was no longer the same. We would never be able to get an accurate reading of both our height and weight at the same time. We reach a certain point at which one part or another of our picture of nature becomes blurred, and there is no way to refocus that part without blurring another part of the picture. Nature is so constructed that we can study one aspect of nature or another, without any possibility of studying both aspects simultaneously.

For example, say we were trying to learn about an Oriental carpet by studying the details of the weave and the overall pattern. To analyze the weave, we would focus closely on it, thereby losing track of the overall pattern. To see the overall pattern, we would have to step back a bit, thereby losing sight of the details of the weave. We could never see both at the same time. The more exact our description, the further away we are from a complete description. In other words, we can give an exact description of something if we limit the description to concern a partial aspect only. This reveals a surprising fundamental structural limitation in our perceptual access to information about our world. To quote Einstein, "As far as the laws of mathematics [the language of physics] refer to reality, they are not certain; and as far as they are certain, they do not refer to reality."[452] Uncertainty is not confined to the microworld; it is embedded throughout nature and is therefore inescapable.

• • •

ANIMAL HOUSE

From their writings, it is obvious (and understandable) that physicists themselves haven't fully comprehended and don't quite know what to make of the great truth that they have unwittingly stumbled upon. They have been forced to wrestle, not just intellectually but emotionally, existentially, and spiritually with their own discoveries in the quantum realm. Quantum theory has pushed its adherents to the very edge of the unknown, both out in the world and within themselves. In the classic book *Quantum Theory and Measurement,* which Wheeler cowrote with Wojciech Zurek, the authors write regarding the quantum, "What else is it but an unfamiliar animal, confined to an animal house? And how else can one better capture its newness than by walking around, looking at it through one window after another, seeking to combine fragmentary views into a total picture?"[453] One of the challenges is how to communicate what this mysterious, unfamiliar creature is like to someone who has no comparable reference point.

In encountering the mysterious realm of the quantum, we are like blind men touching the proverbial elephant, coming to different conclusions about the nature of the entity we are encountering depending upon what part of the elephant we are touching. The fact that the person holding onto the elephant's tail thinks the elephant is like a rope, and the person holding onto the elephant's ear thinks it is like a leaf, points to the idea that things depend upon our point of view. But in the quantum world, not only do we reach different conclusions depending on how we "touch" the system, but the act of making one measurement rules out making a different kind of measurement at the same time. Imagine that when one blind man touches the elephant's trunk, the elephant's leg disappears into a kind of fog, escaping the touch of another blind man. Afterwards, upon returning to the quantum elephant, one of the blind men can touch the elephant's leg, provided they don't try to grasp the trunk at the same time.

In coming across the quantum, physicists, using an aperspectival approach, are illuminating a strange object that is casting various shadows on a wall from as many different angles as they can imagine. With enough of these shadows we can possibly attempt to reconstruct the illuminated object. Though seeming incredibly complicated, which it is on one level, Feynman brings out another perspective when he says, "Perhaps a thing is simple if you can describe it fully in several different ways without immediately knowing that you are describing the same thing."[454] Each of our models for understanding quantum reality has both its utility as well as its limits. The multidimensional nature of quantum reality demands a plurality of perspectives. When all of the various perspectives of the multifaceted quantum reality are combined and looked at together, it gives us a greater resolution and capacity to see what no single vantage point can reveal.

It behooves us to "light up" the quantum from as many different angles as we can imagine. Speaking about quantum theory and the observed phenomena, Bohr says that only "by lighting this relationship up from all sides and bringing out its apparent contradictions, can we hope to effect that change in our thought processes which is a *sine qua non* of any true understanding of quantum theory."[455] Any genuine understanding of quantum physics necessarily involves a change in our thought process.

This confined, unfamiliar quantum animal is like a multidimensional dream figure, an aspect of ourselves that exists deep within us. We can track its presence within the bubble chamber of our own mind. It reminds me of how we can never directly see an archetype—which represents, in Jung's words, "*psychic probability*"[456]—but are only able to circumstantially infer its existence by its effects. Jung writes, "We meet with a similar situation in physics: there the smallest particles are themselves irrepresentable but have effects from the nature of which we can build up a model."[457] Jung continues, "Nobody has ever seen an archetype, and nobody has ever seen an atom either."[458] When the existence of two irrepresentable factors are encountered, there is always the possibility that it may not be a question of two factors but of only one.

An archetype, like the atoms of the quantum realm, underlies, informs, and manifests in our third-dimensional world. It is not identical with, but rather transcendent to its spatiotemporal manifestations. It is Jung's opinion that "Sooner or later nuclear physics and the psychology of the unconscious will draw closer together as both of them, independently of one another and from opposite directions, push forward into transcendental territory, the one with the concept of the atom, the other with that of the archetype."[459]

There is no space-time inside the atom. The subatomic, elementary quantum entities, though never seen directly and not existing in space and time, have a subtle body all their own. Through their subtle body, these quantum entities leave their calling card, so to speak, making an impression on both our experimental apparatus as well as within our minds. The realm of subtle bodies exists in a state "between" matter and spirit, like some sort of intermediate realm (similar to a Tibetan "bardo" i.e., a gap or in-between state). To quote Jung, "There did exist an intermediate realm between mind and matter, i.e., a psychic realm of subtle bodies whose characteristic it is to manifest themselves in a mental as well as a material form."[460] The subtle body of the quantum realm is both an expression of as well as a doorway into the indissoluble unity of "physis" (a Greek word which means "nature") and "psyche" (which is a part of nature). From all appearances it seems as if our world is a lower level reflection or projection of a higher-dimensional reality, much like Plato's shadows on the walls of the proverbial cave.

NO PATH

We are a multiplicity of potential selves, like multifaceted gems with many different aspects, each waiting for its moment to shine. We tend to unconsciously identify with whichever one of our multiple selves "lights up," thereby assuming the driver's seat, in any given moment. Each time an internal or external trigger causes us to quantum jump—in no time at all—from one "self" or identity pattern to another, the whole world around us appears to change also, as if we have entered a parallel universe without knowing it. We tend not to notice this sudden change because, from the viewpoint of the "new" self's perspective, the world that we find ourselves in—which is our new self's projection and a reflection of ourselves—seems perfectly natural. There is often very little associative continuity between these different selves, which is why their abrupt quantum shifts are rarely noticed. The world of the previous self that had moments before been in the driver's seat is typically forgotten about as if it had never existed. It is as though each of our selves comes with its own corresponding world.

The phenomena of quantum jumps is one of the strangest goings-on in the quantum realm. Not only is the "path" of quantum objects unknowable, but quantum objects can't even be conceived of as having a path in the normal sense of the word; the very notion of having a path comes into question. Talking about "the whole idea of following a single path," Wheeler simply writes, "meaningless."[461] Schrödinger writes, "For we are so used to thinking that at every moment between the two observations the first particle must have been *somewhere*, it must have followed a *path*, whether we know it or not."[462] All talk about paths, however, is a hangover from classical thinking and traditional ways of visualizing the universe. Quantum particles follow a "pathless path," which at least on the surface sounds similar to the "pathless path" of spirituality echoed by innumerable mystics throughout the ages. Quantum entities appear and disappear, shifting dimensions as we do when we are born and when we die.

To quote physicist Max Born, "No language that lends itself to visualizability can describe quantum jumps."[463] These quantum objects can be at Point A in one moment and, as if by magic, instantaneously be at Point B, *without having traversed a path between these different locations.* Bohr referred to a "quantum jump" as a process "transcending the frame of space and time."[464] It appears that the quantum entity suddenly flickered out of existence, passed through a limbo of "no time" and "no space," and then reappeared somewhere else with no physical process connecting these two states of being. Quantum leaps characterize the discontinuous movement seemingly intrinsic to nature.

Indulging in an unconscious habit of (classical) thought, we assume that we can ascertain the identity of a subatomic particle by keeping the particle under continuous observation. Addressing this very issue, Schrödinger writes, "This habit of thought we must dismiss. *We must not admit the possibility of continuous observation.* Observations are to be regarded as discrete, disconnected events. Between them there are gaps that we cannot fill in."[465] Nature herself seems to reject continuous description. Wheeler uses the image of the "Great Smoky Dragon" to illustrate this same point. The head and the tail of the dragon can be seen, but its body, which seemingly connects its two ends, is obscured by smoke. It is the same with the photon. For example, we may know about its emission and detection, but what happens in between is a mystery.[466]

We have been trying to follow the motion of individual quantum entities such as photons through time and space while all along these entities have no real existence in time and space; it is time and space that exist through and by virtue of these entities. Outside of and transcendent to space and time, photons (which make up light) embrace all of space and time in one singular, eternal, resonant embrace. This is to say that light (from the perspective of an observer within space-time) appears to be traveling (at the speed of light) through and inside of the very space and time that it is at the same time creating. In so doing, light can be seen as weaving the structure of the space-time continuum from both inside and outside as one inseparable and continuous waveform.

Space-time constitutes the phenomenal realm, whereas quantum objects exist in the noumenal realm outside of the confines of space-time. Space and time originate from a deeper dimension of existence that is not *in* space-time, but rather, *in-forms* space-time. Sir James Jeans writes, "The ultimate processes of nature neither occur in, nor admit of representation in, space and time."[467] Quantum physics compels us to think of the creator of the universe as working from a vantage point outside of time and space, just as the artist is outside their canvas.

Space and time—what Heisenberg refers to as "basic forms of human imagination"—don't exist in and of themselves, at least in the way we have been conditioned to imagine them, but are anthropomorphic concepts, constructs of the mind. This is consonant with Immanuel Kant's description of space and time in his seminal book *The Critique of Pure Reason* as "pure forms of intuition," by which he means that space and time are primarily structures of consciousness that provide the underlying framework for our perceptions, not objectively existing conditions in the outside world. According to Kant, sensory information is organized in our minds via an a priori, preexisting, intuitive representation of space and time that lies deep within our consciousness.

To the extent that we are asleep to the fact that space and time are merely constructions of the mind, we become conditioned to experience these constructs as if they existed outside of ourselves in the seemingly outer world. Relating to the world in this way is to become entranced by a false appearance (arising from the mind itself), which hides the real source of space and time as being within our minds. Wheeler writes, "The concepts of spacetime and time itself are not primary but secondary ideas in the structure of physical theory. These concepts are valid in the classical approximation. . . . " In the world of the quantum, Wheeler continues, these concepts "have neither meaning nor application. . . . There is no spacetime."[468] It should be noted that in creating time and space and then thinking that our creation exists outside of us, we are concurrently creating the illusion that we exist as a separate self within a time and space of our own making.

To quote Einstein, "Time and space are modes by which we think and not conditions in which we live."[469] Space and time are like the elements of a language used by an observer to describe their environment. As Leibniz put it, "Time and space are not things, but orders of things."[470] Our minds employ the categories of space and time to inform and make sense of our experience similar to how our dreaming mind displays images and unfolds its experiences in a space and time of its own making. Our perceptions of space and time in a dream exist only to the extent that we conceive them. As if our universe is an immense dreamspace, all points in space and time are ultimately interconnected to all other points via the dreamer.

Our minds have a mysterious power to shape our world that is not limited by the categories of space and time. Jung writes:

> In man's original view of the world, as we find it among primitives, space and time have a very precarious existence. They become "fixed" concepts only in the course of his mental development, thanks largely to the introduction of measurement. In themselves, space and time consist of nothing. They are hypostatized concepts born of the discriminating activity of the conscious mind, and they form the indispensable coordinates for describing the behavior of bodies in motion. They are, therefore, essentially psychic in origin.[471]

Though a function of and originating within our minds, spacetime is a crucial factor in the equation of our universe. Space and time are not external parameters. They are not determinate givens outside of phenomena but are an integral aspect of phenomena. To quote Wheeler, "Spacetime is an important part of the action, not just the place where action occurs."[472] We have been conditioned to think of space as the stage in which the drama of physics is performed; quantum physics is pointing out that the stage is itself one of the performers.

In essence, the ideas of space and time are creations of the mind that serve as the screen upon which we project the contents of the depths of our mind, both conscious and unconscious. Being constructs of consciousness as well as the receptacles for its projections, space and time serve as consciousness's own way of providing a context for its contents so that they can be revealed, brought to light, reflected upon, and contemplated by a consciousness that is forever getting to know itself in new ways. "Man's coming to awareness," to quote philosopher Jean Gebser, "is inseparably bound to his consciousness of space and time."[473] It is becoming increasingly harder to deny that physics, the science of the nature of the physical world, cannot be distinguished from the study of the structures of our own consciousness. This realization is so hard to see because it is so obvious.

• CHAPTER ELEVEN •

MERLIN

It is not that the deeper reality is veiled and we cannot know it; rather quantum physics is pointing out that there is no deeper, independent reality based on our ordinary conceptions of what that means. Our universe is not a two-decker affair made up of appearances and an underlying reality, like a mask with a face behind it. No one model of reality sits on the throne wearing a crown, existing in royal splendor above all the others. Each model of reality has its own uses and utility in its own appropriate area. How different is our thinking that there is one true model, or one deep reality, from the medieval idea that there is one true religion? The notion that there is one ultimately true model of reality is a residue of our "objective reality hangover," a malaise from which our species is in the process of recovery. Quantum physics continually reminds us that there is no one absolute way that things are, for "things" do not actually exist separate from our own minds. Being attached to our version of reality as not only being correct but being the be-all and end-all of interpretations is as silly as claiming that the thermometer tells more of the truth than the barometer.

Whereas in the mythical land of Oz reality stems from the wizard's conjuring trick, in the quantum realm, Bohr argued, there is no wizard. There is "nothing" behind the curtain; all we see is the formless archetypal play of phenomena itself, a display which is empty of inherent existence and intrinsic meaning, yet is inextricably linked to our consciousness and its various operations. This is both a display "to" our consciousness and an expression "of" our consciousness at the same time, as the distinction between subjective and objective reality dissolves. From all appearances, physics seems utterly incapable of ever unveiling the nature of a quantum "veiled" reality conceived of as existing separately and independently of consciousness. It is only through recognizing that quantum reality is reflecting back to us the essential pure nature of consciousness that we can hope to unveil the true nature of the quantum, thereby realizing it to be one and the same as our own true nature.

Einstein once remarked, "The more one chases after quanta, the better they hide themselves."[474] Wheeler calls the quantum principle the "Merlin principle" because of the way the ever-elusive quantum shape-shifts and, Mercury-like, changes form to continually escape our too-limited and limiting conceptions of it. It is noteworthy that this is precisely the way alchemists' described the philosopher's stone. Merlin was the wizard who guided King Arthur, appearing to Arthur in different guises depending upon circumstances. Merlin was constantly changing his identity in the same way that an electron does. Wheeler recounts, "You remember Merlin the magician; you chased him and he changed into a fox; you chased the fox and it changed to a rabbit; you chased the rabbit and it became a bird fluttering on your shoulder."[475] Just like trying to catch a rainbow or chase after a projection, the quantum always eludes our grasp, as if it had multiple personalities. The mercurial quantum is forever shape-shifting in response to the various modes of thought by which we try to apprehend it. If someone says that quantum theory is completely clear to them, it was Bohr's opinion that they haven't really understood the subject.

As physicists have chased the quantum/Merlin principle, to quote Wheeler, "in each ten years of its history, it's somehow taken on a different color, each time growing more magnificent in plumage, more penetrating in meaning, and more comprehensive in power."[476] The further we descend down the quantum physics rabbit hole, the more magnificent the plumage of this very strange quantum bird. The more we appreciate the quantum realm, the more it appreciates, and the more there is to appreciate, as if it's the gift that never stops giving, a wish-fulfilling jewel beyond belief. As Wheeler reminds us, the quantum, the smallest stuff in the universe—"a marvelous stimulus, hope and driving force"[477]—is the crack in the armor that covers the secret of existence. Big stuff indeed!

It is Wheeler's opinion that in exploring this opening we are at the beginning, not the end. He writes, "All that the 'quantum' means and implies we are *still* far from understanding. Let us therefore not close the door to tomorrow's deeper comprehension by trying to supply a premature definition of 'quantum.'"[478] The current story of our understanding of the quantum realm is simply a prologue of coming attractions. Wheeler writes that quantum theory's "successes are legion. But the last word has not been written on it."[479]

The "unimaginable wonders" that have already been discovered, to quote Wheeler, "take second place to those still waiting to be found. The undiscovered lies around us in every direction . . . [we] live still in the childhood of mankind."[480] Wheeler strongly felt that the greatest discoveries were yet to come, a perspective that greatly inspired all who studied with him. We are, in Isaac Newton's words, like children playing with pebbles on the seashore, while the vast ocean of mystery is spread out before our reach. Wheeler writes in his autobiography that "discoveries of the twenty-first century will astound us, or our heirs, with their novelty . . . the discoveries of the preceding decade had only taken the cap off the bottle."[481] Images of letting the magical (and potentially wish-fulfilling) genie out of the bottle come to mind. To quote Freeman Dyson, "The poetic Wheeler is a prophet, standing like Moses on the top of Mount Pisgah, looking out over the promised land that his people will one day inherit."[482]

When asked in an interview whether he thought we will ever understand why the universe came into being, Wheeler answered, "Or at least how. . . . Why is a trickier thing." When pressed if he thinks that physicists might one day have a clear understanding of the origin of the universe, Wheeler enthusiastically responded, "Absolutely . . . Absolutely." [483] Wheeler dreamed of a future when "as surely as we now know how tangible water forms out of invisible vapor, so surely we will someday know how the universe comes into being."[484] Dyson says, "But it is the young people now starting their careers who will make his dreams come true."[485]

DREAM STUFF

Etymologically the word "science" comes from the Latin word "scire," which means "to know." What the founders of quantum physics realized is that the proper subject matter of science is not what is "out there," but rather what we can "know" about our world and ourselves. At the quantum level science becomes inseparable from epistemology. Quantum physics has realized that it is no longer representing the state of, for example, an objectively existing elementary particle, per se, but rather only our "knowledge" of its apparent behavior—a subtle but very important difference. This knowledge is a state of mind experienced in our subjective sphere of consciousness rather than being a state of some actual, external, material thing. This "failure of thing-ness" is one of the fundamental features of the quantum world. In the quantum realm we never end up with things, but always with interactive relationships. Our thinking mind can't grasp or relate to the simplicity, elegance, and ungraspability, of the quantum realm.[486]

Physicist Nick Herbert, author of *Quantum Reality*, calls the fundamental elements that the quantum realm are composed of "quantumstuff," a (non)substance which, in his words, "combines particle and wave at once in a peculiar quantum style all its own."[487] Wheeler's colleague, physicist Wojciech Zurek, refers to this quantumstuff as "dream stuff." This quantum dream stuff, the underlying fabric out of which what we call reality is made, is what is called "epiontic." The word epiontic is the synthesis of the two terms "epistemic" (the root of the word "epistemology," which has to do with the act of "knowing") and "ontic" (the root of the word "ontology," which has to do with "existence" and "being"). To say something is epiontic is to suggest something whose existence is intrinsically intertwined with the knowledge we have of it. To be epiontic is to imply that the act of knowing creates its being, which is to say that just as within a dream, the act of perception creates the existence of whatever is perceived. At the quantum level, being and knowing, perception and reality, epistemology and ontology are inextricably entangled.[488] The world that appears to be an independent material world is constructed from "quantum epiontic dream stuff" which is of the nature of mind, or consciousness.

This quantum epiontic "dream stuff" is capable of producing the seeming solidity of the material world from out of the process of perception. To quote Graham Smetham, author of the excellent book *Quantum Buddhism: Dancing in Emptiness,* "The appearance of the material world is a matter of deeply etched quantum 'epiontic' memes!"[489] The more often a particular perception takes place, the more deeply it is imprinted into our unconscious, and hence, the more likely it is to reoccur in the future. The laws of physics that are extrapolated and distilled from our experience of the universe are based on regularities in our habits of what we pay attention to.

Note the similarity to biologist Rupert Sheldrake's idea of "morphogenetic fields." Sheldrake is of the opinion that memory is inherent in nature, and that the "laws" of nature are more like ingrained habit patterns which create a well-worn groove or karmic rut in our perceptual system that makes it more probable that we will see things in this way in the future. Once our inner perceptions gain enough momentum, at a certain point they will be perceived as external to and independent of ourselves. Perceptions that subscribe to the inherent existence of the physical world feed back and strengthen the unconscious habitual tendency to perceive the world in this same way in the future, as well as making it more likely that the world will continue to appear "as if" it is inherently existing.

If we buy into the perspective that the world objectively exists in and by itself, we have fallen under a self-created and self-perpetuating spell, evoking evidence that simply confirms our original unexamined assumption. This is a process in which our mind's own genius for cocreating reality is unwittingly turned against us in a way that can severely limit us, stifling the awareness of our options and thus crippling our greater potentials. We can become imprisoned by our belief in the objective truth of our perceptions in such a way that we hypnotize ourselves and literally become blind to our imprisonment, remaining convinced that in our deluded state we are simply "in touch with reality."[490] Unfortunately, this is a grievous error with quite serious consequences that many of us have fallen into.

The persistent appearance of the classical world is generated by innumerable sentient beings through a continuous web of rapidly repeated, habitual perceptions over vast stretches of time, which amounts to a collective intersubjective feedback loop. Once the appearance of an apparently stable material world gains enough momentum it develops a self-sustaining pattern which confers a seeming immutability upon our world, a perception which literally becomes reinforced, inscribed, and embedded into the very quantum ground of being.

Speaking of quantum events in his typically poetic way, Wheeler writes, "flaunting their freedom from formula, they yet fabricate firm form."[491] Solidifying the fluid dreamlike nature of our world, we then create a collective dream that seems by all appearances to be solid and fully classical. Referring to the outside world, Zurek writes that in whatever way it manifests it acts "as a communication channel. . . . It is like a big advertising billboard, which floats multiple copies of the information about our universe all over the place."[492] The more often a perception of an independent, objective world is made, the more potent becomes the classical world's advertising billboard campaign, increasing its broadcasting power as it further proliferates its meme into more people's minds. "The universe," Wheeler writes in his personal journal, is "an enormous construction we all have a part in." We are all collaboratively dreaming up the universe together. To quote Wheeler, "We are tiny patches of the universe looking at itself–and building itself."[493]

ALCHEMICAL IMAGINATION

The viewpoint that is emerging from the cutting edge of quantum physics is that, instead of being an epiphenomenon of matter, consciousness is the ontological ground and driving force of the process of reality itself. Max Planck, commenting on what the new physics was revealing to humanity, famously said, "Mind is the matrix of all matter."[494] Consciousness is in some mysterious fashion creating the "stuff" of the material world. Wheeler goes so far as to say, "In what medium does spacetime itself live and move and have its being? Is there any other answer than to say that consciousness brings all of creation into being, as surely as spacetime and matter brought conscious life into being? Is all this great world that we see around us a work of imagination?"[495] Keep in mind these aren't the words of some ungrounded New Age nutcase, but the words of one of the most brilliant and rigorous physicists of the last century. Like Wheeler conjectures: Is this world simply a "work," as well as the "play," of imagination?

Quantum physics is nature's way of telling us something. It is a living revelation that we are all dreaming up, in typical quantum style, to *potentially* awaken us. Everything depends upon nothing, absolutely nothing other than if we recognize what is being revealed to us or not. Our imaginative, dreaming, and visionary capacity links directly into the quantum realm, sourcing itself from the underlying quantum field (the quantum plenum of pure infinite potential), thereby interfacing with and becoming an open portal for the "divine creative imagination" to potentially transform our world. Our creative imagination is truly divine in that it literally affects, in-forms, and shapes the suprasensory blueprint that underlies this seemingly mundane and "solid" material world of ours.

Interestingly, in alchemy imagination was the key that opened up the door to the discovery of the philosopher's stone. To quote Jung, "The concept of imagination is perhaps the most important key to the understanding of the opus."[496] It is the alchemists' version of imagination—what they called "true imagination" (a creative activity originating out of and expressing the wholeness of the self), as distinct from mere "fantasy" (a repetitive and self-soothing activity of the ego, the fundamental purpose of which is to avoid a relationship with life)—that Einstein was referring to when he famously said, "Imagination is more important than knowledge. For knowledge is limited, whereas imagination embraces the entire world."[497]

Alchemists talk about "the imaginative faculty of the soul," a term by which they are giving a clear indication of the secret essence of their sacred art. Alchemists, to quote Jung, saw imagination as "a concentrated extract of the life forces, both physical and psychic."[498] The imaginative faculty of the soul is not merely a human attribute but a divine activity of the soul in which the human imagination participates and bears witness. The human imagination is enveloped in, interwoven, and suffused with the divine, creative imagination—what Jung calls "the world-creating imagination of God"[499]—which is the imagination that is imagining/creating the whole universe in this very moment. Jung writes, "If a man puts his hand to the opus, he repeats, as the alchemists say, God's work of creation."[500]

In becoming akin to modern day alchemists, quantum physicists are tapping into our God-given role as cocreators of the universe.[501] There is a "self-secret" dimension that is staring us in the face that we experience in each and every moment, but we can only see it if we have the eyes to see. In *consciously* accessing this dimension, we begin to actively interface with the creative powers of the universe, giving us the capability to effect real change in the universe.[502] We are created in the image of our creator, which is to say we are gifted with creative power, we are creator beings. In becoming an intermediary through which the divine, creative power is expressed and made real in time, we are participating in a "re-creation" of the eternal act in the play of creation.

Wheeler "confesses" that, in apparent moments of lucidity, "sometimes I do take 100 percent seriously the idea that the world is a figment of the imagination and, other time, that the world does exist out there independent of us."[503] He seemed to be fully immersed, both personally and professionally, in the challenging internal process of attempting to integrate the modern-day Copernican revolution of the mind that quantum physics portended. He continues, "I subscribe wholeheartedly to those words of Leibniz, 'This world may be a phantasm and existence may be merely a dream.'"[504] Quantum physics was revealing that the world that Wheeler had spent his whole life trying to understand existed nowhere except in the very mind with which he was trying to understand it.

To realize that the world is in our minds is to realize that our minds are in the world, which is to recognize that psyche and matter are not separate from each other. Jung writes that "matter is a thin skin around an enormous cosmos of psychical realities, really the illusory fringe around the real experience, which is psychical."[505] The physical world is the surface or skin of an unknown, vaster world which often may seem remote and otherworldly, and yet is a world that we are in and to which we belong. We are indeed "such stuff as dreams are made on." To quote Jung, "Far, therefore, from being a material world, this is a psychic world, which allows us to make only indirect and hypothetical inferences about the real nature of matter."[506]

We couldn't imagine or "dream up" a more dreamlike physics than quantum physics if we tried. Quantum physics is the physics of the universal dream in the sense that quantum physics is simultaneously pointing to the dreamlike nature of reality while being an expression of the very dreamlike nature at which it is pointing. In this spirit, Bohr himself said, "Well, yes, one could also say that we are not sitting here drinking tea, but that we're dreaming all that."[507]

In the same way, in my role at this very moment as writer, and your role as reader, we are both collaboratively "dreaming" these very words that I am writing in this very moment. It is like you (as the reader) are dreaming me up to write these words, and I (as the writer) am dreaming you up to read them. In this moment that I am writing these words, and in the very moment in the future (from my temporal perspective) that you find yourself reading my words (the present, from your perspective), you and I are mutually sharing an interaction taking place in, outside of, across, beyond, through, and over time. [508] This interchange is made possible through our interconnectedness, which is to say, when you get right down to it, that we are in actuality not separate from each other. The implication: these very words are the manifestation of your own mind appearing—just as if you were within a dream—seemingly outside of yourself.

The discovery of the quantum observership-based nature of reality represents the first rupture in the armor of the classical chrysalis that has long encased the human mind and fettered the human spirit, tightly holding it in a state of slumber, dreaming the limiting dream of a deterministic, clockwork cosmos. Irreversibly awakening out of its somnambulistic trance, humanity is going through an evolutionary metamorphosis in which it is unfurling its iridescent wings of creative imagination as it flies into the open-ended space of previously undreamt possibilities, releasing itself into the luminous imaginal sky of freedom.

SUMMARY OF KEY POINTS IN PART I

- There is no objective reality independent of an observer.
- We live in a participatory universe. The observer affects what is observed by the mere act of observing.
- Quantum entities exist in a multiplicity of simultaneous potential states (called a superposition), hovering in an abstract realm between existence and nonexistence prior to being observed.
- The act of observation is the very act which turns the potentiality of the quantum world into the actuality of the seemingly ordinary world.
- Our act of observation not only changes the present state of the universe, it reaches backwards in time and changes what we can say about the past.
- The questions we ask make a difference.
- The laws of physics are not written in stone, but are mutable.
- The universe is a seamless, undivided, and interconnected whole. An expression of this wholeness is that each part of the universe is in communication with every other part in an immediate and unmediated way.
- Quantum entities can jump from one place to another without traversing the path in between.
- Quantum theory is discovering that mind and matter are not separate but interconnected.
- Quantum physics is revealing that the boundary between the inner and outer domains, however conventionally useful, does not ultimately exist.
- Quantum physics is showing us how we ourselves are moment by moment playing a key role in the creation of our experience.

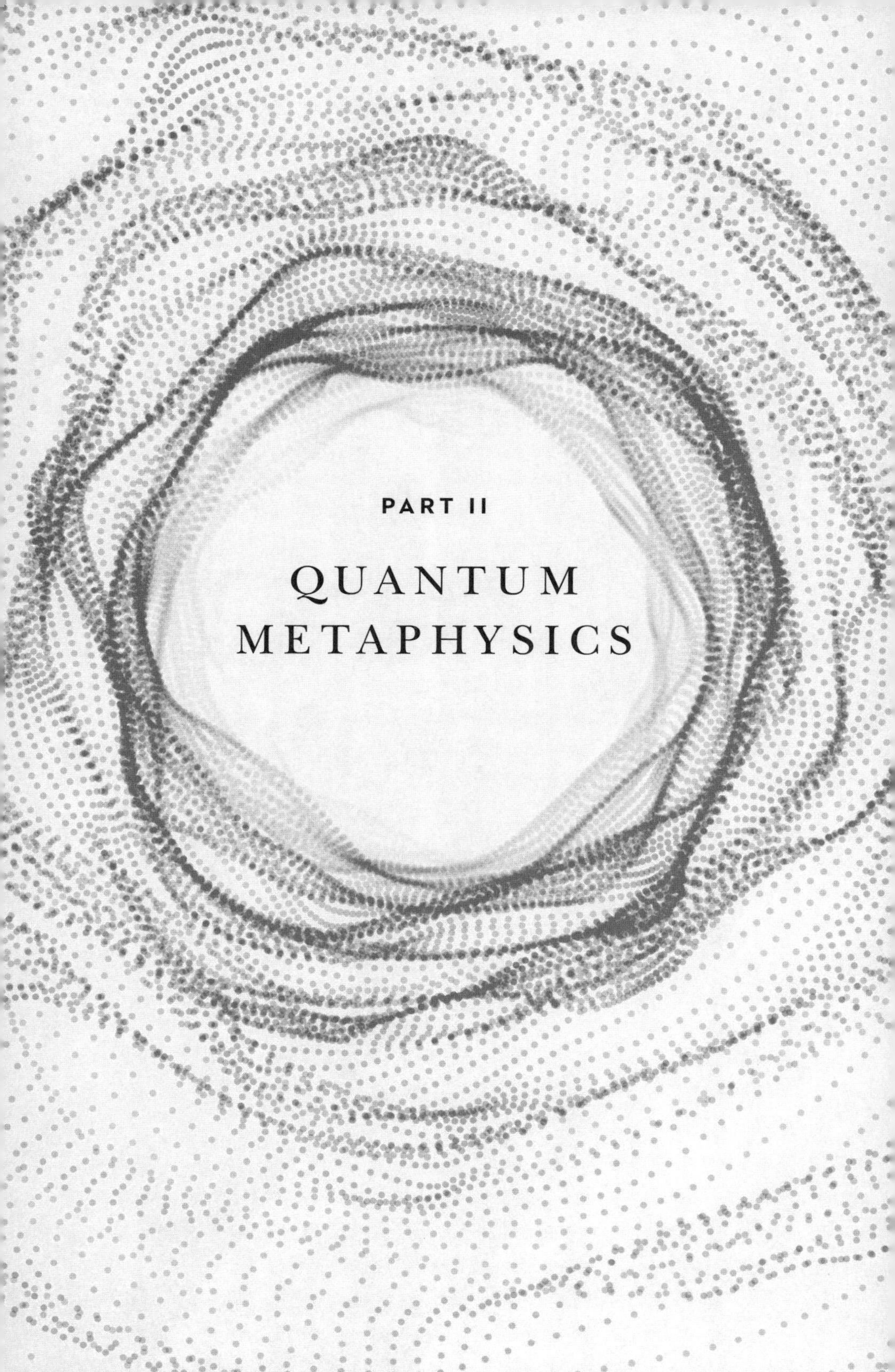

PART II

QUANTUM METAPHYSICS

• CHAPTER TWELVE •

QUANTUM PHYSICS AS SPIRITUAL PATH

I wonder if I am becoming addicted—I can't seem to get enough of quantum physics. It has captured and simultaneously liberated my imagination beyond belief. My appetite for what it is revealing feels insatiable. The more I study it, the more I feel as if I am mutating, metamorphosizing, becoming quantum-physicized into a higher state of coherence in my very soul. Words fail me when I try to describe the realm of pure and utter magic that is quantum theory. Unlike a typical addiction wherein energy gets drained, the more that I invest my attention in the world of quantum physics, the more creative energy makes itself available to me. I feel convinced that what it is revealing to us is of crucial importance for the future of humanity, but, of course, I might be dreaming.

Science is the wisdom tradition par excellence of our modern age. Quantum physics, its crowning jewel—what Wheeler calls "the central principle of every branch of physics"[509]—can be likened to a genuine spiritual path in that its study becomes a doorway beyond both physics and the physical dimension into the realm of metaphysics. Quantum physics' return to metaphysics was inevitable, for physics began with the gnostic search for what Einstein calls the "Old One" behind all phenomena. To contemporary mainstream physics, however, the word "metaphysics" is akin to a swear word, a synonym for "loose thinking" or meaningless nonsense, a code word for unscientific and ungrounded thought. In modern physics as it is commonly practiced today, being "metaphysical" is used as a derogatory euphemism for condemning a theory which doesn't fit into the common, agreed-upon consensus framework. It seems as though contemporary conventional physics wants to "purify" its discipline from the stain of metaphysics.

Mainstream physics claims it is not interested in metaphysics, asserting that it makes no metaphysical assumptions, as it is only interested in seeing reality as it is. Yet, hidden within this very viewpoint is, paradoxically, a tacit form of metaphysics. This metaphysics lies in the unexamined assumptions implicit in the perspective that physics makes about the nature of reality as existing objectively, an assumption so implicit as to be not even recognized as an assumption. The spirit of quantum physics, however, challenges the underlying and unexamined metaphysical assumptions of mainstream physics, at the same time providing the doorway for a radical new form of scientifically grounded metaphysics to emerge.

Metaphysical considerations are unavoidable if we are truly interested in comprehensive knowledge of the whole and not merely in practical, material concerns. Metaphysics is the branch of philosophy that deals with the fundamental structures of reality, and physics is the branch of science that does the same. Metaphysics lies at the origin of the Western intellectual tradition, and is also the foundation of natural philosophy, making it one of the progenitors of science itself. All physics is metaphysics according to Einstein.[510] Metaphysics, according to its most common modern definition, has to do with a transcendent realm "beyond" what is perceptible to the senses, which is precisely what quantum physics points towards. Heisenberg writes, "I began by pointing out that I could see no reason why the prefix 'meta' should be reserved for logic and mathematics [metalogic and metamathematics] . . . and why it was anathema in physics. The prefix, after all, merely suggests that we are asking further questions, i.e., questions bearing on the fundamental concepts of a particular discipline, and why ever should we not be able to ask such questions in physics?"[511]

Going back to the Greek origin of the term, "metaphysics" denotes the exploration and description of the deep, fundamental structure of reality. At the deepest level it has to do with the "stuff" of reality that gives rise to the changing, ephemeral phenomena of our world. The very existence of our universe, as well as the existence of our consciousness itself, calls out for metaphysics in order for it to be more deeply understood and appreciated. The new physics has provided empirical evidence (in distinction from mere philosophical prejudice) for the existence of a different metaphysics than the one underlying classical Newtonian mechanics.

The term metaphysics is related to "mysticism,"[512] which is based on the word "mystery,"[513] implying something hidden. Wheeler writes, "Every man has to search for the answer to a mystery."[514] Even though in his mind "everything" is a mystery, Wheeler is of the opinion that, "The quantum is the greatest mystery we've got."[515] Mysticism is largely a contemplative philosophy that seeks what transcends and unifies the diversity of the world. To quote author and philosopher Aldous Huxley, "The mystics are channels through which a little knowledge of reality filters down into our human universe of ignorance and illusion. A totally unmystical world would be totally blind and insane."[516] Mystics explore their own direct experience to the point where they transcend the subject/object (i.e., observer/observed) duality.

Physicists and metaphysicians are both in the business of wondering about the mysteries of the universe, and humanity and consciousness are an essential part of the mystery they are trying to illumine. Brian Josephson writes, "As to the significance of mystical experience, the natural reaction of the physicist is to debar conscious experience from the framework of discussion altogether. This may be satisfactory as far as the study of matter is concerned (although not even in that case if, as Wigner has suggested, consciousness collapses wave functions), but it becomes a dubious doctrine if the conscious individual himself is our subject of interest."[517] Simply put, the study of the physics of the universe can't help but to include us as well.

Along similar lines, Wolfgang Pauli writes, "Both mystics and scientists have the same aim—to become aware of the unity of knowledge, of man and the universe and to forget our own small ego."[518] Mystics become aware of the unity of the seeming opposites of mind and matter. Eddington writes, "If I were to put into words the essential truth revealed in the mystic experience, it would be that our minds are not apart from the world."[519] Based on this definition, quantum physicists, in their encounter with the strange world of the quantum, are being led to the same conclusion as the mystically inclined. To quote Josephson:

> The question immediately arises whether, in talking about introspective knowledge in general and mysticism in particular, we may not be venturing outside the boundaries of what is legitimate in science. I shall respond to this question by posing another, which is dependent on the fact that, although mystical experience is not an element in current science, mathematics certainly is. The question is, why should the thinking involved in doing mathematics (which is as introspective in character as is meditating) should serve as a legitimate component of science, while mystical experience does not?[520]

Contrary to the pejorative associations that the word mysticism has within the modern scientific community, the genuine mystical path is closely akin to the path of science in that mystics accept only that which is revealed through direct, immediate experience. The word "experiment" is etymologically derived from the word "experience," which is related to the word "empirical." Mystics are those who experiment with their experience and are therefore genuine empiricists, drawing conclusions in a way that is in the true spirit of science.

Coinciding with the advent of classical physics and the notion of the objectively existing world, arose the idea of empiricism as having to do with experience conveyed solely through sensory experiences of a supposedly real, independently existing physical world. This definition of empiricism, which has dominated the scientific enterprise for centuries, narrowed the meaning of the word to refer to data only gathered through the five senses. Quantum physics' dispelling of the myth of an objectively existing independent universe has called into question the unnecessary narrowness of this purely sensory-based definition of empiricism, rooted as it is in the illusion of a solid, material, objective world existing separately from the mind. Quantum physics is helping us to return to the original meaning of what it is to be an empiricist, which has to do with our direct experience—be it inner experiences (dreams, visions, imaginations, intuitions, etc.)[521] or outer, sensory-based experiences. William James used the phrase "radical empiricism" to denote just such an expanded vision. This more expansive meaning of empiricism embraces the vast range of experiences that are naturally available to us within consciousness, experiences which become a rich source of data for building a more accurate vision of reality.

The discoveries of quantum physics make the insights which were once considered mystical "transparent," rendering them readily available for all to see. To quote Einstein, "The most beautiful emotion we can experience is the mystical. It is the source of all true art and science. He to whom this emotion is a stranger, who can no longer wonder and stand rapt in awe, is as good as dead."[522] We ourselves are an essential part of the mystery that is being unveiled through quantum physics. Jung writes, "It has yet to be understood that the *mysterium magnum* [the great mystery] is not only an actuality but is first and foremost rooted in the human psyche."[523] The human psyche is not only the medium through which the mystery reveals itself, but is itself an intrinsic part of the mystery. To quote Feynman, "It does not do harm to the mystery to know a little about it."[524]

To the extent that we are interested in truth, the nature of reality, God, or who we are, we are all metaphysicians. Quantum physics is hinting at something beyond what we normally think of as physics, reaching beyond even what we consider the physical world. Quantum physics is pointing at the very thing that it (and the whole universe, including ourselves) is an expression of.

Once upon a time, physics and philosophy were allied disciplines, and a physicist was equally likely to be referred to as a "natural philosopher," i.e., someone who sought truth in the natural instead of the human world. A lover of wisdom, which is the definition of a "philosopher," Wheeler did work in quantum physics that reached beyond the formalism of physics into the realm of metaphysics in its original philosophical sense. Speaking about philosophy, Wheeler opines, "Maybe philosophy is too important to be left to the philosophers."[525] Quantum physics leads us not only beyond physics into the realm of philosophy, but it also catapults us into the fields of cosmology, psychology, theology, and God knows what else. Quantum physics forces us to reexamine what the concept of existence means. Our conception of the universe's existence affects how we see our place within it, which touches us in our core. The revelation that is quantum physics is simply too profound to ignore. Physics is much too important to be left to the physicists.

The shadow side of science as a modern-day wisdom tradition is that it can, and often does take on the qualities of a religion, with all of its taboos and heresies that violate the open-minded spirit of the scientific method. The tenets of science, if adhered to with unexamined, unconscious, preconceived beliefs about the nature of reality, can easily start to resemble a disguised form of religious dogma, calling for its adherents' intellectual and emotional allegiance in a way that borders on the irrational. People who have been indoctrinated into the dictates of this scientific creed, as if hypnotized or under a spell, can find it difficult or even impossible to imagine that the world can be anything other than the way they have been taught that it is, as if no other way of thinking or knowing about things has ever occurred to them.

The new physics is calling us to free our minds from our unconscious, built-in prejudices, assumptions, and restricted views about our world as well as ourselves. Scientific materialism, with its hidden metaphysical belief in an objectively existing world, can be considered to be the metaphysical illness of our age. The still dominant attitude of scientific materialism has erroneously excluded the subjectively experienced mind from the domain of the natural world to the point that "scientific knowledge" has come to be equated with "objective knowledge." And yet, quantum physics has proven there's no objective anything.

Writer Octavio Paz wonders, "Perhaps tomorrow's metaphysics, should man feel a need to think metaphysically, will begin as a critique of science, just as in classical antiquity it began as a critique of the gods."[526] Quantum entities are a physicist's way of speaking about their experience of reality that gives them a sense of how the world works. How different is this from the function of the Greek gods on Mount Olympus?

Jung writes, "Religion is a symbolic system by which we try to express our most important impressions of unknown things."[527] We should not forget that quantum physics, at bottom, is a purely symbolic procedure in which we are trying to express our "impressions of unknown things." The vision of our universe and our place in it that is revealed by quantum physics is ultimately spiritual—the fundamental interdependence and inseparability of all phenomena is the central pillar in every spiritual wisdom tradition. Is it the worst nightmare of physicists for "quantum physics" and "spiritual" to be mentioned in the same sentence, or is it their deepest, most glorious (though perhaps repressed) dream?

It is difficult to discern where physics ends and metaphysics begins. In our journey into quantum physics, we simply cannot escape metaphysics. In our inquiry into metaphysics, however, we should be careful to neither overindulge, nor have too little. Pauli writes, "In my own view it is only a *narrow* passage of truth (no matter whether scientific or other truth) that passes between the Scylla of a blue fog of mysticism and the Charybdis of a sterile rationalism. This will always be full of pitfalls and one can fall down on both sides."[528] In other words, we don't want to become unreasonably mystical or overly rational in our reaction against mysticism—it is a slippery slope between these two extremes. Heisenberg comments, "We should maintain the tension resulting from these opposites."[529]

SAUSAGE GRINDER

Quantum physics is the most successful scientific theory—as far as its capacity to make accurate theoretical predictions that precisely match with experimental data—of all time. There aren't even any competitors. To quote Wheeler, "Quantum theory in an everyday context is unshakeable, unchallengeable, undefeatable—it's battle tested."[530] As Feynman points out, some say that the only thing that quantum physics has going for it, in fact, is that it is unquestionably correct. The majority of corporately trained and funded physicists, as if having become desensitized to the mind-warping implications of their own theory, are content to take it for granted, just using quantum theory for practical ends, rather than being curious about where it came from and what it indicates about the nature of reality. Describing such physicists, Lederman and Hill write that "they are only clerks taking dictation or workers on an assembly line when it comes to trying to understand why it is this way."[531] Most physicists spend their time calculating amplitudes and probabilities, rarely lifting their heads from the computer monitor to ask what it all means. Practically speaking, there is no problem. This brings to mind Karl Polanyi's notion of "tacit knowing" (knowing in ways that cannot be explicitly denoted or verbalized). For example, we can be satisfied with riding a bicycle without necessarily understanding how the gyroscopic principles that govern its turning works.

Thankfully, physicists such as Wheeler are interested in the deeper philosophical meanings and implications of their mysterious theory. Some of Wheeler's more practically-oriented colleagues have wondered whether he spent too much of his intellectual energy in far-out speculation. But Wheeler himself wondered, on the other hand, whether all the time he invested in mathematics and calculations prevented him from spending more time in contemplative thinking about deeper problems. To quote Freeman Dyson, "The really astounding thing about Wheeler's speculations is that so many of them have turned out in the end to be to be right."[532] The philosophical questions about the meaning of being and reality should, Wheeler felt, regain their rightful place within the discipline of physics, from which they had been banished for centuries. In his later years, Wheeler was more and more investing his attention in the deeper meaning of the quantum, which drew judgment from some of his more practically-oriented colleagues. Setting the record straight, Feynman comments, "Some people think Wheeler's gotten crazy in his later years, but he has always been crazy."[533] Wheeler comments, "I didn't mind that some of my respected colleagues in science thought that I myself had gone a little bit around the bend. They were entitled to remain more conservative, as I tried to be daring."[534]

A dreamer, Wheeler had a need to follow a vision, to look beyond the immediate, to learn something new about what is most fundamental. Commenting on quantum theory, Wheeler says, "It's a sausage grinder. We drop our problems in, and turn the crank, and get out the answers. Where did the sausage grinder come from?"[535] Elsewhere in his writings, due to its seemingly magical powers, he calls it a "magic sausage grinder."[536] It is as though a miraculous object bestowing earthshaking knowledge has fallen from the heavens, helping us to develop undreamed-of technologies; but no one really knows why it works, what it means, where it came from, or what it is ultimately revealing to us.

Wheeler was not just interested in the practical aspects of solving equations, making predictions, developing engineering applications, and building technologies, but was willing and impelled to contemplate what he calls "the Really Big Questions,"[537] such as why is our universe a quantum universe in the first place? He asks, "Why is the quantum there? If you were the Lord, building the universe, what would convince you we couldn't make a go of it without the quantum?"[538] Compare this to Einstein's comment, "When I am judging a theory, I ask myself whether, if I were God, I would have arranged the world in such a way."[539] It gets my attention that great thinkers such as Wheeler and Einstein, in contemplating the really big questions, imagine what God would do. In essence, their questions come down to why we are here in the first place.

Wheeler writes, "Not all of my colleagues consider these questions quite respectable. But if they are not respectable in the twentieth century, they will be in the twenty-first."[540] Wheeler felt that these questions might contain insight into the nature of existence itself. He felt that the quantum was so natural that the universe could not even have come into being without it. "The quantum," he writes, "supplies the machinery by which the world comes into being."[541] Wheeler felt certain that the quantum was a revelation of something deeper that has yet to be found. He writes, "I remain convinced that some deeper reason for quantum mechanics will one day emerge."[542]

NECESSARILY AMATEURS

Wheeler emphasizes that people need to have humility, imagination, and daring in their approach to research.[543] Sometimes the greatest breakthroughs in science happen because someone had the courage to recognize and speak out loud what others have turned a blind eye towards. Wheeler comments, "Here's where it is so important to talk with the young people. Some modest young person comes along with some idea no one else is paying attention to. His idea may just be the central point."[544] Sometimes amateurs are better able to sift the essential from the nonessential.

Not just a world-renowned physicist, Wheeler was also legendary for his teaching style. As impossible it is to list all of his scientific achievements, many physicists consider Wheeler's most substantial contribution to be as a teacher. To quote one of his students, physicist Cheuk-Yin Wong:

> Wheeler's class teaching was in the style of the old masters. He often would come to the class without notes, and would begin his gentle and

> deliberate narration step by step from memory, writing down the concepts and formulas in detail and explaining the thinking that was needed at each step, as if he was thinking aloud. In the process, we students could observe how a great mind was at work, as if pulling out the thread of the silk continuously from a cocoon. On a few occasions, he might get stuck in the middle of the narration, and would stop in front of the blackboard. We students would wait eagerly to see how he would resolve the difficult points in question. A short pause later, he would resume and march on to get the key concepts across. He was well known for drawing pictorial representations of objects that were informative, complex, and often with philosophical implications.[545]

In Wheeler's opinion the real reason universities have students is to educate the professors, rather than the other way around (Wheeler felt that after teaching, writing was the next best way to learn something, which certainly caught my attention!). To quote Wheeler, "In order to be educated by the students, one has to put good questions to them. You try out your questions on the students. If there are questions that the students get interested in, then they start to tell you new things and keep you asking more new questions. Pretty soon you have learned a great deal."[546]

To quote Feynman, who like Wheeler was also a renowned teacher, "In order to make progress, one must leave the door to the unknown ajar."[547] When we encounter the unknown, it touches upon the realm of the unconscious, and is sure to inspire a multiplicity of reactions and mistakes. Anyone who has never made a mistake has never tried anything new. Wheeler thinks of a genius as someone who makes as many mistakes as they can as quickly as possible so as to progress along their path of research. Our capacity to err and make mistakes is intrinsic to the process by which we learn and evolve. Feynman says, "We are trying to prove ourselves as wrong as quickly as possible, because only in that way can we find progress."[548] When asked what equipment he needed when he moved into his new office at Princeton's Institute for Advanced Studies, Einstein replied, "A large wastebasket, so I can throw away all my mistakes." Feynman agreed that "one of the biggest and most important tools of theoretical physics is the wastebasket."[549] If we knew what it was we were doing, it would not be called research. If we avoid error we do not live. Because so many young physicists are forced to specialize in one way or another, Wheeler was of the opinion that the most radical and crazy-seeming research in physics can only be done by "an old fogy who can afford to make a fool of himself." Wheeler was more than willing to take this risk, saying, "If I don't, who will?"[550]

Niels Bohr professed that he always approached every new question from a starting point of total ignorance. Taking his lead from his mentor Bohr, who felt that he was always an amateur, Wheeler thinks of true pioneers as "necessarily amateurs." Bohr and Wheeler's point of view puts into a new perspective what it means to be an expert, authority, or "professional." To quote Feynman, "I never pay any attention to anything by 'experts.'"[551] This sounds similar to what in Zen Buddhism is called "beginner's mind," which is a mind free of the habits of the expert, open to all possibilities. This brings us to a revised definition of an expert, which no longer means someone who knows almost everything about their field, but rather knows from their own bitter experience almost all possible mistakes in their field and how to avoid them.

Wheeler felt that we needed "daring conservatism" in pursuing physics. Revolutionary daring must be balanced by conservative respect for the past, keeping what is valid from the past while boldly extrapolating existing theories to their outer limits. In any case, the ever-emerging discoveries of quantum physics are thirsting for the next generation of daring thinkers to further unfold their deeper meaning. For daring and gallantry are needed in science, as in battle. To quote Dan Goldin, the chief of the National Aeronautics Space Agency, "If we don't dare to dream, we won't find anything. . . . Dreams are how the most exciting science happens."[552]

• CHAPTER THIRTEEN •

PHYSICS IN TRAUMA

Consciousness has insinuated itself into the quantum physics laboratory and mainstream physics has responded, to quote Penrose, "with great reluctance." The physics community has had a most interesting reaction to the advent of the quantum: it has changed definitions, created new forms of logic, introduced superfluous and ad hoc constructs into the theory, and come up with the most ingeniously absurd theories so as to avoid directly dealing with what it has discovered. Banesh Hoffmann writes in his book *The Strange Story of the Quantum*, "Let us not imagine that scientists accepted these new ideas with cries of joy. They fought them and resisted them as much as they could, inventing all sorts of traps and alternative hypotheses in vain attempts to escape them."[553] In their attempts at preserving a material world, however, physicists have been attempting to preserve an illusion.

Seen psychologically, the physics community is in denial about its own unsettling revelations. For example, as I was finishing up this book, I came across a new book by highly esteemed Caltech theoretical physicist Sean Carroll called *The Big Picture.* I opened up to the chapter on quantum physics only to find the following words, "Almost no modern physicist thinks that 'consciousness' has anything whatsoever to do with quantum mechanics."[554] His words were written with the utmost smug authority, mentioning that the "tiny minority" who do think that consciousness has anything to do with quantum physics are the "iconoclastic few." There are innumerable other examples of supposedly cutting-edge physicists summarily dismissing the role that consciousness plays in physics out of hand. From all indications, the physics community is in a serious state of denial regarding their own discoveries.

The standard advice given in quantum physics is "not to worry" about the meaning of its theory. When asked about the metaphysical and philosophical implications of quantum theory, for example, their avoidance is captured in their well-known reaction: "Shut up and calculate." This shut up and calculate school of thinking is more like a school for the avoidance of (critical) thinking. To quote Lederman and Hill, "So bizarre are the consequences of quantum physics that, perhaps to preserve their sanity, the quantum physicist pioneers were driven to denial that they were actually describing a vast new reality, preferring to objectively insist that they had 'merely' invented a new method for making predictions about the results of possible experiments—and nothing more."[555] As fascinating as its new discoveries are, the physics community's unconscious reactions to its discoveries—what has been called an act of "cognitive repression"[556]—are at least as interesting, if not more so. As a student of the psyche, I can't help but wonder what is being revealed by their reactions.

Coming to terms with the revelations of quantum theory challenges us to our very core. To quote Feynman:

> The difficulty really is psychological and exists in the perpetual torment that results in your saying to yourself, "But how can it be like that?" which is a reflection of uncontrolled but utterly vain desire to see it in terms of something familiar. . . . Do not keep saying to yourself, if you can possibly avoid it, "But how can it be like that?" because you will get "down the drain," into a blind alley from which nobody has yet escaped. Nobody knows how it can be like that.[557]

If we persist in demanding to know how the world is, independent of how it appears in experiments, we are goners. In our insistence in this line of questioning we are running up against our ingrained beliefs in a classical world which still tacitly rules our sense of reality at a subconscious level. The difficulty in understanding quantum physics is "psychological," which is to say it touches us within our very psyche. If we try to apprehend the quantum realm with a classical mindset—"How can it be like that?"—we will be entering a psychospiritual black hole with no exit and we will go "down the drain." We approach the quantum at our own risk so we need to make sure that our psyche is sufficiently prepared for the (classically) destabilizing influence of encountering the topsy-turvy world of the quantum.

Niels Bohr famously said, "Anyone who is not shocked by quantum theory has not understood a single word."[558] The worldview emerging from quantum physics has completely and utterly overturned and shattered the old, classical mechanistic ideas of how the universe works. From all appearances, it seems as if the psyche of physics as a whole is having a nervous breakdown; the old structures upon which its view of the world has been based are disintegrating and melting down. To quote physicist Daniel Greenberger, "Einstein said that if quantum mechanics is right, then the world is crazy. Well, Einstein was right. The world is crazy."[559]

When contemplated psychologically the revelations of quantum physics appear so shocking and discontinuous with the previously embraced classical perspective that they have induced a form of trauma in the entire physics community, what I call "Quantum Physics-Induced Trauma" or "QPIT." The physics community's unconscious reactions to its discoveries have the classic features of a trauma that they are in the process of integrating into their conscious awareness. In an unpublished essay, Pauli writes, "I still recall the tremendous shock dealt to me as a student by this state of affairs and its implications. Most of the physicists of my generation and the previous one reacted in the same way."[560] The pioneers who were exploring the cutting edge of physics were as shocked by what they found inside the atom as the voyagers of the starship *Enterprise* from *Star Trek* would have been encountering a bizarre alien civilization unlike anything they had ever encountered before, somewhere at the outer reaches of the universe. Once the quantum revelation is ingested, there is no going back to the classical world, which is recognized to never have existed in the first place. To quote Ramamurti Shankar, author of *Principles of Quantum Mechanics*, "Once we have bitten the quantum apple, our loss of innocence is permanent."[561]

It should be pointed out that it is not just physicists who are in a state of trauma; our entire species is suffering from a form of post-traumatic stress disorder (PTSD), resulting from the unsettling uncertainty that is such a part of our modern world. Jung writes, "I believe I am not exaggerating when I say that modern man has suffered an almost fatal shock, psychologically speaking, and as a result has fallen into profound uncertainty."[562] It is an "uncertainty" that I imagine Heisenberg and the rest of the founding fathers of quantum physics could relate.

It is traumatic to realize that the world that we thought we lived in doesn't exist in the way we thought it did. Abraham Pais, an award-winning physicist who knew Einstein during the last decade of Einstein's life, writes in his biography of the preeminent physicist, "As a personal opinion, it seems to me that making great discoveries can be accompanied by trauma."[563] When we discover something that so completely changes and rocks our world, we can easily find ourselves disoriented, experiencing a shock that needs time to be metabolized, digested, and integrated. The physics community's seemingly unconscious and irrational reactions to the appearance of consciousness in its domain are typical responses to an overwhelming and destabilizing trauma that cannot be integrated in the ordinary way.

The physics community's trauma is a natural reaction, a sane response to a mind-bending discovery that deconstructs the very foundation of the world they thought they had been inhabiting. Capturing the incomprehensibility of quantum physics, Bohr says, "If you think you understand it, you don't."[564] The discovery of the quantum has been like an electroshock to the brains of physicists. To encounter the quantum is a trauma to the classically conditioned mind; to take its revelations into oneself is to drink the transformational nectar of an initially disorienting but ultimately radically liberating gnosis. Speaking about how shocking the insights of quantum physics are, Bohr says, "If you think you can talk about quantum theory without feeling dizzy, you haven't understood the first thing about it."[565] And as with any significant trauma, we are asked to assimilate what has been triggered within us. This is a psychospiritual task that demands real courage, as it involves facing the darker, fearful, and unconscious aspects of the self.

As is often the case in trauma, there emerges an area of experience that is "off-limits" to talk about. Mention the word "consciousness" to corporately trained conventional physicists, and watch their knee-jerk reaction, as if we have just said a dirty word and broken a taboo. This don't-go-there zone in the mind of physicists has its own physics, in that it is a physics that openly interfaces with and directly reflects and reveals the unconscious psyche.

Instead of having a multidisciplinary, holistic vision akin to being modern day renaissance people, the typical physicist of today practices what philosopher José Ortega y Gasset refers to as the "barbarism of specialization."[566] Oftentimes a specialist in one field is incompetent at discussing another. Today's specialists know more and more about less and less. Buckminster Fuller referred to this dynamic as becoming "specialized to death" and felt it was in opposition to life.[567] From this compartmentalized point of view, anyone who tries to synthesize knowledge from different disciplines is denounced as a dilettante and accused of speaking about something they are not qualified or credentialed to discuss.[568] To quote Freeman Dyson, "Fortunately for us, dull specialization has never destroyed the poetry of John Wheeler's imagination."[569]

Feynman was of the opinion that "science is not a specialist business; it is completely universal."[570] This is to say that "doing science," which is basically experimenting through our experience to broaden our knowledge of both the world we live in as well as ourselves, is a field that is open to everyone. In contrast to being a specialist, Fuller's idea was that the world needed more people to become "comprehensive-ists," recognizing that all of the seeming fragmented parts of a system—like all of the different "systems" in the human body—are actually interconnected, interrelated, and inseparable, mutually supportive parts of a greater living whole. Characterizing this all-around, comprehensive perspective, when asked during an interview what his specialty was, Wheeler replied, "My specialty is everything."[571] People with a more comprehensive viewpoint, recognizing the interconnectedness of all that exists, see all areas of knowledge as one multidisciplinary whole system in need of integration and unification. Such people are able to dream the world whole again and truly become the living antibodies and antipsychotic agents for our time.

In any case, the physics community has certainly struggled with trying to come to terms with its discovery of the quantum. To quote Wheeler, "We've had many a hard knock along the way that's driven this quantum idea into our heads, against the greatest difficulties of us all in understanding it. It is quite conceivable that we shall have to have many additional knocks before we get these new things into our heads, but to me the opposite is also conceivable."[572] In other words, Wheeler is of the opinion that we don't necessarily have to go through the school of quantum hard knocks. He feels that the quantum is just waiting to be brought forth into full-bodied realization. It's there for the having. In a sense we already have it—all we have to do is recognize it.

THE POLITICS OF PHYSICS

How amazing that physics, in discovering the miraculous world of the quantum, simultaneously constructs a don't-go-there zone regarding what we are and are not allowed to talk about. This fact alone, a collectively shared defense mechanism, points to an underlying psychological issue within the physics community. From the psychological point of view, the question naturally arises: why is mainstream physics so threatened? It should be pointed out that issues regarding consciousness have not been refuted but merely rejected by those in positions of power and influence, which seems less a scientific process than a (reactionary) political and psychological one. The fact of there being an unspoken elephant in the physics living room, of there being a mysterious secret that cannot be spoken about, are all signs, seen from the family systems theory point of view (which sees the world as a whole, interrelated, and inseparable system of relationships), of a "dysfunction in the family system" of the physics community.

It is not that physicists are merely disinterested in the appearance of consciousness in their experiments; on the contrary, they have become "aggressively/belligerently disinterested" in the metaphysical implications of their own theory. It is as if they have developed a willful ignorance, impervious to the implications of their own data. If we view the physics community as if it were an individual, it has an emotional "charge" (analogous to that of a subatomic particle) and is "reacting" (energetically speaking, in an almost violent way) against something in its own discoveries that is being triggered within itself. Often the greatest discoveries in physics are found by following with unbiased and open-minded curiosity the one anomalous thread in the prevailing theory (in this case, what to do about consciousness) that doesn't seem to fit. Interestingly, the current reaction of the physics community is the polar opposite to this approach: It is actively choosing to look away from what is in its closet, from the thread that is protruding through the cracks in its theory. And yet, if this thread is pulled, it could potentially unravel not only the field of physics' ideas about the world, but physicists' ideas about themselves as well. Jung comments, "New points of view are not as a rule discovered in territory that is already known, but in out-of-the-way places that may even be shunned because of their ill repute."[573]

The lineage holders of corporate/academic physics are like "gatekeepers" who quarantine the radical philosophical implications of quantum theory from the rest of us. To quote Einstein, "Restricting the body of knowledge to a small group deadens the philosophical spirit of a people and leads to spiritual poverty."[574] It is not something solely within the individual psyches of physicists that is resisting the liberating perspectives of quantum theory; it is important to view the physics community within the wider context of the institutional structures in which it operates. From the point of view of the prevailing power structure which funds the overwhelming majority of physics research in the United States, in both corporations and universities, the insights emerging from quantum physics represent a tremendously disruptive new knowledge that could easily threaten the status quo.

Quantum physics is pointing at the primacy of consciousness for how our moment-by-moment experience manifests, thereby illuminating the immensity of our inherent power to create our world more consciously. If recognized and understood by the general population, the revelations of quantum physics would be naturally used for the liberating purposes for which this knowledge is tailor-made. One might even think that the "creator" of quantum physics, the universe itself, designed it in order to free humanity from the shackles of the spiritless, soulless, deadening paradigm of fragmentation (i.e., the Newtonian, classical worldview) that has promulgated limiting and outright false doctrines about the nature of who we are.

The revolution of quantum physics is occurring primarily within the mind, and once its revelations are communicated in readily understood language, metaphors, and symbols so as to be transmitted to an ever-widening circle of people, the possibilities are truly unlimited. The liberating ideas of the newly emerging quantum gnosis are not just catchy but "catching," in that they are contagious. Once sufficiently ignited and set aflame in the psyche of humanity, the revelations of quantum physics can and will spread like wildfire, virally and nonlocally propagating themselves through the collective unconscious of our species. A true "reformation" of the world may be the result.

The overwhelming majority of the physics field, however, has been co-opted by the corporate powers that be to become an instrument for their agenda.[575] For the corporate body politic, the bottom line of generating profits is what's important, after all. Its main priority is focused on whatever activity is the most readily translatable into money.[576] Corporatized physics equates truth and value with utility. Since it is interested in manipulating and gaining control over the seemingly outer world, its focus has to do with issues related to the acquisition of raw power. Professor Ravi Ravindra writes, "We should keep in mind that a majority of all scientists and technologists in the world actually work for the military or for the war machine in one form or another."[577]

Like a compass always pointing north, however, pure physics is solely interested in truth and nothing but the truth, no matter where the quest for truth leads, and is thus deeply grounded in natural philosophy and metaphysics. Speaking of two such truth-seeking physicists (Einstein and Bohr), scientist and Nobel laureate George Wald says, "There were no fences around them, no boundaries beyond which they wouldn't go. They were interested in everything interesting. I thought sometimes of a man walking a puppy. The man walks a straight line, but the puppy's into everything. And they both went like the puppy."[578] Many modern-day corporatized physicists, however, toe the party line. One of the reasons corporatized physicists tend to shy away from the truth of what quantum physics is revealing is because this truth is so alien to the world of everyday physics that they have become used to.

As Feynman points out, "Science is a long history of learning how not to fool ourselves."[579] Unfortunately, a large part of the corporatized physics community has truly fooled itself in dismissing the deeper philosophical implications of quantum theory as being meaningless. The philosopher Søren Kierkegaard wrote, "There are two ways to be fooled. One is to believe what isn't true; the other is to refuse to believe what is true."[580] Mainstream physicists—to the extent they still believe in a (proven to be false) objective reality and refuse to come to terms with the dreamlike nature of reality—have struck out on both counts.

The real heart and soul of physics has become marginalized and devalued by the existing power structure and turned into an alternative, fringe part of physics. This is analogous to what commonly takes place in organized religion, when the radical liberating gnosis of salvation that lies at the esoteric heart of its spiritual doctrines becomes banished as heretical. The original revelations are typically replaced by a distorted version and then monopolized by the powers that be to support the self-preserving interests of the hierarchical institution of the prevailing "church." Speaking about the new ideas emerging in physics concerning the role of consciousness, Josephson writes, "These ideas are not well represented in the standard literature—probably, in the last analysis, because they represent the same kind of threat to current scientific dogmas as scientific discoveries have presented to religious dogmas in the past."[581] To quote Sir John Eccles, "Arrogance is one of the worst diseases of scientists. . . . It is important to realize that dogmatism has now become a disease of scientists rather than theologians."[582]

There is intense pressure in the mainstream, academic (and corporately funded) physics community to remain "politically correct" and not talk about consciousness. Physicists who come out of the closet and "out" themselves as being interested in consciousness seriously endanger their credibility, reputation, funding for their research, and employment options. To quote author Upton Sinclair, "It is difficult to get a man to understand something, when his salary depends on his not understanding it!"[583] Physicists who talk about the mystery of consciousness are condescendingly disparaged, derided as spiritual or superstitious, labeled "unprofessional," seen as New Age hippies, viewed as having psychological hang-ups, and snubbed and treated as pariahs by their own community. The revelations of quantum physics regarding consciousness has created a seeming impasse in the field of physics. To quote Schrödinger, "The urge to find a way out of this impasse ought not to be dampened by the fear of incurring the wise rationalists' mockery."[584]

In breaking the unspoken vow of silence and speaking about what is not supposed to be spoken about (consciousness), genuine adepts of the alchemical art of physics attract the unconscious shadow projections of their colleagues. To quote the esteemed psychologist William James, "By far the most usual way of handling phenomena so novel that they would make for serious rearrangement of our preconceptions is to ignore them altogether, or to abuse those who bear witness for them."[585] James is articulating what seems to be a universal psychological truth; when someone is bearing witness to something that makes others examine their preconceived assumptions about the nature of the world, they do so at their own risk.

Ironically, in their "dogmatic slumber" (to use Kant's phrase), the corporatized physicists are actually blocking much-needed developments in their field. In their unwillingness to look at what is presenting itself in their own theory, they are avoiding their moral responsibility to follow the path towards truth, wherever that may lead. The most intuitive physicists—Einstein, Bohr, Wheeler, and Bohm come to mind—seem to be the ones who have the strongest social concerns. They are able to perceive the wider social and moral context of the problems that they are working on.

There is a mutually reinforcing dynamic between the corporate physicists' personal psychological ego issues regarding confronting within themselves the liberating effects of the newly emerging quantum gnosis and the corporate power structure that they are a part of. These two factors—the internal, unconscious dynamics operating within the psyche of physicists and the corporate and governmental power structures operating in the outside world—collude with and feed into each other in ways both covert and insidious. There is an unconscious incentive-driven blindness intrinsic to remaining part of the global, corporate institutional power structure. This is to say that individual scientists who are embedded in and part of this structure, be it in corporations or academia, have been unconsciously conditioned to avoid inquiring in directions that could threaten the power structure they depend on for their salaries, reputations, and funding for their research. This is a universal phenomenon at work within the human psyche through which power and control are reinforced and maintained at the expense of truth, operating across many different domains throughout the world.

Many corporate scientists are part of an unscientific movement to suppress certain scientific discoveries, aiming to prevent, to quote evolutionary biologist Richard Lewontin, "a Divine Foot in the door." Representatives of this psychologically reactionary and suppressive movement within physics self-righteously envision that they are valiantly defending science from the corrupting forces of subjectivity and irrationality. The irony is that in their unconscious reaction they themselves are driven by the very irrational forces against which they are reacting.[586] Einstein is alleged to have remarked, "It is harder to crack a prejudice than an atom."

It is easy to confuse the map of our prevailing theories with the territory it describes, thinking the two are the same. There is a great proclivity in the physics community to fall under the spell of their prevailing map and believe nothing could possibly be in the territory that is not on the map, all evidence to the contrary. Oftentimes there will be anomalies within physicists' theories, which is to say that the territory (i.e., "reality") will contradict the map, but these differences from the map will simply be rationalized or ignored. Such is the case with how mainstream physics factors out consciousness from their equation of reality, as the evidence that consciousness matters doesn't match with what they have been programmed to think of as true. This behavior seems antithetical to Wheeler's perspective, which was to find the strangest thing that you could and then explore it.

The conviction that the map is unassailably accurate and complete is so strong for many physicists that in numerous cases where there is some discovery of an anomaly that is not predicted by their theory, these physicists refuse to even look at the evidence because they are utterly convinced that it could not be possible because it is not on their map. This is the modern-day equivalent of the church refusing to look through Galileo's telescope to see his evidence for the discovery of Jupiter's moons, which contradicted the church's view of the universe. This is entirely opposed to the open-minded investigative spirit of science, which sees such anomalies as the growing edge of our knowledge that should warrant our most careful attention. And yet this attitude is remarkably common in physics today. The source of this process is to be found within the unconscious psyche.

SCHIZO-PHYSICS

The ultimate goal of science is to come up with an all-embracing "Theory of Everything" (ToE),[587] a single theory which describes the whole universe. It should be pointed out that an unconscious metaphysical assumption about the way the universe exists is implicit in the idea that there can be one theory that covers the whole universe. Wheeler writes, "I, too, dream of an all-encompassing theory, but my dream has quite a different shape than the dream of these particle physicists."[588] In excluding consciousness from their ToE, however, it is as if corporatized physicists are saying that consciousness is not a phenomenon that is part of the whole universe, which is a perspective without any justification whatsoever. To quote Heisenberg, "Yet any science that deals with living organisms must needs cover the phenomenon of consciousness, because consciousness, too, is part of reality."[589] Physics thinks of itself as a discipline that is trying to understand the nature of the universe, and yet, strangely enough, its practitioners' unconscious reactions to the implications of their own discoveries are seen as part of the universe that is not worthy of their attention. Philosopher Jean Baudrillard writes, "It is not enough for theory to describe and analyze, it must itself be an event in the universe it describes."[590]

Interestingly, from one point of view the behavior of the quantum realm itself seems schizophrenic.[591] Speaking about quantum physics, the Dalai Lama asks, "Are we condemned to live with what is apparently a schizophrenic view of the world?"[592] In their reaction to the crazy world of the quantum, physicists are turning a blind eye to something within themselves, as if they are desperately avoiding relating with a part of themselves.[593] In their "schizophrenic" (which literally means "split mind") reaction, they have fallen into a state of cognitive dissonance within their own minds. His Holiness relays a story of how some quantum physicists he knows of have commented on how physicists themselves, "relate to their field in a schizophrenic manner."[594] In the lab, physicists think of subatomic particles as being real things, but once they start philosophizing about the foundation of their theory, they talk about how nothing really exist without the apparatus that defines it. Author Michael Talbot, writing about his "excessive optimism" regarding how the revelations of quantum physics will be ushering in a new paradigm, writes in *Mysticism and the New Physics*:

> One of the reasons for my excessive optimism was that I did not realize how schizophrenic many physicists are when it comes to interpreting some of the new physics' most astounding findings. . . . Time and experience has since taught me that some physicists are oddly schizoid when it comes to extrapolating or expanding beyond their immediate findings. They are not unlike *idiot savants*, individuals who possess a profound genius in one subject, but whose intelligence and vision is merely normal when it comes to looking beyond the narrow focus of their research.[595]

In this state of inner disassociation from a part of themselves, physicists are keeping contradictory viewpoints apart from each other, separated by a watertight partition, a mental firewall. This cognitive dissonance can't help but propagate itself by sending psychic ripples throughout the field of physics. "Deeply rooted convictions, such as the classical empiricism," to quote professor emeritus of physical chemistry Lothar Schäfer, "can brainwash even the most brilliant scholars."[596] From all appearances, many physicists—some of the most educated, intelligent, and influential members of our species—are suffering from a psychological malady. If this is the case among some of the brightest among us, what does this tell us about what is happening among the human race as a whole, as well as deep within the collective psyche of humanity?

Anything that can't be experimentally measured is of no concern to most physicists. But then, how can consciousness, which is "groundless" while simultaneously being the ground of all measurement, directly measure itself? Similar to how elementary particles do not have a precise position and motion, consciousness has no location. Consciousness can never be an object of knowledge, which is to say it can never be objectified and known in an objective manner like other items of our experience. Consciousness, however, is the only tool we have to examine consciousness. Seemingly caught in an endless dilemma with no exit strategy, the situation we find ourselves in is one in which a sentient, self-reflective mirror is reflecting itself as it reflects *upon* itself. Quantum physics is like a cosmic mirror pushing the scientific world right to its edge and reflecting back its blind spot. Etymologically, one of the original meanings of the word "mirror" is "holder of the shadow."

Similar to an individual's personal process, in which the very thing we turn away from is typically where the alchemical "gold" is to be found, what physics is currently turning away from might actually be the most significant clue about the ultimate nature of reality. The very fact that there is something within quantum theory that, when encountered, produces a psychological recoil is a sign of how psycho-activating quantum theory is. In any case, there is definitely something curious and worthy of further contemplation going on within the hallowed halls of physics, particularly within the minds of physicists.

The discoveries of quantum physics throw physicists back upon themselves. To quote Freeman Dyson, "My message is that science is a human activity, and the best way to understand it is to understand the individual human beings who practice it."[597] Quantum physics' realizations about the nature of elementary particles are a magic mirror reflecting something back to us not just about nature, but about *our* nature. To quote Wheeler, "No theory of physics that deals only with physics will ever explain physics. I believe that as we go on trying to understand the universe, we are at the same time trying to understand man."[598] It is in this sense that quantum physics becomes indistinguishable from a form of metaphysics. Schrödinger writes, "I consider science an integrating part of our endeavor to answer the one great philosophical question which embraces all others, the one that Plotinus expressed by his brief: *who are we?* And more than that: I consider this not only one of the tasks, but *the* task, of science, the only one that really counts."[599]

Wheeler consistently emphasizes the importance of asking the right question. Extrapolating on what Schrödinger considers to be the one great philosophical question, Jung writes, "The meaning of my existence is that life has addressed a question to me. Or, conversely, I myself am a question which is addressed to the world, and I must communicate my answer, for otherwise I am dependent upon the world's answer."[600] In other words, instead of unthinkingly taking on others' ideas about who we are, we should, based on our own experience, find out for ourselves. And in finding out who we are, there is no getting around or away from the psyche.

• CHAPTER FOURTEEN •

PHYSICS AND PSYCHOLOGY

By looking away from the implications of their own theory and what it has triggered within themselves, physicists are, ironically, revealing how a psychological factor has entered into the realm of physics. Most animals, including primates such as humans, show a truly staggering ability to ignore certain kinds of information, particularly data which does not fit into their imprinted and conditioned view of the world. Though potentially brilliant in the realm of science, physicists, as much as any other human, exhibit the same "staggering ability" to ignore whatever information doesn't fit with the current theory they subscribe to.

Physics and psychology (*physis* and *psyche*) are meeting through the back door of physicists' unconscious reactions to their discoveries. To quote Jung, "The no-man's land between Physics and the Psychology of the Unconscious [is] the most fascinating yet the darkest hunting grounds of our times."[601] It is as if the psyche and quantum physics are revealing themselves through each other, drawing closer together as both of them, independently of one another and from opposite directions, push forward into transcendental realms. Jung writes, "Microphysics is feeling its way into the unknown side of matter, just as complex psychology is pushing forward into the unknown side of the psyche. Both lines of investigation have yielded findings which . . . display remarkable analogies."[602] This fact certainly seems to point to the possibility that these two seemingly disparate disciplines' subject matters might share an underlying commonality. Wolfgang Pauli writes, "The only acceptable point of view appears to be one that recognizes *both* sides of reality—the quantitative and the qualitative, the physical and the psychical—as compatible with each other, and can embrace them simultaneously."[603]

Both sides of reality—the physical and mental—reciprocally co-arise and mutually reflect and cross-reference each other. Our psyche is set up in accord with the structure of the universe, which is to say that what is happening in the universe at large is in some way happening in the deepest subjective reaches of the psyche. This is to say that what physics is pointing out about the nature of the cosmos is mirrored in the realm of psyche. The unification of psyche and physis—the inside and the outside—demands us to explore the outer world while simultaneously looking within ourselves, as if one eye is turned outwards and the other inwards. Pauli expresses the opinion that "it would be most satisfactory of all if physis and psyche could be seen as complementary aspects of the same reality."[604] It is becoming clear that physis and psyche are two sides of the same coin, a currency that can be used to help our species snap out of our spell of fear and separation.

Pauli further elaborates his thoughts when he says, "It is impossible for me to find this correspondentia between physics and psychology just through intellectual speculation; it can only properly emerge in the course of the individuation process."[605] In other words, the realization of the connectedness between physis and psyche can't be realized through the limited instrument of the intellect, but can only reveal itself as a fruit of an individual's process to become undivided within themselves. This is the result of dedicated and sincere efforts towards personal self-transformation through inner work made over a lengthy period of time. This kind of effort is psychospiritual in nature and is categorically different from the kind of rational, intellectual, and analytical efforts typically made by scientists. It is becoming clear, however, that when it comes to cultivating the deep and penetrating insights and comprehensive understandings that are required in encountering the reality of the quantum, both approaches need to be combined so as to cross-pollinate each other. Jung writes that "only from his wholeness can man create a model of the whole."[606] As the alchemists would say, we can't make the "One" if we are not one ourselves.

In his correspondence with Jung, Pauli makes it clear that he sees the universe as a living symbol. Recognizing the symbolic dimension of the universe allows the inner and the outer aspects of our experience to reveal their reflective nature. In a letter to Jung, Pauli writes, "For me personally, the relationship between physics and psychology is that of a mirror image."[607] The outer physical material of the world and the inner psychic material of our mind are mirrored reflections of each other. Jung writes in a letter to Pauli, "In consequence of the indispensability of the psychic processes, there cannot be just *one* way of access to the secret of Being; there must be at least two—namely, the material occurrence on the one hand and the psychic reflection of it on the other (although it will be hard to determine what is reflecting what!)."[608] Is the outer world reflecting the psyche, or is the psyche reflecting the world?

Jung says, "Modern physics is truly entering the sphere of the invisible and intangible, as it were. It is in reality a field of probabilities, which is exactly the same as the unconscious . . . we [physics and psychology] are both entering a sphere which is unknown. The physicist enters it from without and the psychologist from within."[609] It is as if both physics and psychology are entering the unknown "sphere of the invisible" from opposite directions. Interestingly, both approaches seem to lead to a similar "place"—a mysterious place that is not a place in the ordinary sense. Jung writes that "man's inner life is the secret place," where "the spark of the light of nature" is to be found.[610]

The more we inquire into the essence of matter, the more we are confronted with its incomprehensibility. It is an inescapable fact that we don't know what the bound-energy wave packets we call matter actually are. To quote Jung, "We don't know whether our psyche is material or immaterial, because we don't know what matter is."[611] According to some contemporary theories, 90 to 99 percent of the matter in the cosmos is called "dark matter"; it does not emit, absorb, or reflect light, and we thus cannot see it. It is totally unknown to us, which is to say that we don't know what it is or what it does. Physics has discovered in its own realm an outer correlate to the cosmic unconscious. When asked about dark matter, astrophysicists say that it is not literally dark, rather the word dark is used to indicate that they do not know what it is—they are literally "in the dark." The darkness of dark matter is thus simultaneously existing out in the vastness of the cosmos while being within the minds of physicists, as well as within the darkness of the unconscious of all of us.

Similarly, speaking about the psyche, Jung comments that "when we say 'psyche' we are alluding to the densest darkness it is possible to imagine."[612] In investigating the realms of matter and the psyche, we are encountering the darkness of an unknown mystery. Speaking about the entities that seemingly inhabit the quantum realm, Eddington comments, "Something unknown is doing we don't know what—that is what our theory amounts to."[613]

Every quantum system has an inside and an outside that are inseparably connected. According to quantum physics, reality is not built out of matter, as matter was conceived of in classical physics, but out of psychophysical events—events with certain aspects that are described in the language of psychology and with other aspects that are described in the mathematical language of physics. Jung writes, "Psyche and matter exist in one and the same world, and each partakes of the other."[614] To this end, it is important to develop a unifying psychophysical language that fluently and fluidly partakes and speaks of both physics and the psyche at the same and different times. Jung came up with an image to express their relationship, comparing "the relation of the psychic to the material world with two cones, whose apices, meeting in a point without extension—a real zero point—touch and do not touch."[615]

In Pauli's words, the psychologist is led "from behind" (through the unconscious) "into the world of physics." And the physicist, contemplating what's in front of him or her, can't help but discover the world of the psyche. Somehow, the act of inquiring into the psyche opens into the world of physics, and vice versa. Jung comments, "The psyche, inasmuch as it produces phenomena of a non-spatial or non-temporal character, seems to belong to the microphysical world."[616] And, it should be added, the microphysical world belongs to the psyche. Though on the surface very distinct, the "fields" of physics and psychology—not to mention the "fields" that underlie and inform both disciplines—seem to be inseparably interpenetrating each other so as to become practically indistinguishable.

Jung writes that "there are processes going on in psychology that are absolutely indispensable in physics."[617] These "processes" that Jung is referring to are the human activities of observing, perceiving, and thinking, which ultimately cannot be separated out from the practice of physics. Hence, Jung writes, "It is my conviction that the investigation of the psyche is the science of the future."[618] The psyche is the subject of all knowledge, being the womb in and out of which both art and science are born. Jung writes, "The realm of psyche is immeasurably great and filled with living reality. At its brink lies the secret of matter and of spirit."[619] The investigation of the deeper levels of the psyche brings to light much that we, on the surface, can at most dream about. Psychology—the study of the psyche—is particularly unique because, to quote Jung, its "object is the inside subject of all science."[620]

The synthesis of physics and psychology touches on deep matters of the soul. To quote Pauli, "It seems that there must be very deep connections between soul and matter and, hence, between the physics and the psychology of the future, which are not yet conceptually expressed in modern science. . . . Such deep connections must surely exist, because otherwise the human mind would not be able to discover concepts which fit nature at all."[621] Pauli is inviting us to express new concepts—i.e., new ideas—which will lead us to new ways of looking at our experience and putting it into words, thereby helping us to see the deep connectedness between soul and matter, psyche and physics. Notice the similarity to the words of Einstein, "Body and soul are not two different things, but only two ways of perceiving the same thing."[622]

The processes that physics is discovering in matter are reflections of similar processes in the psyche. To quote Pauli, "The most modern physics, even in the finest details, can be represented symbolically as psychic processes."[623] Compare Pauli's comments with philosopher and mystical theologian Henry Corbin's words that "there is no pure physics, but always the physics of some definite psychic activity."[624] Once again, a physicist and a mystic sound like each other. In quantum physics we are given hints of a possibility of reconstituting psychological processes in another medium, that is, in the microphysics of matter. Physicists continually deconstruct and then reconstruct the matter of the external world in and through the medium of their minds. To quote Mindell, "It seems to me that quantum physics describes not only the material universe but the psychological as well."[625] In the realm of microphysics the operations of the psyche are being reflected through the patterns by which quantum matter emerges into observable form. Lothar Schäfer writes, "In a metaphorical way, you could say that quantum physics is the psychology of the universe."[626]

A psychologist knows that each nervous system creates its own model of the world, just as the modern-day physicist knows that each instrument used in an experiment also creates its own model of the world. In both the fields of psychology and physics we are outgrowing medieval Aristotelian notions of an objective world, and are entering a new, mysterious, and unknown realm where nothing is as it seems. Jung comments, "All things are *as if* they were."[627]

The coming together of physics and psychology certainly seems inevitable, though it has its own timing. Pauli writes, "*Both* physics *and* psychology have still, it appears to me, a long way to go. These 'drawers' are still separate, and only in our dreams is anything like an aurora consurgens[628] of a future unity to be seen. To my mind, however, such a process cannot be rushed, far less dispensed with."[629] Only in our dreams can the future unity of physics and psychology be accomplished. And yet the coming together of physis and psyche is revealing that our waking life is dreamlike in nature, which is to say that dreaming processes are at the roots of both physics and psychology. Are we merely dreaming that physics and psychology are coming together, or is the universe itself dreaming their union? Is there a difference?

SYNCHRONICITY

Jung's closest colleague, Marie-Louise von Franz, writes, "It is a remarkable coincidence that, at approximately the same time as physicists discovered the relativity of time in their field, C. G. Jung came across the same fact in his explorations of the human unconscious."[630] Jung called those moments where an inner situation is expressed through events in the outer world "synchronicities,"[631] which he conceived of as being a "contact point for physics and psychology."[632] It is our classically conditioned minds that impute a sequential, linear order (time, for example) onto the synchronically arising present moment, thereby obscuring our realization of the synchronistic nature of our universe. Jung's idea of synchronicity was the internal corollary in human experience of the seemingly external quantum idea. When he was first bringing out his idea of synchronicity, Jung felt that physicists were, in his words, "the only people nowadays" who would be able to "deal with" and bring "critical understanding" to what he was pointing at.[633]

In a synchronicity, there is a peculiar interdependence of external events with the subjective, psychological state of the observer. Synchronistic experiences reflect back the fact that the human mind does not exist in isolation from the world, nor is it just aware of the world, but is somehow linked to the world in ways few of us are aware of. In synchronistic phenomena, mind and matter reciprocally inform and reflect each other, as if inseparably interconnected at their core. The mental and physical dimensions are two interconnected aspects, like the form and content of something, only separable in thought, not in reality. The factor that connects the inner and outer realities is "meaning."

Jung described his conception of physical reality as "dreamlike"; this correspondence between the inner and outer realities is similar to how events in a dream mirror back the state of the dreamer. If there is no division between mind and matter in the ground of being from which everything emerges (what Bohm calls "the implicate order"), then it becomes more understandable why our world would have infused within it synchronistic traces of this deeper, all-pervasive interconnectivity. Synchronicities can be thought of as momentary fissures in the fabric of reality that allow for a brief glimpse into the underlying unity of our cosmos.

To quote von Franz, "*In a synchronistic event a coniunctio of two cosmic principles, namely, of psyche and matter, takes place,* and in the process a real 'exchange of attributes' occurs as well. In such situations the psyche behaves as if it were material and matter behaves as if it belonged to the psyche."[634] Jung considered psyche to be a quality of matter while at the same time conceiving of matter as a concrete aspect of the psyche.[635] Jung was fervently interested in questions such as, in synchronistic phenomena, "how does it come that even inanimate objects are capable of behaving as if they were acquainted with my thoughts?"[636] In moments of synchronicity, psyche and matter, despite seeming to be opposites, disclose their inseparability, revealing a deeper dimension of existence where they are the same thing. Though synchronistic events reveal them to be interconnected on a deeper level, psyche and matter are, conventionally speaking, what we refer to as inner and outer reality.

These strange, inexplicable coincidences that Jung called synchronicities often consist of a highly unlikely but auspicious and attention-grabbing coincidence between inner states and outer events. These "highly unlikely" events can be referred to as being "probabilistic miracles," in that on the surface they seem like miraculous violations of natural scientific laws but are actually anything but. These synchronistic probabilistic miracles don't contradict the natural sciences, but rather are quite compatible and consistent with (and actually exploit) the probabilistic and observer-dependent nature of quantum mechanics. According to one interpretation of quantum theory, every possible state actually happens in some potential or parallel universe. This is to say that the highly unlikely but possible can be realized "in reality," even if the probability is exceedingly low. There is a world of difference between something being, in Wheeler's words, "incredibly, ridiculously unlikely" compared to being completely impossible. Quantum physics is shedding light on the nature of this boundary between the possible and impossible—it is questioning what is truly possible.

To quote Wheeler, "Quantum mechanics does strange things to what we call causality if we examine it with sufficient care."[637] Jung realized that the *causality* principle—what he calls "one of our most sacred dogmas"[638]—was insufficient to explain certain manifestations of the unconscious. Causality is rooted in the notion of time; abandoning the causal conception pushes pure physics toward the realm of metaphysics. As an explanatory principle, causality is only one possible category of thought for describing the connection between events.

Just as Einstein added time to space to produce the much deeper concept of space-time (in which space and time are inextricably linked in one unbroken continuum), in his idea of synchronicity Jung proposed a new principle in nature, completing the notion of causality by adding a noncausal link. We have to "abandon," in Jung's words, "a causal description of nature in the ordinary spacetime system, and in its place to set up invisible fields of probability in multidimensional spaces."[639] Synchronicities only "make sense" in a holistic, dreamlike, and magic-filled universe, which is to say that they appear utterly nonsensical or impossible from a pre-quantum, mechanistic perspective.

Along a similar vein of thinking, Philip K. Dick refers to the possibility of human beings being able "to discard the modem of causation . . . as the basic ontological structuring category—by which world is ordered, arranged, understood," permitting "a much more accurate and acutely qualitatively different experience of reality. . . . And this radically transformed experience (Dasein) of reality, a way of being-in-the-world, or participating in shaping world (the observer participant) had to wait until such discoveries and realizations as quantum mechanics . . . we're talking about the lifting *for the first time in human history* of a massive perceptual/conceptual occlusion having to do with the ontological structuring factor we call causality."[640] The discoveries of quantum physics are allowing us "for the first time in human history," in Dick's unforgettable phrase, "to discard the modem of causation," which, until removed, acts as an occlusion to seeing "reality" (whatever that is) more clearly. Wheeler sounds like a combination of Dr. Seuss and Yoda when he says, "Why demand of science a cause when cause there is none?"[641]

Quantum physics widened the threesome of classical physics—space, time, and causality—to include the "acausal orderedness"[642] of synchronicity, thereby making it a foursome. Interestingly, Jung considers getting from three to four a "two-thousand-year-old problem." Pauli refers to it as "the main work." Getting from three to four is the age-old problem of alchemy, encapsulated in the axiom of Maria Prophetissa: "Out of the Third comes the One as the Fourth." Viewing numbers as symbolically representing archetypes, shifting from three to four adds a sense of completion, bringing about a unity. The overly rationalistic perspective of physics, losing a holistic view of reality, had fostered the "will to power"[643] of the human shadow; adding the fourth is to embrace the irrational element of nature and of ourselves. In its investigations into the nature of matter, quantum physics is encountering the epistemological boundaries of rational thought. Psychologically speaking, going from three to four symbolizes a stage of inner development known as the individuation process, which is what the magnum opus (the great work) of alchemy is all about.[644]

The alchemical procedure could be seen to represent the individuation process of a single individual. Could quantum physics be the modern-day vehicle for the individuation process of our species on a collective scale? Somehow the revelations of quantum physics are in the service of connecting us with the holistic nature of the universe, which necessarily brings to light and reveals the unified part of ourselves. Whether the realization of our deeper unified holistic nature actually occurs or not is a function of how we wield the creativity inherent in the open-ended freedom of our quantum nature.

To quote Jung, "Synchronicity is no more baffling or mysterious than the discontinuities of physics. It is only the ingrained belief in the sovereign power of causality that creates intellectual difficulties and makes it appear unthinkable that causeless events exist or could ever exist. But if they do, then we must regard them as *creative acts*, as the continuous creation of a pattern that exists from all eternity, repeats itself sporadically, and is not derivable from any known antecedents."[645] This quote by Jung has an interesting footnote in which he adds the following, "Continuous creation is to be thought of not only as a series of successive acts of creation, but also as the eternal presence of the *one* creative act." From this perspective, there is a single, underlying event that appears spread out and elongated throughout time and space.

In a synchronistic event, there is usually a felt sense of participating, as Jung puts it, in "*acts of creation in time*."[646] Interestingly, in quantum physics the act of observation itself is considered to be a unique act of creation in time, what Wheeler calls "an elementary act of creation."[647] The resultant universe arising concomitant with our observation continually unfolds from the singular now moment, generating endlessly unique explications of itself that give rise to the appearance of sequential events happening over linear time. If we are not careful, we can easily become entranced by this display, imagining we live in a linear-sequential world, which would be to fail to recognize our participation in the one creative act in the eternal present.

• CHAPTER FIFTEEN •

THE EVERYDAY WORLD OF QUANTUM REALITY

The idea of an objective reality without an observer is meaningless because it can never be experienced. Human experience is the inescapable basis of science, the empirical raw material from which our ideas and theories about the nature of the universe are constructed. To posit a world that exists independent of human experience (i.e., without an observer) is a mental abstraction that has no traction with the actual experiential world in which we live. A world independent of our experience of it, as seductive as it may be to imagine such a world, would be a world forever inaccessible to us, and thus unavailable to our scientific exploration. Thus, such an imagined world is not a world that could ever be considered real to us, for the ability to experience is a key criterion of reality.

Seen psychologically, being transfixed by (or fixed in) the viewpoint that there is an objectively existing world can be compared to a rigidified complex within the human psyche. This ossified complex repetitively recirculates the same thought into materialization over and over again—in our example, the idea of an objectively existing world. The insights of quantum physics, which can be liked to "living information," are like a metabolic antitoxin designed to dissolve the calcified complex and restore elasticity to the psyche, dissolving the idea of an objective universe in the process. Seen as a whole system, our collective psyche has become one-sided and quantum theory is a compensatory medicine secreted by the psyche in an attempt to return us to a more fluid and balanced state.

What seems to be an independent universe is in actuality a play of appearances, a persistent and persuasive "false imagination,"[648] an unexamined and clearly mistaken metaphysical assumption. The image of an objective world appears to and within the mind, arising from the mind itself. But an actually existing objective world independent of an observing consciousness does not—and quantum physics irrefutably proves *cannot*—exist in reality. Our situation is similar to seeing a mirage of water in the desert and either thinking (and fooling ourselves) that the apparition of water exists as actual water, or seeing through the illusion to realize that the image of water is in fact a magical display of our perception, not separate from our mind itself.

Another example is a kitten reacting to her reflection in a mirror, believing she is observing a second kitten separate from herself. The kitten is reacting to her own projection of herself that appears on the surface of the mirror as if her reflection objectively exists and is other than herself. In becoming conditioned by her own energy, she mistakenly thinks that she has nothing to do with creating and energizing that to which she is reacting. Unless her reflection is recognized as her own face, her reactivity can go on ad infinitum.

Our quantum powers of cocreation play out in so many different areas of our lives that we oftentimes don't even notice them. One example would be the course of events in the financial markets. As observer-participants, how we collectively view the course of events in the markets influence how the markets manifest. And how the course of events in the markets manifest reciprocally influences how we, as observer-participants, view these events. In a circular feedback loop, we are influenced by what we influence. This dynamic is symbolically reflecting the spontaneous creativity inherent in the quantum field, of which we are all a part and in which we all partake (whether knowingly or not). We, as observer-participants, and the events occurring in our world are parts of one unified quantum system.

Though the world of the quantum seems so mysterious and remote, quantum dynamics are actually something that each of us experiences in our ordinary, day-to-day lives far more than classical mechanics. Say, for example, I'm having an experience in my life and am interpreting it in such a way that I am utterly convinced of the real, true, and objective nature of what is happening. I then receive information which proves to me that what I had been thinking was playing out was not actually going on at all, but was only my own projection, my own unfettered imagination filtered through my unconscious and overlaid on the inkblot of the world. This is analogous to thinking that the world exists in an objective way and then snapping out of our fixation on the external world and realizing the subjective nature of our perceptions. Our perceptions are just one arbitrary and limited way of seeing a multitextured, ever-changing reality and not reality itself. If we don't see through this illusory version of reality that our perceptions present to us as being "the way the world is," we have then entranced ourselves through the power of our own mind.

Another everyday life example with which we can probably all relate may further clarify this point. We are all multifaceted beings, having a multiplicity of selves that get evoked and show up at different times depending on life's circumstances. Certain people elicit certain parts of ourselves. For example some people make us feel seen and appreciated, as if they are calling forth a positive, healthy, loving, and creative part of ourselves. Other people might have more of a tendency to see our shadow aspects, the perception of which serves as a hook to attract their shadow projections onto us. When we are around them we might feel ourselves burdened by their darker projections of who they imagine we are. We might even find that when we spend time with them, we find ourselves embodying and acting out the very shadow they are projecting onto us. Whether the shadow within us evokes their projection, or vice versa, is a whole other question that has many similarities to the questions and issues inherent in the observer effect of quantum physics.

Say, for example, someone is holding a negative image of "who we really are" in their mind's eye (which becomes like a lens or filter through which they see us). Projecting this shadow image onto us, they are, in essence, unconsciously calling forth a negative shadow aspect of ourselves to interact with. This shadow isn't "caused" by their projection, it is already within us in potential, but their projection increases the probability that it will manifest. Once we play into and act out this darker aspect of ourselves, this will confirm to them the "objective truth" of their shadow projection, as they now have all of the evidence they need to prove to themselves the rightness of their viewpoint. This will serve to further solidify in their mind's eye their negative image of who we are in a self-reinforcing feedback loop. All the while they will be convinced that they are just relating to objective reality, to who we "really" are, without realizing their own creative participation in calling forth the reality that they imagine exists outside of themselves (a reality which is reflective of an unconscious part of themselves).

This dynamic of concretizing in our imagination that someone exists in a certain fixed, objective way is a reflection of the same deeper process that underlies how we become entrained and entranced into thinking that the universe objectively exists separate from ourselves. Quantum physics is the elixir that can, if properly understood, dissolve our unconsciously ingrained tendency to solidify others, the universe, and ourselves as well.

Taking the insights of quantum physics into the realm of relationship would greatly impact the way we communicate with each other. For example, instead of focusing on the other person and telling them what they are doing, we can simply express what our experience is. Though the difference appears to be subtle, these are radically different ways of expressing ourselves in a relationship. When we tell someone what they are doing, we ourselves are under the illusion that what we see them doing is an objective fact, untainted by our own subjective filters. Expressing ourselves in this way opens up the door for the other person to disagree with us, maybe even resulting in an argument. On the other hand, if we simply express what we are perceiving, we are owning our subjective experience and not talking about the other but about ourselves. How can anyone argue (though some people might try) with what our experience is? This shift in how we express ourselves then empowers the other person to do the same, creating a shared space for both people to freely express what they are experiencing without personalizing the other person's perceptions.

How we see the world and one another affects the world, others, and ourselves. This process can be wielded consciously with a positive intent in such a way that can actually help people. An example would be when we are around people who, though they might see our shadow aspects, focus instead on our wholeness. When we are seen and appreciated in this way, it tends to evoke these more wholesome parts of ourselves to manifest, as if the other person, seeing the good and inviting the light-filled part of ourselves to shine, were creating a bridge to enable us to actually step into the more positive aspects of ourselves.

In spiritual terminology, this can be thought of as a "blessing," in contradistinction to when we project the shadow, which can be thought of as a "curse." It should be pointed out that we are only able to see someone in their wholeness if we are in touch with our own wholeness. In a positive feedback loop, seeing the other person's wholeness further connects us with our own. We can more deeply realize our own light as it is seen reflected through others. In the realm of interpersonal relationships there are ever-present possibilities for us to channel the immense creative power of the quantum for the good of everyone. This quantum dynamic in-forms and infuses all of our relationships with everything, which is a process that is always happening in one way or the other.

Another example: Say there's someone we haven't yet met, but who knows many of the same people we know. If we ask these mutual acquaintances their opinion of this person, they might all describe him or her differently, sometimes in ways that seem totally opposite (for example, one person might describe the person as very open, another person might describe the same person as being very closed). I've seen this in intimate relationships, too. For example, someone breaks up with their partner because, based on their own subjective experience, their partner isn't able to look at their shadow and self-reflect. This supposedly unself-reflective person then begins a new relationship, and their new partner has a totally different experience of them, finding them incredibly self-reflective. Who is seeing correctly? Is the new partner simply blind to what the first partner was seeing? Or is the new partner more enlightened, not needing to dream up their partner's shadow aspects so as to work out some unresolved part of their own unconscious process? Maybe both are in some way true. It is not a question of who is "right"; rather, this situation points to the multidimensionality of who we all are, as well as the subjective nature of our perceptions.

In other words, every person's image of this individual, however contradictory, could be accurate, relatively speaking. Not only is each person seeing potential aspects of the person, but what they are seeing is in some way a function of the filter through which they themselves give meaning to their experience. This is to say that their perceptions could be picking something up in the other person, or it could be a reflection of something within themselves. From the quantum physics point of view, the person's wave function collapses into a particular manifestation depending upon how the observer-participant is interacting with the person in question. Who in actuality is this person they are all experiencing so differently? Does the person exist objectively (as an object)? Are their various subjective experiences of the person merely the other person's projections onto who the person actually is? Or does this person have a multiplicity of potential versions of him or herself in a state of superposition at each and every moment, and which shows up is dependent on the circumstances they are in, who they are with, and how they are being "dreamed up"? This is an example of how the underlying probabilistic reality that quantum physics is pointing at applies to the domain of human interpersonal interactions and continuously informs our day-to-day relationships.

This makes me think of an experience I had many years ago. My longtime girlfriend and I had broken up, but we were still living together. It was early in the morning and I was asleep. She wanted to ask me a question about something, but was afraid of my answer. As if fulfilling her fear, she literally woke up the part of me that indeed did have a problem with what she was asking. To this day I remember my inner experience in that moment: out of the multiple potential aspects of myself (one of which would *not* have had a problem with what she was asking), in the way she asked the question she called forth the part of me that would confirm her negative expectation. The part of me that she had awakened reciprocally evoked and reinforced in her mind that this was indeed who I really was. She, in turn, then reacted to this part of myself, and we were what I call "off to the races," once again re-creating and playing out a problematic aspect of our relationship. Calling forth the part of me that didn't give her the answer she was looking for came with its own corresponding universe that seemed to support the "objective" nature of what was happening, as she now had all the evidence she needed to confirm the truth of her viewpoint. Our whole process was an expression of the quantum nature of the universe in action.

Here's another example. Say someone, while hanging out with a friend, seems to act out their unconscious in a particular moment in a way that seems insensitive. And yet by doing this they actually catalyze their friend to have a deep insight or breakthrough around some issue they were struggling with. The question naturally arises: Was the person who acted out their unconscious solely being insensitive, or was their acting out in this seemingly insensitive way an aspect of how plugged in they are? In other words, were they unknowingly picking up and acting out the unconscious in the other person and/or the field?

Reminiscent of two-valued logic, we are used to thinking that the choice is either one or the other, they were either being unconscious and insensitive, or they were exquisitely tuned in. The four-valued logic of a quantum universe points out that these two seemingly contradictory opposites can both be true at the same time. In a superposition of states, the person could at one and the same time be both insensitive and highly sensitive to what on an unconscious level needed to happen to help their friend have their breakthrough. Most of us who have been conditioned to live in a classical universe might find it a bit challenging to hold both of these mutually exclusive opposites as being true at the same time.[649] Interestingly, one of the chief features of the philosopher's stone—the symbol for an expanded consciousness—is that it represents a conjunction of a vast number of seemingly irreconcilable opposites.

Another example, suppose someone in a family is emotionally upset and maybe even having psychological problems. In the old, classical model of the world, this family member is isolated and might even be seen as having a particular mental illness (localized) inside of their brain. But from the more holistic view that is emerging from quantum theory, the person would not be seen as separate from the set of complex interrelations that comprise the family system, a system (similar to a nonlocal field) which itself is embedded within and a manifestation of the larger culture of society and the world. Instead of being treated as a separately existing objective entity that exists apart from the system (becoming what is known as the "identified patient"), it would be recognized that the person (and their "illness") are expressions of an underlying disturbance in the family system, which includes the other family members. This is analogous to how in quantum physics the experiment (observed) is understood to be not separate from the experimenter (the observer), who, in turn, is not separate from the rest of the universe.

One more example of the isomorphism between quantum physics and our everyday lives: Before an electron is observed, to quote Arnold Mindell, "it is as if the electron were dreaming."[650] In its state of open-ended possibility, we don't seem to have access to the electron's dream. This is parallel to what happens within our own subjective worlds when we are undecided about something and entertain multiple possible perspectives of events, people, or situations that arise within our lives. In our minds we routinely exist in a state of superposition of possibilities when we daydream about a range of possible paths that we could take until we decide on one that we want to actualize. At the moment of decision, the universe that includes the path we have chosen instantaneously materializes, while the other possibilities dissolve as if they had never existed. Some quantum physicists even imagine that all of the potential paths we didn't choose, though not actually existing in this universe, exist in parallel universes. A mental world in which we are holding a superposition of simultaneous possibilities is quite a natural part of the subjective experience of being human.

Our moment-to-moment, ordinary experience is infused with the quantum realm. Like any part of our everyday experience, we have become so thoroughly accustomed to our quantum nature that we oftentimes don't even notice it—it's just "the way things are." This familiarity can mask the rich field of ever-shifting quantum possibilities that underlie the outer surface of our everyday experience. The fact that what we see and experience appears to be so definite and dependable can easily hide the fluid and multidimensional dynamics of quantum creation that inform and give shape to the apparent stability and reliability of our experience of the world.

Since the subjective side of human experience is a domain that scientific exploration has traditionally and habitually eschewed, this profound parallel between fundamental aspects of our inner subjective realm of mind and the superposition of states of a quantum entity prior to observation is rarely pointed out. It would be a most fruitful venture to bring the "out there" and "far-out" realizations of quantum physics "in here"—into our own subjective experience of ourselves. It would be helpful in deepening our understanding of quantum physics, psychology, and ourselves to map the correlations between the quantum realm and our normal states of consciousness, as quantum physics is not only pointing out that the world we live in is quantum through and through, but that we ourselves are quantum entities in the flesh.

Before a quantum entity is observed, it exists in a superposition of states, including any and all possibilities. This seemingly strange situation directly correlates with our day-to-day lives. Say something happens that is very challenging, painful even, to the point that it creates a sense of being wounded. This wound could become a real problem, getting in the way of living our lives, or it could become a doorway through which we can access our deeper gifts and potentials to help others with their wounds, a capability that we didn't know we had prior to being wounded ourselves. Both of these possibilities—having the wound manifest an obstacle/problem or opportunity/initiation—exist in a superposition of states within the experience itself. The experience does not exist objectively one way or the other, just like in the world of the quantum, how things actually turn out depends upon how we dream them.

I will give one final example that I imagine most of us can relate to—I know I can. Over the course of life we have all been traumatized to one degree or another. When we become traumatized, a part of the wholeness of our psyche splits off and, if we aren't able to sufficiently deal with and integrate this dissociated part of our psyche, it can develop a seemingly autonomous life and quasi-independent will of its own. In psychological-speak these split-off parts of the psyche are called "autonomous complexes," which, due to their autonomy, manifest to our minds *as if* they were self-existing entities.[651] These ostensibly autonomous entities are quantum in nature in that they link the subjective and objective dimension of our experience.

When we subjectively experience these seeming entities inside of our mind-stream, depending on our awareness, we might experience them as objectively existing entities that are totally other than ourselves who are attacking us, or our subjective experience might be to realize that we are—or have been up until this moment—*unconsciously* creating them. Whichever of these exclusive viewpoints we hold at any given moment becomes the case, which (from the point of view of two-valued logic) excludes the other possibility from having any reality. The four-valued dream logic of quantum physics, however, points out that both possibilities could be true—existing in a superposition—at the same time. This is to say that we could *experience* these entities as being objective *and* realize that we are creating them at the same time. It all depends upon where we are in all of this. In other words, who are we relative to these seeming entities? How they manifest, or what our subjective experience of them will be, depends on how we dream them.

How these seeming entities manifest to us (do we experience them as happening to us from outside of ourselves or are they our own doing, our own creation which simply need to be recognized as such and reintegrated back into our wholeness?) depends on nothing other than our own awareness, just like in the quantum realm. These two apparently contradictory and mutually exclusive subjective experiences are an expression of the same deeper process that informs the wave/particle duality, in which, for example, how light manifests (as wave or particle) depends on how we observe it. It appears that how everything manifests, be it inner or outer, depends on how we dream it.

Our species is suffering from a form of collective trauma. The signature of trauma is that though the traumatic event happens in time as an actual historical event, we then internalize it. This is to say that we unconsciously reenact the trauma in both our inner and outer lives. Compulsively repeating the trauma (through what is known as "the repetition compulsion"), which is our attempt to heal from it, re-creates the very trauma from which we are trying to heal in an infinitely self-generating feedback loop which can go on forever, both temporally and atemporally (outside of time, in the realm of the unconscious). A primordial form of madness, this is a precise description of the dynamic of the aforementioned wetiko psychosis that our species is collectively acting out, writ large, on the world stage. This infinite regress only stops when we realize the role we are playing in creating our experience of being traumatized. Realizing that we are colluding in our own trauma through our unconscious reactions snaps us out of being a victim while simultaneously introducing us to our intrinsic creative power to shape our experience.

Whether or not we realize our own complicity in reconstituting our trauma is an inner correlate to whether we recognize that the outer world doesn't objectively exist separate from our perceptions of it. To say this differently, just like in the inner world of trauma, do we experience the outer world as completely separate (which it clearly is on one level), or do we realize our own agency, recognizing that we have an active hand in constituting our experience of the world and ourselves at each and every moment?

Ironically, the realization that (as quantum physics explicitly reveals) there is no objective reality is so shattering to our classically-conditioned sense of things that it is itself traumatizing—the aforementioned QPIT: "Quantum Physics-Induced Trauma." This is a special form of trauma, however, in that it can so shake us up such that we irreversibly snap out of the malaise from which many of us have been suffering. In other words, once we sufficiently wake up and see through the illusion that there exists an objective world separate from our own mind, there is no going back to sleep. A door has opened, and it very much behooves us to go through it when it appears. QPIT is a unique form of trauma that—homeopathically speaking (where "like cures like")—heals our existential trauma. Encoded in this particular form of trauma is not only its own medicine but a gift.

The classical world's objective reality delusion simultaneously evokes and is evoked by the subjective experience of the separate self—the two reciprocally co-arise, mutually conditioning and reinforcing each other. The separate self is the context and framework in which all of our previous traumas endlessly recirculate themselves. In rendering transparent the illusion/delusion of the world existing objectively, quantum physics can potentially help us to see through the once convincing mirage of the separate self with which we have been unconsciously identified. The separate self can be considered to be the primordial trauma.

The many new and eye-opening insights of quantum physics run counter to our intellectual habits and are also inconsistent with our implicitly accepted beliefs and understandings of the nature of reality. In the process of taking these new quantum ideas "on board" and installing them into the operating system of our mind, the most important place (or rather time) to try to apply them is our immediate experience in the present moment. The now moment is "where" we can engage with and activate the creativity that is intrinsic to the quantum realm. It is in the now moment when these new quantum understandings can be most powerfully translated from the realm of thought, intellect, and theory into the domain of our present experience of reality. As we continually deepen our understanding of the fluid, creative, and open-ended nature of our present moment experience, we become more of an engaged observer-participant in the moment-to-moment universal dance of quantum creativity.

• CHAPTER SIXTEEN •

LIFTING THE VEIL

To put our discussion in historical context, modern-day physicists are the current lineage holders of a wisdom tradition that has been passed down through the ages. Quantum physicists are the living representatives of a distinguished tradition that has its roots in the very first scientists, theologians, natural philosophers, pagan priests, creative artists, and Paleolithic shamans, all of whom have been involved in one way or another in attempting to understand, illumine, and express reality. The discoverers of quantum physics were deeply spiritual people. They were sincerely interested in truth, wherever it may lead. True trailblazers, they were grappling with the deepest philosophical and metaphysical questions that human beings have ever encountered.

The majority of modern-day practitioners of quantum physics, however, are no more interested in metaphysics than the ordinary person. Speaking about his colleagues, Wheeler said, "They're content to take the theory for granted, rather than to find out where it comes from."[652] Unlike many of today's corporately trained physicists, however, the founding fathers of quantum physics were passionately interested in, and deeply disturbed by, the philosophical implications of their discoveries.[653] Schrödinger, for example, referring to the new physics that he himself was helping to create, famously said, "I don't like it, and I'm sorry I ever had anything to do with it."[654] Similarly, expressing his struggle to take in the implications of quantum theory, Pauli wrote to a colleague that he wishes he "had never heard of physics."[655] Einstein said that if what quantum theory is pointing at is true, he'd "rather be a cobbler or a clerk in a gambling casino than a physicist."[656]

The real movers of history, people who end up being applauded and remembered by later generations, are those who are willing to step aside from the prevailing, mainstream way of thinking to pursue their minority and marginalized point of view. Not being able to find the words to describe the majesty of what they had discovered, the founding fathers of quantum theory "fell into stammering" when asked to discuss the implications of their own theories. There is not a veiled quantum reality that they were uncovering; they were first starting to realize that the very notion of an objectively existing independent reality no longer applied. The whole meaning of reality came into question. These pioneers in physics were beginning to realize that they had stumbled upon an epochal discovery that was destined to change the course of history. Finding the quantum realm is like discovering the Holy Grail; its magic can potentially change everything.

Quantum physics doesn't come with instructions on how to use it. And like the mythical Holy Grail, the powers that quantum physics is unleashing can be used for good or for evil. Pauli, who was considered "the conscience of physics," wrote, "I believe that this proud will to dominate nature does in fact underlie modern science . . . the anxious question presents itself to us whether this power, our Western power over nature, is evil."[657] As with any discovery, there is an obvious shadow side to the revelations of quantum physics. To cite a quote commonly attributed to Nikola Tesla, "It's hard to give unlimited power to limited minds."

There is an underlying and unexamined cultural belief that the solutions to our societal problems can be found in technology, which is a direct consequence of our mechanistic worldview and the blindness that it induces. The legend of Faust selling his soul to the devil for unlimited knowledge and power is an image of a deeper archetypal pattern that informs the mechanistic scientific enterprise. Jung points out that when Eros is absent, a vacuum is created and that vacuum is typically filled by the negative, destructive aspect of the shadow. Connecting with the quantum and realizing our quantum nature bestows such an immense power that this knowledge could only be entrusted upon those with high moral character. Yet many of the new technologies developed out of the quantum revolution are used by some people against others to fulfill their dreams of power and domination. In our modern world, knowledge is power. Clearly knowledge is a powerful currency. Banesh Hoffmann writes:

> And here it was that the curtain fell, a curtain of dreary silence and suffocating secrecy hiding a deathly fear. What of the tremendous new theories. . . . Such things are now military secrets, to be told by spies but not by scientists. Yet a corner of the curtain has been lifted to let some fragments of knowledge escape to the light. . . . The days of the nightmare are upon us, and science is in mortal peril of becoming an occult, unfertile priesthood, passing its mysteries on to chosen novitiates who meet stern tests and take the solemn vow of eternal silence. We can but hope the danger soon will pass, and someday, when the skies are brighter, science will again be free to stride forth boldly, in goodly fellowship, along its enchanted path into the unknown.[658]

As long as power struggles continue on earth—and at the moment their end is not even in sight—we must also fight for knowledge. Our task is to help each other to lift the curtain, thereby overcoming the secrecy and compartmentalization by which the knowledge of quantum physics is held as guild secrets among its "chosen novitiates," who act as guardians of this supreme knowledge. Lifting the veil of secrecy allows the liberating quantum gnosis to "escape to the light," so as to allow these "tremendous new theories" to resume their unfolding along their "enchanted path," extending this knowledge to all humankind, thereby helping all of us.

The emerging quantum gnosis, in placing a divine power within the reach of humanity, places an enormous responsibility in our hands. Jung writes:

> He can no longer wriggle out of it on the plea of his littleness and nothingness, for the dark God has slipped the atom bomb and chemical weapons into his hands and given him the power to empty out the apocalyptic vials of wrath on his fellow creatures. Since he has been granted an almost godlike power, he can no longer remain blind and unconscious. He must know something of God's nature and of metaphysical processes if he is to understand himself and thereby achieve gnosis of the divine.[659]

Because of the practically "godlike power" that quantum physics has bequeathed to humanity, we no longer have the option to remain blind and unconscious. We are being asked—*demanded*—to know something of "metaphysical processes" and deepen our insight ("gnosis") into our nature, particularly how we fit into the grand scheme of things.

Our species is presently confronted with an evolutionary imperative that we ignore at our own peril. Scientific materialism has trapped us into a dangerous evolutionary bottleneck out of which the demand to expand in consciousness or continue our endless self-destruction is bearing down upon us with apocalyptic force. It is noteworthy that quantum physics is playing an "apocalyptic" function within the collective psyche. Psychologically speaking, the archetype of the apocalypse has to do with the coming of the "self"—our true nature—into conscious realization. Etymologically, the word apocalypse has to do with unveiling what was hidden. Quantum physics' role in exposing matter to be an ultimately insubstantial, lower-dimensional, holographic shadow play of the higher-dimensional light of consciousness is a profound instance of just such an unveiling.

Nature always seems to evolve in such a way that enhances an organism's survivability in its environment. As if tailor-made for the multiple crises we find ourselves in, quantum physics is just what the doctor ordered to enhance not only our continual survival, but our full-blown blossoming into our higher potential as well. It is offering us a way out of our dilemma by revealing to us a new direction for our scientific, psychospiritual, and social evolution. This new direction is not a direction in external, third-dimensional space, but instead can only be found hidden within yet to be discovered domains of our own mind. We are already in possession of this medicine, we simply have to recognize this fact and more deeply inquire into its multifaceted curative powers to begin to utilize its benefits.

To quote Stapp, "The re-bonding [between mind and matter] achieved by physicists during the first half of the twentieth century must be seen as a momentous development: a lifting of the veil."[660] The discoverers of the quantum world were aware that momentous changes were afoot, but had as little foreknowledge of the deeper meaning of their discoveries as a caterpillar has of its destiny to become a butterfly. There is a distinction between the act of making a discovery and the act of understanding the discovery that one has made. Physicists have been swept off their feet by the strong currents of their discoveries, and know not where they are being carried. It was years before the survivors began to realize that the maelstrom that had so overwhelmed their science had been the convulsive birth pangs of a new era filled with astonishing possibilities. Speaking of the continually emerging quantum theories, Hoffmann continues:

> Though they be destined to be forsaken by generations to come, they remain a wonderful adventure of the human mind, a wonderful exploration of the works of God. . . . They yet contain within themselves something of the eternal, and to our mortal gaze they stand a dazzling edifice of towering majesty, whose brilliance gladdens the soul and sends forth brave, struggling rays to pierce the murk and gloom that press around. Here in such theories and discoveries is a revelation, all too scant, of the mighty wonder that is the universe.[661]

Quantum physics is one of the greatest all-time discoveries of the human mind. It is a living revelation of that which is most important for us to know. Quantum theory is a prime example of Einstein's remark, "Modern science does supply the mind with an object for contemplative exaltation. Mankind must exalt itself."[662] In contemplating and appreciating the revelation of quantum physics, we exalt ourselves, truly praising the Logos in the process. Our task—a "mission possible"—is to help the brilliant rays of quantum physics "pierce the murk and gloom" that have seemingly enveloped our world and show us the way to our intrinsic freedom. The worst nightmare of the powers that be is a magical (and unstoppable) idea whose time has come.

HYPOTHESIS OF THE REAL WORLD

The belief in an objectively existing, independent universe is a strongly ingrained unconscious assumption that still holds sway deep in the recesses of most people's unconscious minds. The philosopher and mathematician Alfred North Whitehead refers to mistaking an abstraction for a concrete fact as "the fallacy of misplaced concreteness."[663] It is helpful to inquire into how this fallacy of misplacing concreteness onto a universe that is anything but solid can potentially hold sway over and entrance our minds in a way that translates into creating real problems in the world.

The Dalai Lama tells a story of asking his friend and one of his "scientific gurus," physicist David Bohm, what is wrong with the belief in the independent existence of things apart from that it does not accurately represent the true nature of our situation. His Holiness relates: "His response was telling. He said that if we examine the various ideologies that tend to divide humanity, such as racism, extreme nationalism, and the Marxist class struggle, one of the key factors of their origin is the tendency to perceive things as inherently divided and disconnected. From this misconception springs the belief that each of these divisions is essentially independent and self-existent."[664] Bohm is pointing out that having a misconception of a situation leads to a mistaken belief that what we are seeing—in this case, division among people—independently, objectively exists on its own, which is a deluded perception that leads to all sorts of problems. His Holiness commented on Bohm's response, "I wish there were more scientists with his understanding of the interconnectedness of science, its conceptual frameworks, and humanity."[665]

Falling under the spell of our own mind's power to create images, and then thinking that our viewpoint is nonnegotiably true—be it our point of view about a particular issue or about the entire universe—is at the root of much rigid ideology and man-made destruction. By becoming entranced by our own mind we can become self-righteously convinced that we are in possession of the truth. This can easily inspire crusades to convert the unenlightened, as has been tragically evidenced throughout history again and again. It should be noted that the same underlying psychological dynamic that causes us to misconstrue the nature of the apparent physical universe also causes us to divide and polarize among ourselves. This then causes us to create different factions within society with irreconcilable differences that appear to be objectively real, thus creating the preconditions for endless, internecine conflict and war.

Referring to what he calls the "hypothesis of the real world,"[666] Schrödinger writes, "Without being aware of it and without being rigorously systematic about it, we exclude the Subject of Cognizance from the domain of nature that we endeavor to understand. We step with our own person back into the part of an onlooker who does not belong to the world, which by this very procedure becomes an objective world."[667] In excluding from the world the "Subject of Cognizance" we pay, in Schrödinger's words, a "high price," as we are removing life from nature, turning it into a corpse, creating a dead image from a living universe. At the same time we are reducing a part of ourselves to be a simulation of this same inanimate matter that has nothing to do with our essential sentient spirit. In excluding ourselves from the universe, materialist, reductionist science is first destroying the world in theory before proceeding to destroy it in practice.

In writing ourselves out of the script of this world, science is precluding any possibility of experiencing our true nature, thereby negating our power to make any real difference in the world. In excluding ourselves from our image of nature, we are, "by this very procedure," in the same moment conjuring up the appearance of an objective world, which we then take to be both self-existing (mistaking the appearance to be the thing itself) and an unquestioned given. These two processes reciprocally and synchronously co-arise and mutually reinforce and condition each other. This is actually one process with two interrelated and mutually self-reinforcing aspects. We do two things simultaneously: construct the world of objects, and exclude from it the "Subject of Cognizance"—ourselves. Like two sides of the same coin, fabricating the image of an objective world out there and identifying as a separate self in here, reciprocally co-arise and co-generate each other.

The hallucination of a solid, objective world that exists independently of a subjective observer is the projection of a mind entranced by its own intrinsic ability to creatively and effortlessly conjure up how it experiences things. To think of the world as objective is to conceive of the world as being "not I."[668] When we think the world objectively exists, we, as the other half of this process, dream ourselves up to be a separate subject (an "I" separate from the "not I") who encounters the world as an object; the world as object with ourselves as subject reciprocally co-arise, co-evoking, and co-conditioning each other. To say the same thing from the other point of view is just as true—when we think we exist as a subjective reference point in time, we dream up the world as object. This process is nonlinear, synchronic, circular, and takes place outside of time—instantaneously, in no time at all (which makes it hard to see). To quote the Dalai Lama, "Thus, there are no subjects without the objects by which they are defined, there are no objects without subjects to apprehend them."[669]

Scientist and philosopher Francisco Varela reminds us, "This grasping after an inner ground is itself a moment in a larger pattern of grasping that includes our clinging to an outer ground in the form of the idea of a pre-given and independent world. In other words, our grasping after a ground, whether inner or outer, is the deep source of frustration and anxiety."[670] Becoming fixed in a perspective that views humans as subjectively existing apart from the world or the world as objectively existing outside of ourselves are both forms of grasping, which according to the insights of Buddhism, is the root cause of suffering.

This brings to mind what William James calls "the fantastic laws of clinging,"[671] which can be understood to underlie the very process of how we participate in creating both our mental and physical worlds in the way that we do. James is naming an internal process in which a multitude of related ideas and streams of thought intermingle and cling together, in his words, "weaving an endless carpet of themselves."[672] As if having a life of their own, these ideas hang together like a persistent entity. This inner process is a reflection of how we dream up our world to appear as if it is a "persistent entity" that exists in an objective fashion.

As one of its chief features, the universe has built into itself the potentiality for self-awareness. "We cannot escape the fact that the world we know is constructed in order (and thus in such a way as to be able) to see itself," says mathematician G. Spencer-Brown, "but in order to do so, evidently it must first cut itself up into at least one state which sees, and at least one other state which is seen." Out of its wholeness, the world splits itself into subject and object so as to objectify itself to itself in order to be seen as an object and therefore known. Spencer-Brown continues, "In this severed and mutilated condition, whatever it sees is only partially itself. We may take it that the world undoubtedly is itself (i.e., is not distinct from itself), but, in any attempt to see itself as an object, it must, equally undoubtedly, act so as to make itself distinct from, and therefore false to itself. In this condition it will always partially elude itself."[673]

Without a break in the coherent symmetry of simply being fully itself, however, the true nature of "being" would have no way to encounter and become aware of itself. This is similar to light transforming aspects of itself into particles so as to reveal its potential in a new way. Severing itself into subject and object, the true nature of being becomes "false to itself," invariably eluding itself like an ever-receding rainbow until the correlation, interconnection, and ultimate inseparability of the subjective and objective domains of our experience are recognized. Jung writes, "The division into two was necessary in order to bring the 'one' world out of the state of potentiality into reality."[674]

This is analogous to psychological reality. When we are unconsciously identified with a content of the unconscious, the only way we have of becoming conscious of it is to project the unconscious content outside of ourselves—where it gets "dreamed" into form—and see it objectively, outside of ourselves as it appears in the seemingly external world (this is why Jung points out that the unconscious always approaches us from outside of ourselves). Hopefully we can then recognize that what we are seeing "out there" is a reflection of what is "in here," which is when we can take back and own—becoming conscious of—the projected content as ultimately belonging to ourselves.

It should be noted that this dynamic is how the process of dreaming works. In dreams our unconscious projects itself into the seemingly externalized dreamscape where we then interact with these unconscious parts of ourselves. Teleologically, the ultimate aim of this process is to work something out, to *potentially* recognize these unconscious aspects as our own reflections so as to integrate them within ourselves.

Dreaming ourselves up to be "subject" to the world, we truly suffer from a case of mistaken identity, unnecessarily limiting ourselves in the process. We then give away our intrinsic power, having disconnected from our sovereignty and creative potency to call forth our experience of both the world and ourselves. Dissociated from our own creative agency, the world then seems to create our experience for us, with ourselves playing the role of the passive (and oftentimes victimized) witness. Seeing the world in this disempowering way, we then dream up all the evidence we need to prove to us the rightness of our point of view in a self-fulfilling prophecy that further alienates us from one another, our world, and our true selves.

Science, which it should not be forgotten is made by man, always contains implicit statements—be they right or wrong—about the nature of humanity. The contemporary scientific worldview that still reigns throughout the "civilized" world is particularly deadening and lacking in awareness of the sacred world-shaping creative power that lies dormant within the human soul. Additionally, thinking we are seeing the world objectively appears to turn us into objects as well, an operation which ultimately immobilizes our human potential, hinders our compassion, and kills the soul.

When we think the world objectively exists independent of ourselves, we are distorting our image of the world, which is a process by which we can't help but distort our image of ourselves at the same time, since we are inescapably part of the world. Speaking about this process, Schrödinger writes, "I conclude that I myself also form part of this real material world around me. I so to speak put my own sentient self (which had constructed this world as a mental product) back into it."[675] We construct the world out of our sensations, perceptions, and memories, and then, as if making a collage, glue our made-up (and ultimately unreal) image of ourselves (which is similarly a function of our creative imagination) into the seemingly real scene that surrounds us. We have then reduced our idea of ourselves to a similarly objectified concept that corresponds in type to the nature of the rigid concepts by which we frame our understanding of the "objects" of the seemingly objective physical world.

The illusion of external reality is so convincingly real that it produces a strongly held, concurrent belief that there is an inherently real psychological center of operations, a subjectively existing reference point and center of volitional action within us (our ego, our sense of individual self) which then reciprocally feeds back into the illusion of an inherently existing outer reality in a potentially infinitely self-perpetuating feedback loop. We have then fallen prey to what David Bohm calls "an illusion-generating illusion." To quote Bohm, "The key to sanity is to see through this basic illusion-generating illusion."[676] If we don't see through this process, we can then easily remain unaware of the creative power within us, a power that is being blindly misused to create the very illusion in which we become trapped. Our creative power, wherein lies our potential gift to the world, is nevertheless always present within us. But if it is not seen and consciously engaged with, it can be used against us to generate the illusion that we have little creative power.

Expressing the essence of quantum physics, as well as a central tenet of Buddhism, Niels Bohr commented that "an independent reality, in the ordinary physical sense, can neither be ascribed to the phenomena nor to the agencies of observation."[677] What Bohr says is so profound that it warrants highlighting. Quantum physics is showing us that we can't ascribe an independently existing reality—not to the outer world or to the "agencies of observation" (us). In other words, we don't exist in the way we have been imagining we do, if we have been imagining we exist as independent, objectively existing agents separate from the universe. Jung writes, "One is *oneself* the biggest of all one's assumptions."[678] Quantum physics thus not only challenges the nature of what we call reality, it calls into question our very sense of who we are. Wheeler comments in his own inimitable style, "We may someday have to enlarge the scope of what we mean by a 'who.'"[679]

It is one thing to recognize that this universe doesn't exist in the way we've been imagining it does; it is quite something else to recognize the inner correlate of this realization—we ourselves don't exist in the way we've been imagining. To the extent that we're unconsciously identifying with a self-constructed model for who we are instead of recognizing and simply being who we are, we are living a lie. We are then negating the truth of our existential situation, which leads to a state of delusion. Quantum physics, when contemplated deeply enough, can completely unravel our illusory sense of self in a way that, to the ego, can feel like the most frightening thing of all, like some sort of death experience. This is the "edge" that quantum physics is forcing its practitioners to confront within themselves, an edge which is at the bottom of the unconscious creation of the aforementioned "don't-go-there zone" in physics.

Realizing that the world doesn't exist objectively and is not made up of separate things has another corresponding half. The realization of the absence of the objective world occasions the realization that we, as alleged subjects of this realization, have nothing to be subjects in relation to. This is to say that the idea (which is just a thought in our mind) of ourselves as subjects separate from and encountering the world as an object spontaneously re-visions itself. This re-visioning entails the discovery of ourselves to be a field of pure immaterial awareness without any edges, borders, boundaries, or features. Freed of its unnecessary and limiting identifications, this awareness naturally recognizes itself in everything that it beholds.

This process of objectifying both the world and ourselves veils the real subject within us, which is something other than our self-fabricated ego. Jung writes, "Every science is a function of the psyche, and all knowledge is rooted in it. The psyche is the greatest of all cosmic wonders *and the sine qua non* [indispensable condition] of the world as an object. It is in the highest degree odd that Western man, with but very few—and ever fewer—exceptions, apparently pays so little regard to this fact. Swamped by the knowledge of external objects, the subject of all knowledge has been temporarily eclipsed to the point of seeming non-existence."[680] The "subject of all cognizance," the part of us that is perceiving the world, is who we truly are. In Schrödinger's words, "Our perceiving self is nowhere to be found within the world-picture, because it itself is the world-picture."[681] Our picture of the world reflects ourselves.

The root meaning of the word "psychology" is the study of the psyche and the soul. We can conceive of the soul as a vital, animating core of luminosity, sentience, and aliveness, the very thing that links us to the divine, to each other, and to the part of us that is most ourselves and most human. Finding our soul has to do with becoming conscious of our true identity and discovering our purpose for living. The soul can't be explained—it is the soul itself which is the principle informing every explanation—but we know it when we experience it, as when we meet people who are connected with their own soul. We can conceive of the soul as being more like a perspective than a substance; it informs our way of seeing rather than being something seen. The soul can never have knowledge of "objective" reality—for it, and we, are not objects. The soul can only know what it *is*.

To realize that we do not exist in the way we have been conditioned to believe is to have a radical phase shift in our sense of reality and identity, crossing an event horizon in our mind in which figure and ground reverse themselves. This is not only a realization which takes place in the psyche; it necessarily involves finding ourselves within and enveloped by psyche. Jung comments, "The psychical is no longer a content in us, but we become contents of it."[682] From all appearances the psyche has spilled outside of our skull and is synchronistically in-forming and giving shape to the outside world. Instead of the psyche being within our brains, just like in a dream, we discover ourselves to be inside of the psyche; we, and our world, thus become "en-souled."

To quote writer Jorge Luis Borges, "We (that indivisible divinity that operates in us) have dreamed the world."[683] This physical world is, as Sir Arthur Stanley Eddington calls it, "mind-stuff,"[684] which is to say that, just as within a dream, the "stuff" of this world is inseparable from the mind of the dreamer, which is us. In other words, to see that the world doesn't exist as an object out there, combined with seeing that we don't exist as an objective subject in here, is the doorway to the realization of the dreamlike nature of reality, which is the very realization that quantum physics is ultimately revealing to us.

GETTING IN TOUCH

Schrödinger writes, "We cannot make any factual statement about a given natural object (or physical system) without 'getting in touch' with it. This 'touch' is a real physical interaction. Even if it consists only in 'looking at the object.'"[685] We get "in touch" with an object when, like an artist, we are "touched" by it, which is to say when we experience the object within ourselves. Schrödinger comments, "Matter is an image in our mind."[686] Quantum physics is linking the subjective and objective domains into a higher, more coherent synthesis. In Schrödinger's words, "In perception and observation subject and object" are "inextricably interwoven," their influence being unavoidably "mutual," their relationship a true "*inter*-action."[687]

It is not just that we affect the universe through our act of observing it; we ourselves are reciprocally affected. To quote Wheeler, "Anything that affects something else must, in turn, be affected *by* that something else."[688] To say this differently, we can't touch without being touched. This reminds me how in the realm of the psyche it is impossible to see the unconscious—be it inside or outside of us—and not have that seeing affect the unconscious, which, as if coming full circle, simultaneously impacts our consciousness.

In the act of observation, the subjective and objective domains reciprocally cocreate each other as their difference becomes blurred. To quote the Dalai Lama, "Once you do away with any possibility of grounding epistemology in a truly existing external world, or internal world, for that matter, then the only option you have is to develop an epistemological system where there is a mutual dependence between subject and object."[689] The subject and object, like any parts of the universe, are interconnected, interdependent, and reciprocally co-arise; one doesn't exist without the other. As we go down the quantum physics rabbit hole, the mysterious boundary between the subjective and objective becomes fuzzier and more uncertain. When we slowly take off our eyeglasses, for example, how far must we move them before they are an object rather than part of the observer? Where does the observer begin and end?

Schrödinger comments, "The world is given to me only once, not one existing and one perceived. Subject and object are only one. The barrier between them cannot be said to have broken down as a result of recent experience in the physical sciences, for this barrier does not exist."[690] In our ordinary state of consciousness, we have developed a chronic, habitual, and unconscious pattern of actively erecting and maintaining a seeming boundary between subject and object that does not actually exist. We don't have to get rid of the barrier between the subject and object because, as Schrödinger reminds us, this barrier is illusory; not only that, we ourselves are creating and then falling for this illusion. In this process we are tricking ourselves out of our own mind. Heisenberg writes that "a complete separation of the observer from the phenomenon to be observed is no longer possible."[691] We are simply asked to "see through" and recognize the nature of our situation in which the observer is the observed. We are invited to recognize ourselves in what is being observed.

Upon closer inspection, for example, atoms dissolve into a mysterious, little-known deeper structure that ultimately merges with the field of the whole universe. Even the instruments we use to try and observe atoms and their effects are themselves composed of atoms, which similarly merge into the universal field. Going further down the rabbit hole, we ourselves, with our brains and nervous systems, have a similar constitution. So if we step into a deep enough point of view, we, in our act of observation, are in a sense like that which we observe. Atoms are made out of and are crystallizations of the very awareness that is observing them.

Instead of falling prey to Whitehead's fallacy of misplaced concreteness and superimposing an imaginary solidity onto a fluid universe that is continuously in flux, which will conjure up the universe to simply reflect back to us this seeming concreteness, we can allow the universe to reveal and glorify its dreamlike, synchronistic nature. The more we see the dreamlike nature of the universe, the more dreamlike the universe will reveal itself to be. This is a creative and creativity-generating feedback loop, the activation of which brings forth what can be thought of as a higher technology of mind.

The more I deepen my research into quantum physics, the more indistinguishable it becomes from a spiritual path. To quote Einstein, "I am of the opinion that all the finer speculations in the realm of science spring from a deep religious feeling, and that without such feeling they would not be fruitful."[692] And like spiritual adepts, the true practitioners of physics single-mindedly—practically religious in their devotion—focus their attention on their discipline. Commenting on his contemplations about physics, Feynman simply says, "I can't stop." Similarly Wheeler, who refers to the universe as "our museum of wonder and beauty, our cathedral,"[693] let's his mind "run free over the nature of space and time."[694] He confesses, "I have to admit that I never stop thinking about physics."[695] Physics clearly took hold of and, in Wheeler's words, "fired up" his imagination, using him as one of the instruments of its realization in our world of space and time. Reminiscent of Jung's process of "active imagination," after Einstein's death Wheeler even published an imaginary dialogue that he had with the esteemed physicist.[696]

Every spiritual wisdom tradition from time immemorial has pointed out in its own creative way that grasping onto the idea of intrinsic, independent existence—both in the seemingly objective outer world and within the subjective domain of our own selves—is the fundamental mental affliction, the root cause of our self-created delusion with all of its concomitant suffering. Clinging onto the idea that we exist in a way that we simply do not is a deeply entrenched unconscious disposition, a habitual pattern that at a certain point gains enough momentum to develop a seeming autonomy such that it regenerates itself, as we invest our life force into an illusory identity and unconsciously re-create (and defend) it, moment by moment. It should be pointed out that this is a widespread form of madness that afflicts our species.[697]

These same spiritual wisdom traditions point out that the realization of what in Buddhism is called "emptiness" (the lack of intrinsic, independent objective existence of both the outer world as well as ourselves) is the fundamental cure for our psychic "dis-ease."[698] In Buddhism, "emptiness" denotes the ultimate ground and nature of reality. Once the delusion of an objectively existing world is seen through and overcome, we are much more capable of generating great compassion for all beings, as there is a deeper sense of the interconnectedness of all of life. In discovering that there is no objective world out there and no objective subject in here, quantum physics is discovering the medicine or fundamental cure for the psychospiritual illness that ails our species. In so doing, quantum physics is promoting itself to the ranks of a spiritual wisdom tradition.

• CHAPTER SEVENTEEN •

QUALIA

Quantum physics shows that the world appears in one way and exists in another. To quote Stephen Hawking, from his book about quantum physics called *The Dreams that Stuff Is Made Of* (an interesting choice of title, I might add), "We are reminded of Bertrand Russell's words, 'We all start from "naïve realism," i.e., the doctrine that things are what they seem. We think that grass is green, that stones are hard, and that snow is cold. But physics assures us that the greenness of grass, the hardness of stones, and the coldness of snow are not the greenness, hardness, and coldness that we know in our experience, but something very different. . . .' It is these dreams that stuff is made of."[699] The greenness, hardness and coldness of the world are subjectively experienced *qualia* (the Latin word from which we get the word "quality") created in and by our consciousness, using our brain, nervous system, and sense organs as processing facilities. Jung writes:

> We can make only the dimmest theoretical guesses about the nature of matter, and these guesses are nothing but images created by our minds. . . . It is my mind, with its store of images, that gives the world color and sound; and that supremely real and rational certainty which I call "experience" is, in its most simple form, an exceedingly complicated structure of mental images. Thus there is, in a certain sense, nothing that is directly experienced except the mind itself. Everything is mediated through the mind, translated, filtered, allegorized,

> twisted, even falsified by it. . . . We live immediately only in the world of images.[700]

Compare Jung's words to Schrödinger's: "On the one hand I undoubtedly form part of nature, while on the other hand objective nature is known to me as a phenomenon of my mind only."[701] Are Jung and Schrödinger's words the words of a psychologist describing the world of physics, or of a physicist describing the world of psyche?

Russell comments, "Everything that we can directly observe of the physical world happens inside our heads, and consists of *mental* events. . . . The stuff of the world may be called physical or mental or both or neither as we please; in fact the words serve no purpose."[702] Russell, a logician par excellence, is succinctly expressing the aforementioned four-valued logic. His words also express that the distinction between mind and matter is illusory.

Scientific materialism leaves consciousness out of its picture of the world and thus falsifies the most important fact about reality: We only *experience* it. To quote Schrödinger, "'The world of science' has become so horribly objective as to leave no room for the mind and its immediate sensations."[703] Previous to the advent of quantum physics, physics had become so "horribly objective" that there was no living space for the subject of experience. This has developed into a taboo against subjectivity that still pervades science and constrains the practice of the scientific method with unnecessary restrictions on its freedom to explore and investigate anything and everything, be it subjective or objective. Schrödinger writes, "The scientific picture of the real world around me is very deficient. It gives me a lot of factual information, puts all our experience in a magnificently consistent order, but is ghastly silent about all and sundry that is really dear to our heart, that really matters to us."[704]

Mechanistic/deterministic explanations of life ignore the actual *experience* of living. The felt presence of the immediate experience of the individual, which is the medium through which all that we know and learn is transmitted into our beings, is strangely ignored and even devalued by this type of scientific mindset. Philosopher David Chalmers suggests that consciousness, or "experience," in his terminology, should be seen as a "fundamental feature of the world, alongside mass, charge and spacetime."[705] All experience, however, is made of qualia; the theory of qualia gets at reality through directly lived experience. Rooted in consciousness, the only reality that we can ever know is qualia; we live in a *qualia-verse.* Hopefully physics is returning to its authentic purpose, which is not only to comprehend the nature of the measurable world, but the nature of the experienced world as well.

What physicists describe as elementary particles are actually patterns of activity momentarily flickering into seeming existence, the result of insubstantial quantum fields interacting with each other as well as with our own consciousness. Elementary particles only exist for us as transitory traces within the field of our experience. Science fiction writer Philip K. Dick writes, "How real is a Beethoven symphony without one of us? We are part of the equation with it, and essential to it; half is on the record, but we are part of the playback equipment."[706] Similar to how a rainbow doesn't ultimately exist without a consciousness that is experiencing it, we are the other half of the equation that brings a Beethoven symphony—not to mention the universe—to life.

Wheeler writes, "There is not a single sight, not a single sound, not a single sense impression which does not derive in the last analysis from one or more elementary quantum phenomena."[707] There is simply no way to know that reality exists outside qualia, which are the quantum building blocks of creation. Our sensory qualia are often mistaken as being perceptions of a real, objective, independently existing world "out there." We must remember that our experience of the world is brought to us through our five senses and then neurologically remixed within our minds into a convincing experience of a seemingly solid, external, and material world. Upon closer inspection, in actual fact all we have is the mysterious immediacy of our firsthand phenomenological experience; the idea of a real physical external world is an unwarranted presumption that we are overlaying onto our direct experience of qualia. When carefully contemplated, it becomes obvious that the appearance of a solid, external world arises within our awareness; it is not nearly as external as we tend to think.

Schrödinger writes that "the stuff from which our world picture is built is yielded exclusively from the sense organs as organs of the mind, so that every man's world picture is and always remains a construct of his mind and cannot be proved to have any other existence."[708] Instead of thinking that the physical world is primary and our psyche is derivative from the "real" physical world, Jung points out, "It is an almost absurd prejudice to suppose that existence can only be physical. As a matter of fact, the only form of existence of which we have immediate knowledge is psychic. We might well say, on the contrary, that physical existence is a mere inference, since we know of matter only in so far as we perceive psychic images mediated by the senses."[709] Since we are never able to describe anything "as it is" but only "as it appears to our minds," we can never have a pure physics, but only "neuro-physics"—i.e., physics as known only through the mediation of the human nervous system.

According to quantum theory, the greenness of grass, the hardness of stones, the coldness of snow, in fact the entire "outside world," does not exist "out there," independently of and separate from ourselves, but rather exists nowhere except within our own minds. To realize this is, in Henry Corbin's words, "to be delivered of the fiction of an autonomous datum."[710] Wheeler writes in his journal, "Aren't we mistaken in making this separation between 'the universe' and 'life and mind'?" In other words, is our thinking that the universe is separate from our mind a grave error? In an interview in which he was asked about this idea, Wheeler responded with his typical humility, "I'm one of the most baffled men in the world on this subject."[711]

THE REALITY OF THE PSYCHE

There is not one universe that exists and another one that is perceived; the way our universe exists is inextricably linked and inseparable from how it is perceived. Our knowledge of the world begins not with matter but with perceptions. Everything we know and can ever know about the universe is conveyed to us via our perceptions. Perception does not consist of passive reception of and reaction to signals from the outside world, but is an active filtering and interpretation of the world's signals, which is to say perception is a creative transaction with the world. Nothing is perceived except the perceptions themselves. To quote Richard Conn Henry, "The only reality is mind and observations, but observations are not of things."[712] Speaking about the "benefit of seeing the world as quantum mechanical," Henry continues, "someone who has learned to accept that nothing exists but observations is far ahead of peers who stumble through physics hoping to find out 'what things are.'"[713]

Our perception of the universe is a creative part of the universe happening through us that actually influences how the universe manifests. To quote physicist Andrei Linde, "What if our perceptions are as real as (or maybe, in a certain sense, are even more real than) material objects?"[714] Our perceptions have a fundamental ontological reality of their own. They are something in and of themselves, reflecting a reality that *is* itself, and are not merely secondary reflections of some supposedly existing material world. Due to the fact that our social institutions are still organized around an outdated materialist, reductionist, Newtonian worldview, many of our collective problems today are essentially crises of perception, just as was true for physicists in the 1920s. Our (classically-induced) habits of perception tend to freeze and concretize the fluidity of the world (not to mention our mind) thereby limiting and restricting the richness of the immediacy of what is available to us at any given moment of our experience.

Jung simply refers to the ontological reality of our thoughts, perceptions, beliefs, and projections as the "reality of the psyche."[715] With this phrase, Jung means that the psyche exists in its own right, sui generis, having its own category of existence per se. Speaking of the psyche, Jung writes, "It is not there where a near-sighted mind seeks it. It exists, but not in physical form. . . . Not only does the psyche exist, it is existence itself."[716] Pointing out how "all important" the psyche is, Jung refers to it as "man's greatest instrument," "the essence of man," "the indispensable sine qua non of all existence" as well as, ominously, "the greatest danger which threatens man." "The life of the psyche," Jung writes, "is the life of mankind. Welling up from the depths of the unconscious, its springs gush forth from the root of the whole human race."[717]

Though formless, immaterial, and unquantifiable, the psyche has a unique reality with its own open-ended sphere of seemingly unlimited influence, molding us, our experiences, and our world in an ever-changing array of forms we can only imagine. Jung was of the opinion that "the idea of psychic reality" (the reality of the psyche) was "the most important achievement of modern psychology if it were recognized as such."[718] To quote Jung, "Between the unknown essences of spirit and matter stands the reality of the psychic—psychic reality, the only reality we can experience immediately."[719]

Jung also refers to the "reality" of the psyche as the "objective psyche." He refers to the psyche as real or "objective" (an unfortunate choice of words in my opinion) because the psyche, due to its archetypal foundations, has a universal substratum that exists *a priori*. Interestingly, while physics has had to incorporate the subjective element of the observer into its considerations of the seemingly objective world, psychology, in its explorations into the nature of the psyche and our subjective experiences, has been led to assert an objective element.

Quantum reality is not subjective just as it is not objective. The quantum dimension is the bridge, the intermediate realm in between the subjective mental realm "in here" and the seemingly objective world "out there," somehow coupling the two. In *The Holographic Universe*, author Michael Talbot has coined the term "omnijective" to refer to this indivisibility between the psychological and physical dimensions of our experience. The reciprocally co-arising nature of both the objective and subjective worlds is reflected by quantum theory, which is an instrument for the realization of the ultimate inseparability of these realms. The quantum occupies an intermediate domain between the subjective and the objective, between the inner and the outer, sharing in the attributes of both but being the same as neither.

A prototypical example of this state is an image in a mirror, which acts as a bridge or isthmus between the object reflected and the mirror, partaking of the qualities of both the object it is reflecting and the mirror. On the one hand, the image that resides in the mirror is indivisible from the mirror, while at the same time being different from it. In other words, the reflections in the mirror are inseparable from the mirror but are themselves not the mirror. The image in the mirror also cannot be separated from the object it is reflecting, while at the same time it is not identical with it. A mirror is merely the place of the appearance of the image it reflects, but the substance, the thing-in-itself that the image represents, doesn't reside in the mirror.

We currently do not have a theoretical framework for psychophysical phenomena, just as the early nineteenth century did not have electrodynamics. Just as electromagnetic phenomena are neither solely electric nor magnetic, psychophysical phenomena are neither solely mental nor physical but a combination of both that is greater than the sum of its parts. The psychophysical problem has to be recast in a way that finds a radically new viewpoint characterized not by the two-valued logic of either/or, but by the previously mentioned four-valued (dream) logic of both/and that is clearly articulated in Buddhism. A third viewpoint is needed which can unite the opposites of the physical and psychological realms. This third viewpoint is Jung's idea of the "reality of the psyche"[720]—a perspective which introduces us to the dreamlike nature of our universe. Ontologically speaking, the psyche is not only as real as the physical world; it is as real as we, as psychic entities, are real. From our point of view, it doesn't get more real than that.

AS VIEWED, SO APPEARS

Quantum physics indicates that our universe is arising exactly as a dream arises—as an immediate and unmediated reflection of the observing consciousness. It should therefore not be surprising that there is a similarity with other wisdom traditions that are also creatively articulating and illuminating the dreamlike nature of our universe. For example, one of the most succinct teachings of Tibetan Buddhism describing the dreamlike nature of reality—while at the same time being an expression of the very dreamlike nature it is pointing at—can be essentialized in four words: *As Viewed, So Appears.*[721] As Viewed, So Appears can be thought of as an equation or formulation for how we cocreate reality within this dreamlike universe of ours. As Viewed, So Appears is as all-pervasive and universal a law in the realm of the light of consciousness as gravity is in the physical dimension. Contemplating As Viewed, So Appears can be a catalyst to awaken in us the realization of how, in this very moment, we are dreaming up the universe into materialization out of the infinite field of unmanifest quantum potential.

In a dream, our inner process projects itself seemingly outside of ourselves so as to be experienced and encountered as though it objectively exists. This process helps us get in touch with unconscious parts of the subject—ourselves. Because the appearances within a dream are not separate from how we view it, if we change the way we view our dream while having it, the dream has no choice but to spontaneously shape-shift and mirror back this change in perception—changing the way it appears. This is because a dream *is* nothing other than our own consciousness externalizing itself, taking on forms that embody the state of the dreamer, appearing as immediate reflections of the mind that is observing it. We, as percipient beings, are in fact generating the very dreamworld we are experiencing; our perception of the dream (in a very real sense) produces it. The dreamscape is a reflection and an instantaneous reflex ("reflex-ion") of the way we are viewing it—As Viewed, So Appears.

One way to think about this is to consider that a dream is a projection of the mind (I am not talking about the conceptual mind, but Mind with a capital "M." This Mind is the featureless "subject of cognizance" which bears witness to all the forms of our inner and outer experiences. This Mind is the dreamer of the dream, what I call the "deeper, dreaming self"). A projection, whose sponsor is the mind itself, is an inkblot reflecting back to us what is happening within ourselves. As soon as we connect the dots on an inkblot and interpret it in a particular way, the inkblot instantaneously shape-shifts and mirrors back our projection. It is not as if one moment we view the inkblot one way and the next moment it appears that way. The very moment we view the inkblot a certain way is the very same moment it appears that way—As Viewed, So Appears. The inkblot just reflects back to us our own interpretation. This process doesn't happen in time or over time, it happens outside of time, faster than the twinkling of an eye, faster than we can think or blink, faster than the speed of light. Once we project onto the inkblot, the inkblot will provide all of the necessary justification and convincing evidence to prove the rightness of our projection, thereby compelling our assent to its "reality," in a self-confirming and never-ending instantaneous feedback loop that is completely self-generated by our own minds. Because this process happens in no time, we typically don't see it and get fooled by the reality-creating power of our own mind into thinking that what we are seeing objectively exists independently of the mind.

Even though the viewing and appearing are simultaneous, the viewing is more primary in that it is where the real power or leverage lies in effecting a change in how our universe manifests. *Our interpretation of our experience—investing our universe with a certain meaning—is the part of the universe through which we can change the universe.* We are the generators of meaning. When we inquire into the dreamlike nature of our reality, we begin to touch what I call the "meaning of meaning," discovering that there's no intrinsic *meaning* embedded in our waking dream separate from our own mind's *interpretation* in the present moment. This brings to mind the philosopher Friedrich Nietzsche's words, "There are no facts, only interpretations."[722]

By connecting the dots in the inkblot, so to speak, we are superimposing, or mapping a meaning pattern (projecting or dreaming) onto the screen of our experience, and being that the seemingly outer dreamscape is nothing other than our own projection, it will simply reflect back our interpretation. The universe mirrors back to us our point of view in such a way so as to confirm our perspective in a self-validating closed feedback loop that endlessly pours back into itself. The meaning appears to be inherent in the outer inkblot, whereas in reality the origin of the meaning is our own mind.

We project our inner state onto the omnifaceted outer world. The outer world responds in no time, simulating our projection by supplying the details that fill in the picture, and then generates our inner state back to us in enriched synchronization so as to confirm the "objective truth" of our projection. The outer world is not just a passive mirror, but rather a dynamic and amplifying one. Over time this process, what Philip K. Dick calls a "push-pull feedback loop," keeps on "mutually generating (creating) *a more and more articulated hologram-like reality.*"[723]

This waking dream we are living in, however, being of a denser vibration than a night dream, is more solidified into materialized form, and is hence "slower," having more lag time, in the way it is a function of our mind's creative imagination. Due to the apparent solidity of this waking dream, the effects of the creative imagination on how this universe of ours actually gets dreamed up are visible only to those with a much more subtle, penetrating, and rarefied vision. The cutting-edge experimental instruments that quantum physicists are using to explore the microworld are helping our species refine our vision and gain insight into how the way we perceive our universe affects the way it manifests. Our physical universe very convincingly appears to have the continuity of being something that seems solid and exists objectively, but we shouldn't get fooled or entranced by the seeming concreteness of the universe's dreamlike display. The universe's intrinsic reflective nature oftentimes is adroitly concealed in plain sight. It is sentient, playful, and alive, just like we are.

As Viewed, So Appears is such a profound articulation of how we create our reality that if we think As Viewed, So Appears is *not* true, the entire universe will shape-shift and reflect this back to us. This apparent negation, however, paradoxically demonstrates As Viewed, So Appears' all-pervading validity, further proving its profundity as a meta-principle par excellence. Our universe is dreamlike in nature; quantum physics is the physics of the dreamlike nature of reality. If physicists think that quantum physics is *not* the physics of the dream, then due to the magic of As Viewed, So Appears, quantum physics will manifest so as to supply all the needed evidence to prove that it is not the physics of the dream. Based on the principle of As Viewed, So Appears, reductionist, materialist scientists who don't see the dreamlike nature, but think that the outer world objectively exists and has nothing to do with how they observe it, will experience a world that reflects back their viewpoint. Ironically, this perception confirms to them the rightness of their viewpoint that the world exists separate from their observation through the very quantum principle of genesis by observership that they are denying.

We are like magicians who, entranced by our God-given power of evoking reality, have unknowingly entranced ourselves and have become enchanted by our own creation. The revelations of quantum physics can potentially help us to snap out of our self-created spell. Everything depends on if we recognize what it is revealing to us; this cannot be repeated often enough.

• CHAPTER EIGHTEEN •

COMPLEMENTARITY

Quantum entities are simultaneously waves and particles. This is completely impossible from the conventional point of view, as waves and particles are polar opposites that mutually exclude each other. By definition, to be a wave is not to be a particle, and vice versa. Waves spread out and oscillate, they are disturbances in some medium or field, whereas a particle is a localized, concentrated, bullet-like object with a certain mass. They are phenomena of totally different kinds, and it would be hard to conceive of two more contradictory possibilities. No more contrary entities exist in nature. One is not more fundamental than the other, nor can we reduce one to the other. This distressing conundrum deeply troubled the soul of many physicists. It was intolerable for science to harbor such an unresolved, contradictory dualism gnawing at its vital parts. "Physicists," to quote Hoffmann, "could but make the best of it, and went around with woebegone faces sadly complaining that on Mondays, Wednesdays, and Fridays they must look on light as a wave; on Tuesdays, Thursdays, and Saturdays, as a particle. On Sundays they simply prayed."[724]

How can the impossible be happening? And what does it mean that it is? This dilemma is, to quote Richard Feynman, "impossible, *absolutely* impossible, to explain in any classical way, and which has in it the heart of quantum mechanics. In reality it contains the *only* mystery. We cannot make the mystery go away by 'explaining' how it works."[725] Clearly, when we label what is actually happening as "impossible," something is being reflected back to us about the limited way we are viewing the world. This mystery is calling for a novel, radical, and (r)evolutionary way of thinking about things, as well as new and more conscious ways of feeling, sensing, and experiencing our world—a real "re-visioning" of our moment-by-moment experience.

It should be noted that revolutions are sometimes born from conceiving the so-called impossible. If there is one unifying principle applicable in every field of human knowledge, Wheeler feels that it is the idea that the unknown can be found out; in Wheeler's words, "Every darkness can be lighted."[726] Wheeler writes, "The unknown is knowable, the impossible is possible."[727] We should be careful at subscribing and buying into "the venom of the impossible."[728]

In confronting the deeper paradox at the heart of the wave–particle duality, Bohr came up with the idea of "complementarity," which Wheeler calls "the central idea of the quantum."[729] Bohr's idea was that the incompatible and seemingly contradictory opposites of, for example, waves and particles were not just contradictory but also complementary and necessary descriptions of the same underlying reality. Two descriptions of a thing are complementary only if each by itself is incapable of providing a complete description of the thing in question while both together provide a more complete description. Defying a unique description, quantum reality demands several diverse, mutually exclusive, contradictory, and paradoxical perspectives, which, when seen together, form a more complete picture of the underlying state of things. Waves and particles are two aspects of the same thing, which makes no sense as long as we are entrenched in the dualistic viewpoint of classical reality.

Neither of these two descriptions, wave or particle, is exhaustive; the very quest for a single model has to be given up.[730] Each description is only partially correct and has a limited range of application. Though we can consider only one of these aspects at a time, they are alternative and complementary images, different faces of the same thing. It is easy to fall under the assumption that there is an objectively existing object that these descriptions are referring to, but as quantum physics points out, this is not the case.

The complementary aspects of quantum reality complete each other, in a sense bringing each other to realization by virtue of their oppositional mutuality, as if deep down they are cooperative adversaries. Speaking about waves and particles, Banesh Hoffmann writes that the new physics had discovered that "they were not enemies. Their whole battle had been a sham. Their persistent warfare had been one long fraud, a superb example of classical propaganda. . . . If we try to regard the wave and particle as two distinct entities, we must think of them not as implacable feudists but as professional wrestlers putting on a show. But they are really not distinct. They are alternative, partial images of the selfsame thing."[731] Eddington proposed the name "wavicle" for this higher-dimensional paradoxical entity.

Hoffmann continues, "Like the little girl with the curl, the electron sometimes shows one side of its nature, and sometimes the other. . . . It would not be an electron did it not display a well-rounded personality."[732] The complementary aspect of particle and wave is a central feature of the new physics and a reflection of the "well-rounded," whole fabric of both our world and ourselves. It should be noted that when a person is described as "well-rounded," it means that they are not stuck in a fixed viewpoint but can see the world from various points of view. They can view the world through the different lenses of both art and science, and by integrating these perspectives, can arrive at a deeper understanding of reality. This brings to mind novelist F. Scott Fitzgerald's idea that the mark of a true artist is being able to hold two mutually exclusive viewpoints at the same time.

The wave–particle duality offers us a new model for seeing ourselves both as distinct, autonomous, and sovereign individuals, and at the same time as members of a greater body, the collective social web of interrelationships, in which we are all contained and through which we forge further identity as well as a wider capacity for creative relationship. If human consciousness is quantum mechanical in its origins, then each of us has a more personal "particle" aspect and a more impersonal "wave" aspect, both of which, when seen together, make up the greater totality of who we are.

At any moment quantum entities play a certain role, be it wave or particle, but each of these is only a role, and not definitive or absolute. The required change in our conception of these quantum entities can be likened to, in astrophysicist Piet Hut's words, "a change from 'is' to 'as.'" To quote Hut, "An electron *is* not a piece of absolute substance. But an electron can appear *as* a particle or *as* a wave. It can play a certain role."[733] Just as we shouldn't identify electrons with their momentary role, we shouldn't identify or concretize ourselves—or each other—in whatever particular role we tend to be playing at any given moment.

Some people even consider John Wheeler himself as a living embodiment of the principle of complementarity. On the one hand, Wheeler is at home in the prosaic world of practical calculations. A master craftsman, he always had his feet on the ground. On the other hand, Wheeler was also very much at home in the poetic world of following his intuition, imagination, and dreams. Feynman tells a story of having worked on a physics problem and bringing it to Wheeler, who immediately saw the solution. To quote Feynman, "I only realized later that a man like Wheeler could immediately *see* all that stuff when you give him the problem. I had to calculate, but he could see."[734] The poetic Wheeler was unafraid to ask outrageous questions and took nothing for granted. To quote Freeman Dyson, "The prosaic Wheeler and the poetic Wheeler are equally essential. They are the two complementary characters that together make up the John Wheeler that we know and love."[735]

Wheeler can also be seen to be exemplifying the complementarity principle in another way. One of the major themes found throughout his work is how consciousness, at the quantum level, is involved in bringing the physical world into being. Again and again he talks about how consciousness, through innumerable acts of observer-participancy, is the central factor in the materialization of the world into form. And yet, he also writes, "'Consciousness' has nothing whatsoever to do with the quantum process."[736] He is simultaneously asserting that consciousness both does and doesn't have to do with the quantum process, which is a completely contradictory, paradoxical, and nonsensical statement that precisely fits with the four-valued logic of the quantum world. Notice the similarity to Robert Oppenheimer's comment on the electron, "If we ask, for instance, whether the position of the electron remains the same, we must say 'no;' if we ask whether the electron's position changes with time, we must say 'no'. . . ."[737]

We live in a universe that appears to have two complementary aspects. One part of our universe obeys locality and is large, old, expanding, and seems to be mechanical. Another aspect of our universe is nonlocal, built on forms of space and time that are unfamiliar and seemingly incomprehensible to us, and is everywhere interconnected. The local and nonlocal aspects together comprise the warp and woof of the universe. The nonlocal aspects of our universe are enfolded or interwoven into the local aspects in such a way that these two aspects are simultaneously different yet ultimately inseparably the same.[738] Similarly, to use Bohm's terminology, the explicate and implicate orders are complementary aspects that, ultimately, are not separate. Just as underlying and in-forming the manifestations of the explicate order lies the implicate order, the phenomena of quantum physics lies beneath, is coupled with, and encompasses those of classical physics.

QUANTUM PHYSICS AS A SPIRITUAL TREASURE

In another instance where quantum reality and the psyche mirror each other, the unconscious itself behaves in a compensatory or complementary manner towards the conscious mind. In a letter to Jung, Pauli writes, "The epistemological situation regarding the concepts of 'consciousness' and the 'unconscious' seems to offer a close analogy to the situation of 'complementarity' in physics."[739] Similar to how we can glean more insight into a person's process by contemplating their dreams, if our species is seen as one individual, we can understand more about ourselves by viewing quantum physics as the dream of humanity and interpreting it in such a light. Just like a dream compensates a one-sidedness and false attitude in the dreamer, quantum physics can only be fully appreciated when it is seen in context, relative to the classical physics mind-set (with its overly rational, mechanistic, deterministic, materialistic, and reductionist way of viewing things), for which its theory is a compensation and from which it arose. Quantum physics is truly a product of our time.

Quantum physics is a dreaming phenomenon, and like a dream, it is also a cipher of information in need of being deciphered. Something that is thirsting to be known is striving to be born out of the unconscious of humanity into the light of the world in the form of quantum physics. This brings to mind the allegedly divine saying, "I was a hidden Treasure, I yearned to be known. That is why I produced creatures, in order to be known in them."[740] The quantum can be likened to a "hidden treasure" existing deep within the recesses of both the universe and the human mind that "produced creatures" (us, for instance) in order to be known and made conscious.

In Tibetan Buddhism there is a wisdom tradition that insures both its purity and longevity over the course of time by continually revealing itself in a never-ending series of hidden spiritual treasures.[741] These treasures (be they sacred objects or liberating teachings) are discovered not only within the physical world, but within individual practitioners' minds as well. Even if discovered by or within one person's mind-stream, these spiritual treasures are of immense benefit to the whole community. These treasures typically get discovered during times of great need within the community; the specific form that the spiritual treasure takes on speaks precisely to this particular need. These spiritual treasures don't exist objectively in isolation from the person and the community in which they appear, their appearance is a collective dreaming process in which the treasure precipitates out of and into the field of the shared consciousness of the community. Similarly, quantum physics can itself be likened to a "modern-day spiritual treasure" that our species as a whole is "dreaming up" to address our need for healing from the one-sided spell of reductive, materialistic science that we have fallen under.

Typically, these hidden treasures are discovered when, in an auspicious coinciding of inner and outer factors, we encounter what Philip K. Dick calls a "dis-inhibiting signal" in the seemingly outer environment that resonates with and thereby activates something within our minds. These dis-inhibiting signals (what in previous writings I have called "lucidity stimulators") are seemingly random occurrences within the universe that can take various forms—be it a synchronistic event, something someone says, a phrase in a book we are reading, or even a mere syllable—that unlocks something deep within us and helps us to discover the treasure that was hidden within our minds. It's as if the dis-inhibiting signal is helping us to remember something that our soul once knew, what is known as *anamnesis*, an "unforgetting" which is the opposite of and antidote to amnesia. The emergence of quantum physics in our world could be seen to be the disinhibiting signal that helps us to remember our quantum nature—but in typical quantum style, "potentially." How quantum physics actually manifests and the effect it has upon us depends on if we recognize what it is revealing to us. In a sense, anything can be a dis-inhibiting signal if it stimulates our lucidity, even these very words.

Quantum physics can be viewed as a potent and living symbol crystallizing out of and into our shared waking dream that is speaking to something deep within us. Quantum theory can be likened to a force of nature that has blossomed and borne mysterious fruit within and out of the human mind. An expression of the self-regulating capacity of the psyche, when the fruit of quantum physics is ingested by a sufficient number of people, it will restore the balance to a badly disoriented human psyche. Spiritual wisdom traditions typically represent the archetypal figure of the teacher who lives within us in a personalized form that is always using whatever is happening in our lives as a means to teach us. The quantum can be conceived of as being our inner teacher appearing through the medium of science.

Both the unconscious and conscious are inseparable aspects of the greater world of psyche. The building blocks of quantum physics are the result of the interaction between the measuring instrument, the measured object, and the observer's consciousness. In the same way, the symbolic products of the unconscious (such as the symbols in our dreams) are psychological phenomena resulting from an interaction between consciousness and the unconscious psyche. As if comprising a deeper unity, the conscious and the unconscious reciprocally inform and reflect each other, crystallizing in and as the symbolic reality of the dream. Similarly, the observer and the observed in quantum physics form an inseparable unity, joined together within (and expressions of) a deeper wholeness. When physicists "look at" quantum reality, they are engaged in a purely symbolic procedure. It should get our attention that "symbols" are the language of dreams.

Dreams, the unmediated expression of the unconscious, are pure nature. They are the part of nature that is concerned with the survival of the species, a compensation for a one-sided, unadapted attitude in the dreamer. The type of knowing that dreaming is concerned with is not in opposition to, but rather is complementary to scientific knowledge, which is knowledge of the world as an object. One of the functions of scientific knowledge is to separate, compartmentalize, fragment, and analyze the world into bits and pieces small enough for us to understand so that we can handle it and use it for our own ends. Dreaming, on the other hand, serving humanity's need for unity and connectedness, is a complementary and compensatory way of expressing and apprehending the nature of our existence. Our dreams help us to transcend ourselves and experience ourselves as part of a larger whole. Dreaming, to quote Dr. Montague Ullman, one of the leading researchers in the field, "is the wave counterpart to the particulate notions of science."[742]

It is important to embrace a plurality of different epistemological approaches rather than elevating one above all the others. "It should not be forgotten that science," according to Jung, is "only one of the forms of human thought."[743] He elaborates, "Science is not indeed a perfect instrument, but it is a superb and invaluable tool that works harm only when it is taken as an end in itself. Science must serve, it errs when it usurps the throne . . . it obscures our insight only when it holds that the understanding given by it is the only kind there is."[744] Unfortunately, science in our world today has a tendency to not only arrogantly presume that its specific way of knowing things is superior to other ways of knowing, but forgets that its deeper purpose is to serve the higher good of humanity. Jung comments, "Science must prove her value for life; it is not enough that she be the mistress, she must also be the maid. By so serving she in no way dishonours herself."[745]

UNUS MUNDUS

Just like the conscious and the unconscious are complementary aspects of psyche, Jung and Pauli were convinced that mind and matter were complementary aspects of quantum reality. In the quantum world, all mental and physical phenomena are complementary aspects of the same transcendental reality. Jung writes, "But this much we do know beyond all doubt, that empirical reality has a transcendental background—a fact which, as Sir James Jeans has shown, can be expressed by Plato's parable of the cave. The common background of microphysics and depth-psychology is as much physical as psychic and therefore neither, but rather a third thing, a neutral nature which can at most be grasped in hints since in essence it is transcendental."[746] Quantum physics can be seen as simultaneously the expression of this transcendental background while being the very portal through which we become introduced to the universe's transcendental nature.

The transcendental background of our empirical world is what Jung refers to by the term *unus mundus* (one world), which is the unitary ground underlying both psyche and matter. For Jung, the unus mundus is not metaphorical but a real world which subsumes, underlies, and exists prior to the world of our everyday experience. He describes the unus mundus as "the potential world of the first day of creation, when nothing was yet 'in actu,' i.e., divided into two and many, but was still one . . . a potential world, the eternal Ground of all empirical being."[747] Notice the similarity of this "potential world, the eternal Ground of all"[748] to the quantum wave function with its built in infinite potentiality prior to measurement. Based on ever-increasing empirical evidence, Jung writes that "we have every reason to suppose that there is only one world [the unus mundus], where matter and psyche are the same thing."[749] Psyche and matter are in continuous contact with one another and ultimately rest on irrepresentable, transcendental factors. To quote Schrödinger, "It is the same elements that go to compose my mind and the world."[750]

Though existing outside of space and time, certain dynamic manifestations of the unus mundus break through into our ordinary temporal dimension in the form of synchronistic phenomena. Through his idea of synchronicity, Jung was attempting to reenvision the complementary realms of psyche and matter as one undivided reality. Jung writes, "That even the psychic world, which is so extraordinarily different from the physical world, does not have its roots outside the one cosmos is evident from the undeniable fact that causal connections exist between the psyche and the body which point to their underlying unitary nature."[751] The idea of the unus mundus is founded on the assumption that the multiplicity of the empirical world rests on an underlying unity, which is to say that all of the different things in the world belong to one and the same field of potential. This very same underlying unity is what quantum theory is revealing to us.

Jung comments, "But if a union is to take place between opposites like spirit and matter, conscious and unconscious, light and dark, and so on, it will happen in a third thing, which represents not a compromise but something new."[752] The result of this conjunction is theoretically inconceivable, as a known quantity is combined with an unknown one. This "third thing" (*tertium comparationis*) partakes in the common qualities of both of the opposites, but isn't the same as either. Being able to see "the equivalence of psychic and physical processes," Jung writes, "the observer is in the fortunate position of being able to recognize the *tertium comparationis*."[753]

This "something new" is the self—our indescribable and incorruptible true nature. Psychologically, the self unites and is itself a union of the opposites. Bohr saw complementarity as an expression of embracing the opposites that are built into nature. As humans we are partly empirical, partly transcendental, partly conditional, and partly unconditional. Jung refers to the self as our "super-empirical totality."

The self, our super-empirical totality, is something that has always existed, but came into consciousness via alchemical operations. This is to say that the self is created, at least in part, by humanity.[754] Stepping into the role of creator re-creates the alchemist (or dare we say, the quantum physicist) in the process. Becoming aware of the quantum nature of our universe instantaneously affects the underlying quantum field, changing us in the process. There is no getting around it—becoming aware of the quantum and ever-more deeply contemplating its implications transforms us within our core.

Contemplating the meaning of quantum physics mirrors the very observer effect that is at the root of its theory. Studying the quantum realm is not a passive act in which we are detached, sitting in the audience, unaffected by what we are seeing. It is impossible to contemplate the quantum without changing not only what we are looking at, but ourselves as well. No one remains the same upon encountering the quantum realm—it changes us forevermore. All of our senses, emotions, thoughts, and somatic and perceptual systems operate through quantum processes. The quantum is not objective in that it is not separate from ourselves as subject. It is a magic mirror that, if held just right, reflects our true nature.

• CHAPTER NINETEEN •

NO SAMENESS

Schrödinger asks, "What is matter? How are we to picture matter in our mind?"[755] Quantum events are in a constant state of change, never staying the same for an instant. Quantum entities are processes rather than things, just as the ring of light created by rapidly moving a flashlight in a circle is not really an object but an appearance in the mind, an artifact of our perceptual system. We then mistake the appearance for the thing itself, as if relating to the shadow as primary instead of the thing casting the shadow. Another example is the phenomenon of a water wave, in which the up and down movement of the water particles make us believe that a "piece" of water moves over the surface. In a water wave, the water particles do not move along the wave but move up and down as the wave passes by. What is transported along the wave is the energy that caused (i.e., initially disturbed) it, not any material particle.

Quantum theory has discovered, however, that, when observed, each quantum event is a discrete happening, utterly unique and distinct from all other quantum events. The continuous, persistent endurance of things in nature is only apparent, the impression of continuity being due to the similarity of different entities succeeding one another with incredible rapidity in and over time. We perceive matter as solid simply because the oscillations occur so rapidly. There is no single unchanging "entity" that stays identical from one moment to the next. We have the impression of identity persisting over time simply because new, nearly identical entities keep appearing so as to create similar patterns. Quantum entities, however, have no thread of identity connecting one another between one moment and the next; though appearing similar, they are not the same entity.

Speaking of, for example, a man returning to his childhood home after many years, Schrödinger points out, "Indeed, the body he wore as a child has in the most literal sense 'gone with the wind.'"[756] The actual material that makes up the entity has disappeared many times, and the pattern has been completely filled with new matter. Philip K. Dick expresses this same realization when he says, "The reality which exists *now* cannot be the reality which existed a nanosecond ago—*despite our memories.*'"[757] Dick is shedding light on the fact that our sense of continuity between moments is a function of our memory. In other words, when an object seems to resemble itself from moment to moment, this feeds into our past memory in such a way that we become convinced that it is the same object. The same is true for ourselves as well. Not only do we think that the universe we exist in now is the same universe that we inhabited a moment ago, but we don't notice that we are totally refreshed as well.

The material world is composed of myriad elementary quantum events incessantly flashing in and out of existence, pulsating in and out of the underlying field of infinite potentiality every microfraction of a second. Physicist Nick Herbert describes the quantum, microscopic structure of an ordinary coffee cup as "an assembly of *events* rather than of *things*. These events (called *quanta*) last only for an instant, then fade away. Imagine a trillion trillion fireflies flashing in the space of your coffee cup. The cup is a never-still scintillating network of quantum events . . . it is full of *dots*, and the dots are constantly changing. The old fashioned notion of the cup as made up of atoms is just one frozen frame of the microscopic light show."[758] It is not the same coffee cup from moment to moment; appearances to the contrary, the coffee cup, as well as the whole universe, is continuously reborn anew in each instant.

These quantum entities are what you and I are made of—not to mention the rocks, the trees, and the stars. To again quote Schrödinger:

> We have . . . been compelled to dismiss the idea that such a particle is an individual entity which in principle retains its "sameness" forever. On the contrary, we are now obliged to assert that the ultimate constit-

> uents of matter have no "sameness" at all. When you observe a particle of a certain type, say an electron, now and here, this is to be regarded in principle as an *isolated event.* Even if you do observe a similar particle a very short time later at a spot very near to the first, and even if you have every reason to assume a *causal connection* between the first and second observation, there is no true unambiguous meaning in the assertion that it is the same particle you have observed in the two cases. The circumstances may be such that they render it highly desirable and convenient to express oneself so, but it is only an abbreviation of speech. . . . It is beyond doubt that the question of "sameness" of identity, really and truly has no meaning.[759]

Quantum processes are not causally connected from one moment to the next; their connection is acausal, atemporal, nonlinear, and synchronistic. In other words, what appears to be the same quantum entity traveling through space and time is actually a new and unique entity at each and every moment. To quote Hans-Peter Dürr, "There are no objects which are temporally identical with themselves."[760] In other words, there are no objects that over the course of time are identical with a previous version of themselves.

Imagine a strip of lights timed to turn on and then off, one after another in just the right way so as to create the illusion of a continuous movement along the strip. In a magical display, the particle appears to move across space-time as it creates the illusion of continuity. Schrödinger writes, "Atoms—our modern atoms, the ultimate particles—must no longer be regarded as identifiable individuals. This is a stronger deviation from the original idea of an atom than anybody had ever contemplated. We must be prepared for anything."[761]

Attempting to describe and make sense of the situation that was revealing itself in the new physics, David Bohm came up with the idea of the "implicate order," which, interestingly, he refers to as a "new form of imagination." The implicate order is the higher-dimensional substrate out of which our physical world emerges moment by moment. Enfolded in and unfolding out of the implicate order is the materialized universe, which, with each moment, unfolds back into the underlying implicate order only to be replaced by a newer version. To quote Bohm, "The implicate order can be thought of as a ground beyond time, a totality out of which each moment is projected into the explicate order. For every moment that is projected out into the explicate there would be another movement in which that moment would be injected or 'introjected' back into the implicate order."[762] The universe is recurrently creating itself and being created anew out of this implicate, unmanifest, yet all-pervading multidimensional plenum of infinite potential.

Because it is so counterintuitive, Schrödinger reiterates, "It is better not to view a particle as a permanent entity but as an instantaneous event. Sometimes these events link together to create the illusion of permanent entities."[763] Physics tells us that matter is composed of more than 99.9999999 percent empty space. In Eddington's words, "Matter is mostly ghostly empty space."[764] How do we wrap our mind around this? The new physics has discovered that matter is a pulsation of energy temporarily emerging out of a deeper substratum of boundless, unmanifest potential that creates the illusion of solid objects in three-dimensional space. This illusion is fabricated within our brain and nervous system in such a way that a physical world appears to be really there outside of us, when in fact its real basis is a neurologically generated *standing wave*[765] holographic pattern that is witnessed by consciousness in such a way as to trick us into seeing it as a solid, external world of physical objects.

Our physical world can be likened to clouds in the sky. Seen from the ground, clouds look like substantial objects, but if we "enter" the clouds while flying in an airplane, we discover that there is no hard-and-fast boundary around them. They simply dissolve into a fine mist. Quantum physics reveals that the same is true for our world, which appears substantial, and yet at bottom is just an endless series of unpredictable fluctuations whose ultimate existence is transitory and insubstantial. Wheeler invented the idea of "quantum foam" to point to this level of quantum reality where space-time "churns itself into a lather," into a "roiling chaos" where the very notion of space and time cease to have meaning. It should also be noted that, in our example, the clouds are an unmediated expression of the underlying sky. This is to say that the clouds and the sky are not two separate entities, but are indivisible. Form is emptiness. Emptiness is form.

In modern-day physics, the notion of matter has been refined into immaterial fields and forces; matter can be thought of as a defunct idea, a non-concept. As philosopher of science Karl Popper once put it, in the new quantum universe, "matter has transcended itself." In the quantum world, there is no such thing as matter, only "matter-like" entities, namely, entities that appear like matter but aren't made of matter. Schrödinger writes, "Our conceptions of matter have turned out to be 'much less materialistic' than they were in the second half of the nineteenth century."[766]

The world that quantum physics is disclosing is a magic mirror in that this lack of sameness, this lack of a continual thread of identity from one moment to the next, is true not just of elementary quantum entities, but of ourselves as well. This is to say that physics' quantum reflections can help us to get over and "transcend" ourselves. For if we live in a thoroughly quantum universe, what we are discovering about quantum entities is true about us—we ourselves are quantum entities as well. Embracing our quantum nature helps us to dispel the belief that we possess a continual solid thread of identity that exists over time, thereby liberating us into a much vaster, more open-ended and unencumbered sense of self that is free to creatively redesign itself in each moment.

This is such a staggering point that it bears repeating: The lack of any thread of identity of quantum entities from one moment to the next is reflecting back to us this same quality in ourselves. The discoveries of quantum physics are revelatory of the inner world, as though through our explorations into the outer world nature is reflecting back to us our own quantum nature. Quantum physics has revealed that all seemingly solid, objectively existing forms that appear to have continuity over time, including our sense of self, are bereft of substantial, intrinsic existence and are merely our imaginary projection. The question naturally arises: Who is the entity doing the projecting? This, indeed, is THE question. As Schrödinger reminds us, discovering who we are is not just one of the tasks of science, but "the only one that really counts."

In any case, we are new, novel, completely refreshed each and every moment, re-created and re-creating ourselves anew every nanosecond. We are being asked to simply recognize that this is the actual nature of our situation. This very recognition effects a liberating transformation simultaneously in our sense of reality and our sense of self. Quantum physics has gone beyond simply being about our ideas of physics and the so-called physical world and is actually holding up a mirror to us reflecting back our open-ended nature. We are thus invited to de-solidify ourselves, recognize ourselves anew, and discover our intrinsic freedom in the emptiness of what is being revealed.

It should be noted that this discovery of quantum physics, this lack of "sameness" from one moment to the next, is not new but has been expressed in various spiritually informed wisdom traditions over many centuries. In these traditions, the universe is seen as being created and passing away at each and every moment. At the instant of passing away, something like what has passed away immediately takes its place. Henry Corbin writes, "At every moment the world puts on a 'new creation,' which veils our consciousness because we do not perceive the incessant renewal."[767] Corbin continues, "In the realm of the manifest, there is only a succession of likes from instant to instant."[768] The cosmos is a recurrent and recurring creation, refreshing itself at each and every moment. To quote Zen master and scholar D. T. Suzuki, "My solemn proclamation is that a new universe is created every moment."[769]

According to Buddhism, the world is seen to be an indefinite "series of flickering events," comparable to the flame of a butter lamp. These flashes of energy are constituted of a rapid succession of instantaneous events, like frames in a movie. From the Buddhist point of view we are ever-changing conglomerates of processes that take form in self-organizing patterns. The "problem" comes in when we reify our idea of ourselves as truly existing in concrete form. We are then creating a seemingly problematic situation for (and as) ourselves from one that is not ultimately problematic. The idea of an underlying material substratum that exists from one moment to the next is a figment of our imagination; nothing corresponds to it in reality. The very notion of the existence of a continuous identity is just a thought in our minds. Identity is in the mind of the beholder.

Philip K. Dick also articulates this endlessly recurrent creation in his own unique way. He writes:

> The universe is destroyed "every day" (actually every trillionth of the second) "and re-created." . . . Introduced as a totally *new* factor is an apperception of the flicker pulsation in which the system (reality) switches on and off [and on again]. . . . When it pulse-phases to its off position it ceases to exist; when it comes back to its on position it is

> slightly different. . . . Thus in a certain real sense it abolishes and then re-creates itself at a very rapid rate, a sort of flicker. Each time it re-creates itself it is different, hence in a real sense new.[770]

Quantum physics, mystical Islam, Buddhism, Philip K. Dick, and many other spiritual wisdom traditions are all pointing at the same "lack of sameness" of our universe from one moment to the next. Every moment our universe is completely refreshed anew. As Schrödinger reminds us, "We must be prepared for anything."

ONLY SAMENESS

For the sake of completeness, and to show the utterly paradoxical character of the quantum realm, I should mention that there is also an alternative theoretical perspective in physics which is diametrically opposed to the "no sameness" point of view. This other point of view is a coherent explanation for the universe as we perceive it that has its own self-consistent internal logic, claiming that the entire universe is the seamless manifestation of a singular indivisible field. From this point of view (seeing the universe as a singularity) the universe is never divided, for all division is only apparent division and everything is simply an expression of a radical sameness. From this perspective, all quantum entities are expressions of this universal sameness. We could call this perspective "only sameness" and it provides a complementary perspective to that of "no sameness."

One of the things that makes quantum entities such as electrons different from everyday objects is their indistinguishability—specifically, there are absolutely no intrinsic features by which we can distinguish one electron from another, which is to say that every electron is the same as every other. Wheeler and Feynman came up with a radical idea why electrons are utterly indistinguishable: because, according to their "out there" hypothesis, there is only one electron in the whole universe. According to their theory, this one electron weaves backward and forward in time like a thread going back and forth through a tapestry. We see the multitude of places where the thread goes through the fabric of the tapestry and mistakenly attribute each to a separate electron. Their idea seems in alignment with the "only sameness" point of view. This implies that the appearance of innumerable separate electrons is an illusion caused by the structure of space-time. This perspective would say that each new emergence of a quantum entity is actually a recurrence of the same quantum entity in a different guise.

The question naturally arises: Which perspective is "true"? "No sameness" or "only sameness"? They both provide a satisfactory, coherent, and self-consistent description of the observable universe within their particular framework. They are complementary perspectives that, when taken together, add depth and give us a fuller appreciation of the deeper undivided wholeness of the universe. Instead of thinking that the paradoxical nature of our situation is illogical, this point of view embraces the apparent contradiction and synthesizes the opposites into a higher unity rather than simply affirming one or the other in a way that is perfectly parallel to the "complementarity" of the wave–particle duality. This could be called the no sameness–only sameness duality. To be able to hold and embrace this paradoxical and seemingly contradictory point of view involves a higher form of logic that, instead of using an "either/or" mode of thinking, demands a "both/and" mode of thinking. This form of thinking is a reflection of the deeper wholeness that quantum physics is revealing, a wholeness that is fundamental to the universe and intrinsically exists within us.

WHIRLPOOLS

A whirlpool or vortex in a river has a definite location in space and time, and yet it has no independent existence separate from the river that generates and supports it. The whirlpool is constantly being re-created, refreshing itself moment by moment, taking on a self-perpetuating pattern that persists over time. A local property or concentration of the underlying field, it is as if the whirlpool, whose substance is never the same, is continually dying and being reborn every moment. The water flowing through the whirlpool is constantly new and ever-changing, but the self-reinforcing pattern of the whirlpool remains the same. The apparent entity of the whirlpool is "abstracted" from the underlying flowing movement, arising and vanishing with the total process of the flow.

Bohm refers to such seemingly self-existing entities as whirlpools that exist embedded within a larger process, "relatively autonomous sub-totalities." Speaking of the nature of these phenomena, Bohm writes, "Such transitory subsistence as may be possessed by these abstracted forms implies only a relative independence or autonomy of behavior, rather than absolutely independent existence as ultimate substances."[771] The relative autonomy and seeming stability of the whirlpool is not because of its separateness, but derives from the whole motion of the underlying river, which is to say that the part (i.e., the whirlpool) "implicates" the whole. It should be pointed out that another example of a "relatively autonomous sub-totality" that we take for existing independently is the idea of our very selves. Like whirlpools arising out of the underlying river, we arise out of the highly energetic and dynamic quantum plenum of space.

It is easy to look at the "relatively autonomous" and seemingly stable form of the whirlpool and think that it has an independent existence. In actuality there is no such self-existing entity as a whirlpool; it is but an aspect and expression of the whole. It is impossible to determine where the whirlpool ends and the river begins. To say that one whirlpool is "separated" from another by the water "between" them is a metaphoric way of talking which has some usefulness, but we should be careful not to entrance ourselves into thinking that we are dealing with two separate entities. Because each seemingly separate whirlpool is indistinguishable from the same river, each whirlpool is ultimately indistinguishable and inseparable from one another.

In the words of mathematician Norbert Wiener, "We are but whirlpools in a river of ever-flowing water. We are not stuff that abides, but patterns that perpetuate themselves."[772] This is reminiscent of Buckminster Fuller's description of a human being as what he referred to as a "pattern integrity," by which he means that a person's self or identity is not a thing but a continually created stable pattern that appears to exist over time.

Embracing the notion of wholeness and non-separation doesn't eradicate or negate distinctions—there is a certain convenience and practical utility of thinking that there are two separate whirlpools. There is a mutual interdependence between the parts and the whole—without parts there can be no whole, and without a whole it makes no sense to speak of parts. Without intimately understanding wholeness it is easy, however, to confound the relationships between parts and whole. The key is to be able to hold both viewpoints at the same time, seeing the interconnected wholeness without losing sight of the distinctions. When viewed as separate, it is important to realize that the whirlpool is only "relatively" so, with a wholeness that derives from its inseparable connection to the larger whole in which it is contained. As Wiener reminds us, "We are but whirlpools . . ."

• CHAPTER TWENTY •

LANGUAGE

For Niels Bohr, a key part of "understanding" quantum phenomena involved the ability to describe our observations. In Bohr's words, we had to solve "the problem of using the right words."[773] To quote Schrödinger, "If you cannot—in the long run—tell everyone what you are doing, your doing has been worthless."[774] Every considerable advance in science involves a "crisis in communication." The world of the quantum doesn't easily lend itself, however, to ordinary language.

It is challenging to find language for a world that is beyond the reach of our imagination. Language itself is a tool of the imagination. One way or another, imagination becomes embodied in and through language while at the same time language takes on form through imagination. Every language carries within it a prevailing worldview that informs not only our thinking and perception, but our imagination as well. Language shapes our ability to imagine the possibilities inherent in our world. Our relationship to our creative imagination not only gives shape and form to progress in the arts and sciences, but to our daily lives as well. Though defying the standard forms our imagination takes, quantum physics at the same time opens new possibilities to the imagination.

Speaking about the quantum dimension of reality is as much a problem of language as of physics. Our language, however, is based on daily experience, and atoms are not. To quote Bohr, "When it comes to atoms, language can be used only as in poetry. The poet, too, is not nearly so concerned with describing facts as with creating images."[775] The mind-expanding insights of quantum physics cannot be contained within the bounds of ordinary technical scientific discourse and, in Heisenberg's words, its "precision-orientated language." The need for poetic language to describe the behavior of the quantum realm signifies a breakdown of the traditional metaphor of a clockwork or mechanical universe. Particles do things we would never dream of them doing in pre-quantum or classical physics. Plato, in dealing with the "limitations of precise language," according to Heisenberg, "switched to the language of poetry, which evokes in the hearer images conveying understanding of an altogether different kind."[776]

Language falls short and only goes so far; many of the revelations of quantum physics can only be pointed at by images and parables. The new physics demands an imprecise language that instead of simply describing facts, evokes pictures in our mind along with the notion that these pictures allude to and have only a vague connection with reality. Expressing the probabilistic and indeterminate nature of the quantum world, these inner mental pictures represent only a "tendency" toward reality. Reflecting Bohr's notion of complementarity, physics has found itself using ambiguous language which reflects the principle of uncertainty that characterizes the quantum world.

Misunderstandings can be engendered by a careless use of language. To quote Wheeler, "We humans have limitless power to confuse ourselves with words."[777] Modern physics deeply suffers from the tension between the demand for complete clarity and precise language, on the one hand, and the impossibility of expressing quantum phenomena in unambiguous language, on the other. We cannot, for example, speak unambiguously about the behavior of an electron in an atom. Though the imprecise and ambiguous language and images used in quantum physics can never fully correspond to the meanings they are trying to express, they help to draw us closer to the truth of their meaning.[778] As poet and philosopher Friedrich Schiller wrote, "Truth dwells in the deeps." Nature has its own language, and we are being asked to learn it. The quantum renders and makes "reality" into language. Theologically speaking, it's not that the gods have fallen silent; rather, we have become unable to see or hear them.

Though quantum theory has given us new understanding, we still use the old words. Our current language is steeped in the pre-quantum, classical world of objects. Our language is thus a potential agent for reifying the classical world model. Through our unconscious use of language we are unwittingly putting ourselves under the "spell" of a false world image. Our language not only shapes our worldview but our world and ourselves as well. In a sense it defines us. Most physicists still speak and think, with an utter conviction of truth, in terms that regard the universe as being constituted of aggregates of separately existing building blocks. As Wittgenstein writes, "A *picture* held us captive. And we could not get outside of it, for it lay in our language, and language seemed to repeat it to us inexorably."[779] Our language is potentially, when used unconsciously, an instrument to hypnotize ourselves.

Language is like a net spread out between people in which our knowledge, thoughts, and beliefs are inextricably enmeshed and reinforced. To quote Bohr, "We are suspended in language so that we don't know which way is up and which is down."[780] Language is an all but invisible hand that structures thought and modes of thinking. "Linguistics," to quote linguist Benjamin Lee Whorf, "is fundamental to the theory of thinking, and in the last analysis to all human sciences."[781] It is helpful to remember to exercise our awareness of the quantum nature of the world so as to overcome the trance-inducing influence our language has over us. Language and thought are bound together, and both can exert an undertow towards the classical world via forces that are as strong as they are unconscious.

Language can easily entrance us into a particular way of viewing the world. For example, the word *matter* has an etymological connotation with the word *mother* (mater), as if to invoke the idea that matter is the mother, the source of everything. When we say that something "doesn't matter," or is "immaterial," what we mean is that it's not important. This is to unconsciously conflate something being material and having importance, as if matter is seen as sacred, and materialism is its religion. Another example: when we say "that doesn't make sense," we are unconsciously making an equivalence between being in our senses and being meaningful, as if when something isn't in our senses, it is meaningless. And yet quantum physics tells us that the most fundamental part of the world is invisible, outside the realm of the senses. Our language has encoded within it a hidden built-in ideology that unconsciously brainwashes us in ways that many of us don't notice.

Language is a pool of meaning in which we all drink as well as contribute. It is the cardinal instrument by which our individual worldviews are linked so that a shared, common reality is constructed. What is actually subjective becomes seemingly objective, particularly when it is agreed upon by groups of people. Language is not a fixed or static thing but an ever-evolving medium always in need of being updated and refreshed so as to keep pace with our constantly evolving consciousness. Language is never a finished phenomenon but a dynamic and creative activity that requires our words and ourselves to be born anew.

Wheeler created numerous new names for things that would help other physicists expand their viewpoints on fundamental concepts in physics, the term "black hole" being a case in point.[782] Instead of telling others that he was coining a new phrase, he would typically just start using it as though it had always been used, and almost overnight all other researchers in the field of "black hole physics" would start using Wheeler's newly minted phrase. Showing the power of finding the right name, by naming it, he helped to create the *idea* of black holes in countless minds. Over time he embellished the phrase, using the pithy slogan, "A black hole has no hair."[783] Wheeler was thinking about a room full of bald people and how hard it is to tell them apart because they show no differences in hair length, style, or color. Similarly black holes lack the "hair" that more conventional objects possess that give them their individuality. To quote Wheeler, "No hair stylist can arrange for a black hole to have a certain color or shape. It is bald."[784]

Language is not merely a means of communicating ideas about the world, but an instrument for bringing the world into existence in the first place. What we find in the world is the result of the way we talk about the world; when we create a new way of talking about the world, we virtually—and literally—create a new world. Language is truly creative. Our experience of reality is not merely expressed in language, but is actually produced by language.

The need for updating new forms of language is not just a psychological need but a social one as well. The power of a new way of looking at the world and the power of its effective dissemination into that very world are interconnected. A new type of language needs to be created that is qualitative rather than exclusively quantitative. Conventional scientific language is based on verifiable measurements and is hence descriptive. Based on the participatory nature of the new physics, the new form of language needs to be more "depictive," somehow evoking its qualitative character. To quote Feynman, "The next great era of awakening of human intellect may well produce a method of understanding the qualitative content of equations."[785] We have to create a new language that calls forth and transmits novel images in and through the medium of the shared collective psyche. We have to create a new language that evokes and helps us see and understand the interconnectedness which underlies and in-forms all phenomena. Interestingly, at the same time of the advent of quantum physics, modern art was becoming more "evocative" and less "reproductive."

In using language to describe the quantum, we run up against the limits imposed by the logic and grammar built into the language we are using. The English language, relying on nouns (separate things) interacting, does not coincide with the actuality of the underlying quantum "process" informing our world. The more we develop intellectual concepts and the more abstract our thinking becomes, the more we run the risk of having our language become dissociated from the fundamental ground of being. A language composed of living words that discloses reality can only develop and organically emerge from intimate, direct experience of the reality of the Logos, which is the ground of language itself.

The quantum is a transcendental realm that cannot be adequately expressed in terms of language or our Western philosophical views, which are contained within categories of space and time. "The real problem," to quote Heisenberg, "was the fact that no language existed in which one could speak consistently about the new situation. The ordinary language was based upon the old concepts of space and time."[786] The challenge is to create a new language that partakes of as well as unites the seemingly opposite realms of the physical and the psychical. Pauli writes in a letter:

> For the invisible reality, of which we have small pieces of evidence in both quantum physics and the psychology of the unconscious, a sym-

> bolic psychophysical unitary language must ultimately be adequate, and this is the far goal which I actually aspire. I am quite confident that the final objective is the same, independent of whether one starts from the psyche (ideas) or from physis (matter). Therefore, I consider the old distinction between materialism and idealism as obsolete.[787]

Elsewhere, Pauli writes, "I suspect that the alchemical attempt at a unitary psychophysical language miscarried only because it was related to a visible concrete reality. But in physics today we have an invisible reality (of atomic objects) in which the observer intervenes."[788] As Pauli points out, our task is to create a "unitary language" that relates to, evokes, and reveals the "deeper invisible reality."

Everyday language is infused with the notion of time. We cannot use a verb without choosing a tense. The linguistic rule built into the fabric of our language is that verbs have to "do" something to the nouns, as if the verbs and nouns are two separate entities that are being combined and engaging in a certain way. In the quantum world there are only verbs; there is only process. This is why Buckminster Fuller wrote a book called *I Seem to Be a Verb*. In the realm of the quantum, there is no distinction between the actor and the action. The subject-verb-object structure of the English language (which is also common to the grammar and syntax of many other modern languages) can easily sustain, propagate, and reflect fragmentation through being bound by the logic of space and time. The underlying structure of our language—and of our thought—tends to divide things into seemingly separate and static entities. The divisions implied in the structure of our language then appear projected, as if they are actually existing fragments in the outer world. Category errors abound in our language, which only reinforces the tendency to make this kind of error in thought.

Language has trouble enough describing what is in front of us, not to mention behind us or within us. In light of the new dynamic worldview emerging from quantum physics, however, we have to develop a new form of language to describe the heretofore unnavigated realms that the new physics has stumbled upon. To quote Wheeler, "How can we hope to move forward with no solid ground at all under our feet? Then we remember that Einstein had to perform the same miracle. He had to re-express all of physics in a new language."[789] Based on the process-oriented nature of the quantum realm, our new language should be based more on verbs than on nouns, and more on action, events, and movements than on static things.

We cannot speak about atoms in ordinary language or in the typical concepts we use to describe ordinary physical objects. To quote Wheeler, "The kind of physics that goes on does not adjust itself to the available terminology: the terminology has to adjust in accordance with the kind of physics that goes on."[790] If we are up to the task of somehow translating the revelations of quantum physics into communicable language, its effects can become contagious, virally spreading like psychic wildfire through the human psyche such that it liberates and unleashes a latent, creative energy lying dormant in the collective unconscious of humanity. Being symbolic, languages, mathematics, and physics are meant to stimulate higher brain functions, which can potentially awaken within us the experience of our interconnectedness with all that lives. Human language is a remarkable form of information processing, capable of expressing anything that can be put into words.

SHAPE

Let's go back to Schrödinger's example of a man returning to his childhood home after many years. He finds the place unchanged—the same stream flows through the same meadows. However, the material that makes up the stream and the meadows has completely changed many times. To quote Schrödinger, "It is clearly the peculiar *form* or *shape* [German: *Gestalt*] that raises the identity beyond doubt, not the material content."[791] According to quantum theory, a trans-empirical domain of reality exists that does not consist of material things but of transmaterial ideal forms. Schrödinger writes:

> The *old* idea about them was that *their* individuality was based on the identity of matter in them. . . . The *new* idea is that what is permanent in these ultimate particles or small aggregates is their shape and organization. The habit of everyday language deceives us and seems to require, whenever we hear the word "shape" or form pronounced, that it must be the shape or form of *something*, that a material substratum is

> required to take on a shape. . . . But when you come to the ultimate particles constituting matter, there seems to be no point in thinking of them again as consisting of some material. They are, as it were, *pure shape*, nothing but shape; what turns out again and again in successive observations is this shape, not an individual speck of material.[792]

This "pure shape" is reminiscent of the Platonic idea of transcendental "Forms" *beyond* physics (hence "metaphysics"). Heisenberg writes, "Material things are the copies, the shadow images, of ideal shapes in reality."[793] It is as if the physical world is a shadow emanation of a higher-dimensional reality. To quote Sir James Jeans, "Plato maintained that the forms possessed a higher degree of reality than the material objects which exemplify them, so that the world was primarily a world of ideas and only secondarily a world of material objects."[794] In a similar vein, physicist Hans-Peter Dürr writes, "The material world which is so tangible to us increasingly proves to be an apparition and dissolves into a reality where it is no longer things and matter, but forms and shapes, which dominate."[795] This idea of pure shape sounds similar to the idea of the primordial archetypal image which in-forms all of the various specific manifestations of the underlying archetype. Hearing the word "shape" and immediately assuming it must be the shape of "something" is an expression of our classically conditioned mind, which thinks in terms of objects having definite form in an objective world.

Quantum physics' discovery that matter and the whole material world are literally lower-level shadow patterns on the walls of the cave of the space-time continuum being cast by a higher-dimensional reality outside the cave represents a modern scientific realization of a philosophical framework established by Plato (called "Platonic Idealism," "Plato's Theory of Forms," or more simply "Platonism"). This insight is famously expressed in Plato's allegory of the shadows on the wall of the cave. In essence, Plato's point of view is that nonphysical, higher-dimensional forms, ideas, shapes, and archetypes are the primary, essential reality that in-form and give shape to the physical world. The material forms in our universe are the particular instantiations that are derived from these ideal forms and stand in relation to them precisely as shadows stand in relation to the higher-dimensional forms that cast them.

Heisenberg writes, "The smallest units of matter are, in fact, not physical objects in the ordinary sense of the word, they are forms, structures, or—in Plato's sense—Ideas."[796] The implication of these insights is that the universe is similar in nature to an idea, a collective thought form becoming materialized into living form. French physicist Louis de Broglie comments, "The great wonder in the progress of science is that it has revealed to us a certain agreement between our thoughts and things."[797] Thought itself appears to be the basic building block of the universe. Is the universe not only made by thought, but is itself one big thought? To quote Jeans, "The universe begins to look more like a great thought than a great machine."[798] Thinking along similar lines, Eddington wrote, "The universe is of the nature of 'a thought or sensation in a universal Mind.'"[799]

The quantum world is not comprised of objects, rather it is an endlessly unfolding, ever-changing dynamic process that is in continuous movement. Similarly, it is easy to assume that if there is movement, there is something that is "doing the movement." In the quantum realm, however, we are never able to find any such entity or substance; it is only encountered in thought, as an idea within our minds. Given that mind is being revealed by quantum physics to be the foundation of the physical world, movement itself may ultimately be the movement of our mind.

When the universe manifests in its wavelike aspect, there is no separate entity that is doing the waving. At the quantum level, the dancer and the dance are inseparably one—there are no separate dancers, there is only the dance. In the quantum realm there is motion but ultimately no moving objects. There are no entities—things-in-themselves—existing behind or as the cause of phenomena, rather we can only talk about "things" as an abstraction existing *in* and not separate from phenomena. The idea that objects exist apart from processes is a residue of our outdated classical conception of the world, and is at the root of our seemingly inescapable sense of separateness from the universe.

To say the same thing differently, relations are not secondarily derived from objectively existing *relata* (independent entities), but it is the relations that are themselves, ontologically speaking, primary. Phenomena are composed of relations without preexisting relata doing the relating; there is only relata within relations. This is to say that things with seemingly determinate boundaries and specific properties emerge through the process of relationship rather than existing prior to, informing, and determining the relationship. This insight turns our traditional ideas and understanding of causality on its head. It is only through the ongoing process of phenomena being reconstituted via relations (without preexisting relata doing the relating) that our concepts about the world become meaningful. It behooves us to remember that this is a process which ultimately takes place within and is not separate from our minds.

If we focus our attention on individual quantum entities we find ourselves unable to account for the seemingly impossible behavior of these entities. Many, maybe all, of the paradoxes inherent in quantum physics are a result of the mistaken assumption that there are individually distinct, determinate, and bounded entities that underlie phenomena. This is itself an externalized reflection of the fact that we ourselves are not a discrete reference point or an "I" separate from the process of the endless becoming of the universe. The universe, like a never-ending work of art, is a work-in-progress in which we ourselves are participating every moment, dreaming it up while simultaneously being dreamed up by it.

MATHEMATICS

It's not just that we have to create a new language to express the new breakthroughs in physics; physics itself is a language, and mathematics is the underlying technical language of physics. A similar question that arises in the field of physics also emerges in mathematics: Is the abstract realm of mathematics not separable from, but rather an expression of our own minds? Is mathematics merely a capacity of human thought? This raises the important issue of the nature of human thought itself—does human thought partake in a higher-dimensional realm beyond space-time where the origin of both thought and the physical world resides? Are the mathematical laws we deduce merely statements about the structure of human thinking, or even more fundamentally, about the structure of human consciousness itself?

In his book *Across the Frontiers*, Heisenberg quotes mathematician and philosopher Bertrand Russell, who said, "Mathematics may be defined as the subject in which we never know what we are talking about, nor whether what we are saying is true." After Russell's quote, Heisenberg adds a commentary, "To explain the second part of this statement, we know only that our propositions are formally correct, not whether there are objects in reality to which they could be related."[800] In other words, does abstract mathematics have any correlation with the real world, whatever that is?

Mathematics is the underlying "symbolic procedure" that is the basis of quantum physics. Just as in quantum physics, Platonic idealism is considered to be the predominant worldview that informs the philosophy of mathematics. This is to say that, according to the prevailing worldview, the realm of mathematics is conceived of as existing in an ideal, timeless Platonic realm, which, rather than being invented by mathematicians, can be discovered within the human mind. From this point of view, the human mind is the screen or "wall" upon which the higher-dimensional mathematical order projects itself as a shadow. The abstractions of mathematics are like the shadows on the wall of the cave, but taking place within the subjective domain of the mind. This higher dimension does not exist "objectively" in the usual "outer" sense since it can only be found within us and subjectively accessed within our own minds.

Quantum physics, in turn, is revealing that physical matter is also like shadows on the walls of the cave, but these shadows are appearing outside of ourselves in the "objective" world. In other words, mathematics, the shadows on the wall inside the physicist's mind, is revealing that matter is the shadows on the wall of space-time. It is noteworthy that these shadows on the wall of the cave are simultaneously manifesting both subjectively and objectively, within and without, as if these are two different vantage points of the same deeper reality that is casting the shadows.

Though easy to take for granted, it is quite mysterious that mathematics, which comes from the human mind, is such a precisely accurate language for describing the behavior of the physical world. In his famous article "On the Unreasonable Effectiveness of Mathematics in the Natural Sciences," physicist Eugene Wigner addresses this issue in detail. He concluded his article by writing, "The miracle of the appropriateness of the language of mathematics to the formulation of the laws of physics is a wonderful gift which we neither understand nor deserve." Once we recognize that physical matter and the mathematics that provide such an effective language for describing its dynamics are both projections from the same higher-dimensional source, we can see that the reason the effectiveness of mathematics in the physical sciences seems to be "unreasonable" is that this deep connection between mathematics and matter had not yet been recognized. This clear understanding also enables us to realize that we do in fact deserve this "wonderful gift," because this correlation is an innate expression of our true multidimensional nature. Both physical matter and the mathematics which describe it are shadowlike reflections appearing within the mind, which itself is a shadow reflection of a higher-dimensional meta-mind.

And yet, there is another school of thought that is of the opinion that mathematics, instead of being discovered, is a creation of the human mind. As an example of this other viewpoint, Harvard physicist Percy Bridgman, a twentieth century Nobel laureate, comments, "It is the merest truism, evident at once to unsophisticated observation, that mathematics is a human invention."[801]

The question naturally arises: What is the "reality" which quantum theory and its corresponding mathematics are describing? Are we discovering this reality? Or creating it? To invent is not merely to create something but to discover it as well. Etymologically, the word "invent" comes from the Latin *invenire,* "to come upon," which points to the connection between creation and revelation. The process of invention, just like the Latin root of the word implies, may in fact be a disguised form of discovery, or "coming upon" something that already exists. Could what we call "invention" be a kind of direct mental perception of a preexisting order that is already present in the higher-dimensional spheres of mind, namely, in the ideal realm that Plato described? Could things that are discovered by the human mind also subsequently be creatively modified in a way that could be called inventive? Could the two seemingly different processes of invention and discovery merely be two different perspectives on one indivisible process that is taking place within the human mind?

Could the blurred distinction between invention and discovery be mirrored by a dynamic within our very mind, such as the process of "thinking?" In our subjective experience of thinking, it can be challenging to differentiate between when "we" are (actively) thinking compared to when we are not identified with our thoughts and we experience that our thinking process is just naturally occurring on its own accord. Are we creating our thoughts? Or simply noticing that they've arrived in our mind? This naturally brings up the question of who we are in this whole process.

Getting back to the math of things, is mathematics invented or discovered? In inquiring into the mathematics of things, this feels like the right question to ask. The majority of the adherents to these different schools of the philosophy of mathematics are firmly entrenched in the perspective that their viewpoint is exclusively true while the other point of view is false. These two perspectives seem mutually exclusive and in opposition to each other. From the point of view emerging from the new physics, however, could both of these perspectives have some measure of truth to them? Could these seemingly contradictory perspectives be an example of a heretofore unrecognized instance of complementarity? When these two perspectives are seen together as composing a complementary relationship, it can potentially give us a more accurate picture of a deeper, mysterious, underlying reality, a reality unlike anything we have previously imagined.

• CHAPTER TWENTY-ONE •

THINKING

In his classic textbook *Quantum Theory,* David Bohm points out that there is a certain analogy between our thought processes and quantum systems. Bohm writes that "the analogy between thought and quantum processes can still be helpful in giving us a better 'feeling' for quantum theory."[802] For example, if we try to observe what we are thinking about at the very moment we are reflecting upon a particular subject, the way our thought process proceeds immediately changes. Jung writes, "Nowhere does the observer disturb the experiment more than in psychology."[803] Just like a quantum system, our thought process and inner experience becomes different from what it was the moment we try to observe and reflect upon it. Observing our own psyche changes the psyche, both the part of it that is being observed as well as the part of it that is observing. As with the observer and observed in quantum theory, the thought and thinker can never be separated—the thinker is the thought; they form a totality. To quote Philip K. Dick, "The universe, then, is thinker and thought, and since we are part of it, we as humans are, in the final analysis, thoughts of and thinkers of those thoughts."[804]

This is similar to when we, as a conscious ego, observe the unconscious—our interaction reciprocally affects both the unconscious as well as ourselves. In other words, the process of becoming conscious inevitably affects the unconscious, which instantaneously loops back so as to complete the circle, changing our consciousness.[805] To quote Pauli, "The mere apprehension of the dream has already, so to speak, altered the state of the unconscious, and thereby, in analogy with a measuring observation in quantum physics [has] created a new phenomenon."[806] The psyche is the means by which we observe the psyche; it is in the peculiar position of being simultaneously subject and object of its own contemplation. In the domain of the psyche, the observer is truly the observed. Is the nature of the psyche mirroring the quantum nature of reality?

Is the dynamic—in which the thinker reflecting upon thought changes the thoughts being reflected upon—a process within the inner light of consciousness that is analogous to what quantum physics has discovered taking place in the outer light of electromagnetism when we observe a quantum particle? The laws of the outer light of electromagnetism may indeed be a lower-level reflection of the laws of the inner or higher-dimensional light of consciousness, thereby demonstrating a powerful example of the alchemical dictum: "As above, so below."[807]

The properties of both our thought processes and quantum systems do not exist separately from and independently of their surrounding environment but are, instead, properties that arise and take on meaning in connection to and relation with other elements. This is similar to how a word takes on its meaning relative to and within the context of the other words it is used with; by itself a single word loses its nuanced meaning. Similarly, we can't understand a musical composition by analyzing its separate parts; it takes on its meaning only when experienced as a whole. Due to their indivisibility with other elements, both our thought process and quantum systems "hang together" with the deeper environment they are a part of, and therefore cannot be analyzed too much in terms of distinct elements. The "intrinsic" nature of each element doesn't exist in an isolated fashion on its own, but is context dependent upon the greater environment in which it takes on its nuanced existence.

Speaking of our thought process, Jung suggests that if "we wished to form a vivid picture of a non-spatial being of the fourth dimension, we should do well to take thought, as a being, for our model."[808] A new idea often comes suddenly, falling into our head out of the blue, many times without any apparent direct cause. Bohm points out that the production of new ideas, which are like inspirations from the beyond, presents a strong analogy to a quantum jump.

Bohm concludes his section on the analogy between our thoughts and the world of the quantum by writing that "the behavior of our thought processes may perhaps reflect in an indirect way some of the quantum-mechanical aspects of the matter of which we are composed."[809] The fact that our thought process and quantum reality are so analogous makes one wonder whether this is not mere coincidence but is revealing something about the thought-like nature of our world. Interestingly, the first words of the Buddha in *The Dhammapada* (a collection of the sayings of the Buddha) are: "All that we are is the result of what we have thought: it is founded on our thoughts, it is made up of our thoughts."[810] Recognizing the subtle, illusory but reality-creating power of our thoughts allows us to create *with* our thoughts rather than being created *by* them.

There is a type of thinking that not only creates problems, but is the very problem itself. This type of thinking creates an apparent problem and then tries to solve the problem, all the while forgetting that it is creating the very problem it is trying to solve. The more it thinks in this way, the more problems it creates. Once our mode of thinking creates problems, it then has all the evidence it needs to confirm its point of view that our situation is indeed problematic—a viewpoint that attains a self-generating, seemingly autonomous life of its own. Thinking can then become a cyclic loop that continually feeds back upon itself as it endlessly creates and feeds on more problems by its very activity. One of our greatest talents as human beings is our incredible ability to deceive ourselves and then forget that we have done so. Thought is a perfect accomplice in this process of pulling the wool over our own eyes.

Bohm was of the opinion that practically all of the problems of the human race are due to the fact that thought is not "proprioceptive," i.e., aware of what it is doing. The body is proprioceptive in that it has a self-perception, it knows when it is moving, and it knows what it is doing. Unlike the body, however, thought is not proprioceptive in the sense that thought creates something, forgets that it did so, relates to its creation as if it didn't create it, and then reacts to its own self-creation as if it existed objectively.

Is an analogous process reflected in quantum physics to the extent that the physicists *think* the world and their experiments exists objectively, separately from themselves? At the moment that physicists think that the world exists objectively, it is effectively true for them, via the aforementioned process of As Viewed, So Appears. They find themselves living within an experiential domain in which whatever they believe to be true appears convincingly to be the case, even if it isn't. We can't get around the fact, as quantum physics continually reminds us, that we are participating in creating our experience of the universe.

Speaking of the physical world, what Bohm refers to as "the whole field of the finite," he writes, "It has the appearance of independent existence, but that appearance is merely the result of an abstraction of our thought."[811] Bohm is implicating thinking as being at the bottom of our having fallen under the persistent illusion of imagining the world to exist separate from ourselves. What we take to be reality is based on what we believe, which is a function of our perceptions. What we perceive depends upon what we look for, which is related to what we think. What we think depends on what we perceive, which determines what we believe, which (coming full circle) affects what we assume to be reality. As Bohm points out, thinking is at the root of this process.

Thinking can easily produce a thought which conceives of or implicitly presumes a separate thinker that it imagines is thinking it. In attributing its origin to this thinker, which it experiences as if it were real and separate from itself, thought then behaves as if it were produced by this thinker, which further serves to entrench the illusory delusion of a separate thinker that has produced it. All the while, the truth of the situation is actually the other way around—the idea of a thinker is itself produced by thought and therefore cannot be separated from the process of thinking. Thought thinks itself.

Like matter, our thoughts are in a continual process of movement, flux, and flow. As Bohm writes in his book *Wholeness and the Implicate Order:*

> Whenever one *thinks* of anything, it seems to be apprehended either as static, or as a series of static images. Yet, in the actual experience of movement, one *senses* an unbroken, undivided process of flow, to which the series of static images in thought is related to a series of "still" photographs might be related to the actuality of a speeding car. . . . Then there is the further question of what is the relationship of think-

> ing to reality. As careful attention shows, thought itself is in an actual process of movement. That is to say, one can feel a sense of flow in the "stream of consciousness" not dissimilar to the sense of flow in the movement of matter in general. May not thought itself thus be a part of reality as a whole? But then, what could it mean for one part of reality to "know" another, and to what extent would this be possible?[812]

Bohm asks a good question: What could it mean for one part of *reality*, which is seamlessly and inseparably whole, to "know" another part? And he presents a radical idea—that thought itself is a part of reality as a whole, which is to say that thought is ultimately not separate from, but interrelated to the physical world. Thought is not just passively reflecting or representing the physical world, but is somehow connected to the physical world in such a way that it affects matter while concurrently being affected by matter. This is to say that both thought and matter are inseparable aspects of one undivided process or "holomovement," as Bohm calls their dynamic indivisibility. To quote Bohm, "In this flow, mind and matter are not separate substances. Rather, they are different aspects of one whole and unbroken movement."[813]

This insight makes an important break from the long-entrenched implicit assumption of Western Cartesian dualism that views the domain of thought (res cogitans) as being immaterial and thereby having nothing to do with the realm of matter (res extensa). To quote quantum theoretician Henry Stapp, "The new physics presents prima facie evidence that our human thoughts are linked to nature by nonlocal connections: what a person chooses to do in one region seems immediately to affect what is true elsewhere in the universe. . . . Our thoughts . . . *do* something."[814] Our thoughts and intentions appear to have the power to influence, change, and transform our world.[815]

Quantum physics, in uncovering the nature of nature, might at the same time be unveiling the nature of our own minds, which are clearly a part of the very nature that they are seeking to understand. According to Buddhist teachings, in understanding the nature of outer phenomena we can understand that its nature is not different from our own minds, which is to see the true nature of everything. Heisenberg writes that "modern physics has perhaps opened the door to a wider outlook on the relation between the human mind and reality."[816] It is becoming glaringly obvious that the structure of nature and the structure of our minds are reflections of each other, comprising a deeper unity that embraces both in an underlying wholeness.

Quantum physics is discovering that the basic building blocks of our physical world seem to be of the nature of ideas, images, and thoughts. It is not hard to imagine that the thought-like foundation and nature of the universe affects and interfaces with our own thinking. The very process of thinking can concretize the continually changing, impermanent, and flowing movement of our dreamlike world—or not. Being "a part of reality as a whole," the process of thinking itself appears to be reflecting back the "unbroken, undivided process of flow," which is the nature of our universe. This underlying flow exists prior to the thought forms which arise and dissolve in the flux, similar to how the movement of a flowing river in-forms its whirlpools.

In any case, the new physics is reflecting back to us that thought needs to be a factor, perhaps a major factor, in our equation of giving meaning and measure to our experience. By describing and creating models for reality, thought is an essential part of the reality it is trying to represent. Thus thought itself must be included in our ever-evolving model of the universe.

To quote Percy Bridgman, "By far the most important consequence of the conceptual revolution brought about in physics by relativity and quantum theory lies not in such details as that meter sticks shorten when they move and that simultaneous position and momentum have no meaning, but in the insight that we had not been using our minds properly and that it is important to find out how to do so."[817] How amazing that the cutting edge of physics is potentially giving us insights into new ways of thinking about things as well as into the nature of thought itself. Nobel Prize-winning physicist William Lawrence Bragg writes, "The important thing in science is not so much to obtain new facts as to discover new ways of thinking about them."[818]

THOUGHT EXPERIMENT

One of the most important modes of exploration in quantum physics is what are called "thought experiments." These are laboratory experiments of the mind in which physicists explore the imagination so as to tease information out of nature. Thought experiments are experiments that are thought about rather than performed, although sometimes they can be performed, too. In a thought experiment no hands get dirty, but minds can become clarified. Einstein himself often used thought experiments as a way of attaining insight into the nature of things. Wheeler himself was considered to be a master of thought experiments, extrapolating them in the most extreme fashion imaginable.

In a typical thought experiment we take an accepted idea and extrapolate it to the ultimate extreme so as to see what happens: Does it break down? Where and why does it break down? What is it revealing to us? We all entertain thought experiments throughout our lives. "Should I do this or do that? What will happen if I do this?" Physicists use this mode of inquiry to deepen their understanding of the universe. The very fact that physics, which is generally seen to be all about the functioning and operations of the material world" (seen as separate from the mind), conducts a large part of its experiments purely in the mind and considers the results of these experiments to be credible contributions to the field of physics, is a clue to the mind-like nature of the physical world. The trust that physicists place in the value of thought experiments is an expression that the underlying order of thinking can faithfully reflect and congruently represent the order of the physical world. Thought experiments are expressions of the profundity and power of our imagination to help us find our place in the universe, and they indicate that the nature of the universe is more thought-like than is generally acknowledged.

What is reflected in the magic mirror of physics can precipitate a Copernican shift in how we conceive of ourselves in relation to the universe. For example, imagine bathing in the sun's rays on a hot summer day. It is a scientifically accepted "fact" that the sun is "out there," ninety-three million miles away from earth. And yet the rays of the sun are the sun's unmediated expression, which is to say that they are indivisible and not separate from the sun by one iota. This is analogous to how the waves of the ocean are not separate from the ocean but are the unmediated manifestation of the ocean. This realization instantaneously helps us to change our perspective and understand that the sun isn't outside of us, but rather as we are enveloped in its rays and awash in its life-giving warmth we are "inside" of the sun. Not only do we find ourselves within the sun, we can further realize that we are not separate from the sun. This is to simultaneously realize that it makes just as much sense to think of the sun being inside of us, which is an expression of our identity expanding to even larger degrees.

Continuing our thought experiment, we realize that we are not a separate entity from the sun momentarily sharing the same space—we *are* the sun. We are the light! In an instant we go from thinking we are far away from the sun to feeling our oneness with it. Once we have this shift in perspective, we can no longer think of the sun in the same old way as an object outside of ourselves. Not just our image of the sun and our relationship to its image have changed, but our image and experience of ourselves relative to the universe have changed as well. In the physical world, nothing has actually changed except our mind's perspective. We have simply recognized something we didn't recognize before.

After our shift in perspective, the age-old idea that we are composed of (crystallized) light, that we are stardust, makes more sense. Our essential being isn't simply made of light, it *is* light. The calcium in our bones and iron in our blood are literally forged in the stars. To quote Nobel laureate Ilya Prigogine, "Matter is just a minor pollutant in a Universe made of light."[819] Interestingly, seeing and being in and of the light is a perennial gnostic theme. The Nag Hammadi Gospel According to Thomas, to use one of many examples, refers to a Gnostic (one who "knows") as one who both sees and is "in the light." The Gnostic Christ is described as "the light which is in the light." A true gnostic is considered to be a light to this world, one who sheds light on the darkness so as to dispel it. Astrophysicist Bernard Haisch writes, "Our world of matter is like the visible foam atop a very deep ocean of light."[820]

Quantum physics reveals that it is a mistake in our thinking to imagine that two separate entities, such as the sun and ourselves, are interacting; the emphasis in the quantum world is on undivided wholeness. The two seemingly separate entities are in actuality inseparable parts of a more inclusive entity or field that includes and unites them both. This is similar to when we see a pattern in a carpet. It has no meaning to say that different parts of the pattern are separate objects in interaction. The seemingly separate parts of the pattern are merely abstracted from the deeper wholeness of the underlying carpet that connects them.

Another example is to consider different views about space. If, for example, we view the skin as the boundary of ourselves, then there's the space within (the separate self) and the space without (which separates the separate selves). But when we see that space as a whole is the very ground of existence in which we are all contained, then space is recognized to not separate us but unite us. Nothing has changed except our perspective.

Seemingly separate objects in our world of everyday things are not truly separate; their apparent independent existence is an abstraction with a certain utility. Speaking of this situation at a conference on quantum physics, the Dalai Lama said that "the notion of reality tends to disappear. So we are in fact almost compelled to refer to objects as this so-called table or this so-called microphone."[821] Another example: if there are two points on a line, we can think of the line as illustrating the relationship between the two separate points, or we can think of the line as being real and the points on the line merely abstractions from that.

Similarly, in the quantum context we can regard terms like "experimental conditions," "observer," and "observed object" as aspects of a single overall "pattern" that are, in effect, abstracted or pointed out by our mode of description. Thus in quantum physics, to quote Bohm, "to think of an 'observing instrument' interacting with a separately existent 'observed particle' has no meaning."[822] In a quantum universe, the observer is the observed. Both observer and the observed are manifestations of the same underlying indivisible process. They flow into and out of each other like a river flows through its whirlpools. The observer is not "causing" the observed any more than our inner state is "causing" a synchronistic event to occur in the outer environment.

In a reciprocally co-arising, nonlinear, and acausal feedback loop, the observer and the observed are in a sense inter-causing each other (in my language, "dreaming each other up") while at the same time being caused by the underlying movement of the whole in which they are both contained and of which they are both expressions. Every aspect of the undivided process of being causes every other aspect in a seamless and ceaseless nonlocal dance of omnidirectional causality that, taken as a whole, is acausal. The seeming distinction between the observer and the observed is a convenient abstraction on the level of relative (as compared to absolute) reality that has a certain practical utility, but ultimately has no reality. In the realm of the quantum there are no separate or separable parts; the universe is an undividable whole. A true singularity, the quantum universe is "singular."

Another example: it is easy to think of a tree as a "part" of nature. The tree, however, is not a part but is coextensive with the whole universe. In its interchange with the environment, it is impossible to say at just what point a molecule of carbon dioxide or oxygen crossing the cell membrane into or out of a leaf stops/starts being air and becomes/is no longer the tree (the same question can be asked in reverse about humans). The tree, with its roots, trunk, branches, and leaves, threads out into the whole environment; it is part of and an expression of a larger ecosystem, which itself is interconnected with and inseparable from the whole universe. Where is the boundary between the tree and the rest of the universe that is supposedly separate from the tree? If this fact is ignored and forests are cut down with only a view toward short-term profits in mind, dangerous consequences can arise, which will affect the whole ecology of the land and spirit. The viewpoint of quantum physics is an expression of the universe as a whole system, as if the universe is one indivisible unit, a singularity.

SINGULARITY

Our entire universe/multiverse may itself be arising within (and as an expression of) a cosmic singularity. Singularities in physics are generally considered to be regions in the geometry of space-time where the mass/energy density becomes so extreme that the gravitational forces reach a critical intensity such that the curvature of space-time becomes infinite and the normal laws of physics collapse and no longer apply (notice the similarity to Wheeler's notion of the laws of physics being "mutable"). If our universe is indeed a singularity, this state of affairs has significant implications for our understanding of what might be possible within our universe, opening up a vast range of previously unimagined possibilities beyond what our conventional ways of thinking allow.

The known laws of physics may not be the ultimate arbiters of what is possible or impossible. For example, there could be an infinitude of sub-singularities arising within a singularity, each hosting an entire open-ended, infinitely creative universe within it. This possibility opens up the limits to our conceptions of both the micro and macroworlds, which is to say that the upper and lower limits of our universe are potentially infinite in both directions. Simply entertaining this possibility, whether it's the actual state of affairs or not, frees our imagination from any arbitrary, unnecessary, or limiting frameworks and belief systems.

Due to their "singular" nature, singularities are conventionally thought of as being featureless, having no structure, distinctions, or separation within their indivisible and unified existence. Having structure implies differences, which in a singularity is supposedly impossible, for in a singularity there are no differences—everything is the same. Yet, this mainstream notion of singularity partakes of the aforementioned two-valued logic (where, for example, structure and structurelessness are mutually exclusive), and is thus "one-sided" in a multisided universe. Two-valued logic is a residue of the ingrained habit of dualistic thinking, which, when applied within physics, creates unnecessary dualistic polarizations which propagate through the field like a virus of the mind (the aforementioned wetiko virus).

Yet the conventional dichotomy between the formless quality of a singularity and the endless multiplicity of our universe is actually an unrecognized complementarity, in which the infinite forms of the universe arise within and are expressions of the underlying featureless emptiness. The singular, indivisible nature of our universe would thus also be able to host apparent (and ultimately illusory) differences within itself. So even though on the surface the universe as a singularity seems to negate multiplicity, it actually implies and embraces within it an endless diversity of forms. Notice the similarity to the Buddhist *Heart Sutra*'s idea of "Form is emptiness. Emptiness is form." Forms and emptiness reciprocally co-arise and are not separate or opposed to each other, rather they interdependently co-originate.

From the limited perspective of two-valued logic, the idea that the featurelessness of our singular universe implies diversity is impossible and makes no sense. Yet it's the one "impossible" thing that makes everything else possible. It is an "impossibility" that cracks open the limited eggshell of the constricted mental domain of two-valued logic, thereby making the existence of a much more open-ended, creative, and creativity-generating universe like ours possible. It is noteworthy that the world's nondual mystical traditions all agree that separation is an illusion, which is exactly what quantum physics is pointing at, and that what fundamentally exists is ultimately indivisible and singular in nature. Ultimately admitting no division or separation, our universe as a singularity is paradoxically able to simultaneously display itself as an apparent (and very convincing) multiplicity with infinite diversity, which itself is an expression of its singular and whole nature.

Putting the same idea in theological/psychological terms Jung writes, "Where would God's wholeness be if he could not be the 'wholly other'?"[823] Substitute the words "singularity" for "God," and "multiplicity" for "wholly other" and you get the same idea. The singular, empty, and formless underlying structure of our universe and the dynamic diversity of the apparent forms arising within it only seem contradictory and mutually exclusive within our divided (and dividing) minds. Both states actually imply and require each other to simultaneously coexist, in distinction to the limited condition of either one or the other exclusively being true.

Bringing our discussion back around to the nature of our own minds, it is easy to assume that thinking is problematic in that it is obscuring the true nature of our mind. This perspective, however, is itself just another thought, and if we invest this thought with reality (think of As Viewed, So Appears) our thinking process will indeed manifest as an obscuration to our true nature. Yet, if we recognize that our thoughts are an unmediated natural manifestation of our true nature rather than obscuring the nature of our mind, our thoughts can be seen as one of its myriad expressions that are actually revealing it. In other words, if we relate to our thoughts as a revelation of our deeper nature, then this will be how our very process of thinking will manifest.

Think of a mirror and its reflections. In this example, the reflections symbolize our thoughts and the mirror represents the open-ended spaciousness of our true nature. From one point of view, the reflections in the mirror are obscuring the silvered surface of the mirror. And yet due to the fact that we would never notice the mirror without the reflections, from another point of view the reflections are actually revealing the mirror. It all depends upon our perspective. The reflections in the mirror are actually inseparable from—and an expression of—the mirror.

All of the great spiritual wisdom traditions point out that our existential situation as human beings is not inherently problematic; the only "problem" is a self-created one—our addiction to dualistic thinking. All genuine, spiritually-inspired traditions throughout the ages have born witness and proclaimed that dualistic thinking, the very flaw in our thinking process that quantum physics is revealing to us, does not correspond to the deeper nature of reality. This is to say that dualistic thinking, which informs our conception of the universe as an object, existing separately from us as a subject, is the generative source of our problems, both individually and collectively. By dualistically thinking that the universe exists objectively, we create the seemingly real and convincing illusion that we are an alien within our own home.

• CHAPTER TWENTY-TWO •

PHYSICS OR THEOLOGY?

The founding fathers of quantum physics were beginning to realize that nature herself and the structure of our own minds are not merely interrelated reflections/reflex-ions of each other but an inseparable unity. Nature isn't outside and separate from the mind; it is an unmediated expression of it. The mind *is* pure nature. Instead of thinking that the outer world was different from the inner world, they were beginning to realize that if something was happening within themselves, it was simultaneously happening within the universe as well. This led to the startlingly obvious realization that human beings—inclusive of their inner lives—are *in* the universe rather than (as the old Cartesian dualism conceived of things) their inner subjective worlds somehow standing apart from the universe. Coinciding with the collapse of the boundary between the subject and object, just as within a dream, the demarcation between the inner and the outer was becoming harder to find as well. In the holistic world that the new physics describes in which separation between the parts doesn't exist, the innermost processes of the psyche spill out and become as much a part of the seemingly external world as the rocks, trees, and stars, as if reality itself were a mass shared dream.

When these brilliant scientists began to metabolize and assimilate within themselves what they had discovered, it was like they had "come to their senses," waking up from a centuries-long slumber. We can tell from their writings that their discoveries truly changed the way they envisioned life itself. As if remembering something they knew long ago, they became (to varying degrees) inwardly transformed. This realization of the dreamlike nature of reality is itself the very expansion of consciousness which galvanized them to realize that consciousness plays the primary role in both physics and the creation of the universe. From the theological point of view, quantum physics is healing the "fall" of mankind. To quote Philip K. Dick, "We did not fall because we sinned; our error—which caused our fall—was an intellectual one: we took the phenomenal world . . . to be real."[824]

Trying to put his inner realization into words, Schrödinger says, "Mind has erected the objective outside world . . . out of its own stuff."[825] Just as our deeper, dreaming mind is the source of our dreams at night, we have a deeper part of ourselves, our divine, creative imagination, that is dreaming up this universe into fully materialized existence. Along similar lines, Schrödinger says, "Consciousness cannot be accounted for in physical terms. For consciousness is absolutely fundamental. It cannot be accounted for in terms of anything else."[826] Being fundamental, consciousness can't be reduced to other features of the universe such as energy or matter. Thinking that the source of consciousness is in the brain is like looking in the radio for the announcer.

Instead of consciousness arising from the brain, the brain and all of matter arise out of, because of, within, and as a dynamic modification of consciousness. Max Planck says, "I regard consciousness as fundamental. I regard matter as derivative from consciousness. We cannot get behind consciousness. Everything that we talk about, everything that we regard as existing, postulates consciousness."[827] The brain is just matter, but consciousness is something wholly other, an entirely different order of reality that includes matter (and the brain) as a subset of itself. As the philosopher Colin McGinn puts it, "You might as well assert that numbers emerge from biscuits or ethics from rhubarb"[828] as suggest that the "soggy clump of matter" that is the brain produces consciousness. The brain does not "produce" consciousness, it is an instrument that tunes into and transmits it. Rather than generating consciousness, the brain may simply be a transducer that acts as a filter as it mediates consciousness at the physical level. From the quantum physics point of view, the universe is of the nature of an appearance to the mind. D'Espagnat wonders how "could mere 'appearances to consciousness' generate consciousness?"[829]

Speaking of how "the Mind" (with a capital "M") "erected the objective outside world," Schrödinger says, "Mind could not cope with this gigantic task otherwise than by the simplifying device of excluding itself—withdrawing from its conceptual creation. Hence the latter does not contain its creator."[830] In its stead, the Mind sends a "stand-in," identifying with its made-up model of itself—the skin-encapsulated ego—instead of the real thing.

Through the revelations of quantum physics, the human mind is assuming its rightful place in the universe. Jeans writes, "Mind no longer appears to be an accidental intruder into the realm of matter; we ought rather hail it as the governor of the realm of matter."[831] Like a lamp that illumines itself, the self-luminous nature of consciousness, changeless in its essence, is completely altering and radically reconfiguring the field of physics from within in previously undreamed-of ways. Hans-Peter Dürr gets right to the point regarding one of the main implications of quantum physics when he says, "Matter is not made up of matter; basically there is only spirit."[832] Are these the words of a physicist or a theologian?

Jung reminds us, "Nature is not matter only, she is also spirit."[833] Matter and spirit are usually conceived of as being polar opposites; quantum physics is disclosing a world where the opposites are not as opposed as we have been imagining, but are secretly allied with each other. Jung comments, "Soil is just matter, the absolute opposite of the spirit, yet it contains the spirit. Without encountering the soil one would never realize the spirit; it needs the resistance of matter in order to reveal itself."[834] If we place a seed in our hand it doesn't grow, but when that same seed is buried underneath the soil it is inspired by the soil's resistance to grow toward the light and actualize its potential spirit in form. From this point of view the matter of the soil is not separate from the spirit of matter. The spirit has no way of revealing itself except through its opposite—matter—which is to say that the opposites of matter and spirit are inseparable aspects of a deeper unified field.

This is the *coincidentia oppositorum* of alchemy, which is a symbol for the goal of the alchemical opus. Uniting the opposites, alchemists discovered that within the darkness of nature a light is hidden[835]—what is referred to as the "lumen naturae."[836] To quote Pauli, "I consider the ambition of overcoming opposites, including also a synthesis embracing both rational understanding and the mystical experience of unity, to be the mythos, spoken or unspoken, of our present day and age."[837] It is as if the insights of the alchemists were a preconfiguration of the quantum gnosis, both of which, through their inquiry into matter, lead us (as if returning us to a place we once knew) to a world where the opposites are united and the physical world reveals itself to not be separate from spirit.

To quote theologian Sallie McFague, "The picture of reality coming to us from contemporary science is so attractive to theology that we would be fools not to use it."[838] As noted by Isaac Newton's biographer Richard Westfall, "The ultimate cause of atheism, Newton asserted, is 'this notion of bodies having as it were, a complete, absolute and independent reality in themselves.'"[839] The implication is that to see through the illusion of reality having an independent existence is somehow related to an experience of a higher power. Rather than disproving the existence of a higher, ostensibly divine intelligence, the new science has helped to demonstrate it. When asked if he believed in a personal god, Wheeler replied, "The idea is a little too concrete for me. I think of divinity as being present everywhere."[840]

In this materialistic age of ours, true scientists are becoming indistinguishable from deeply religious people.[841] "Religious teachers," to quote Einstein, "will surely recognize with joy that true religion has been ennobled and made more profound by scientific knowledge."[842] Physicist Edward Neville de Costa Andrade pronounced in a radio interview that "the electron leads us to the doorway of religion."[843] Who would have thought that microphysics might be the gateway to religion? In a related vein, Bohr writes of the "inseparability of materialistic and spiritualistic views," since "materialism and spiritualism, which are defined only by concepts taken from each other, are two aspects of the same thing."[844] "Like the meridians as they approach the poles," to quote Pierre Teilhard de Chardin, "science, philosophy, and religion are bound to converge as they draw nearer to the whole."[845]

Quantum theory, seen through a theological lens, is concordant with the idea of a powerful god that creates the universe to get things started but then bequeaths part of this power to beings created in his or her image. These beings then have power through their choices to cocreate with the very universe that created them. To quote scientist

Carl Sagan, "Science is not only compatible with spirituality; it is a profound source of spirituality."[846] Though seemingly polar opposites, science and religion can mutually complement and support each other. As Einstein famously said, "Science without religion is lame. Religion without science is blind."[847]

Quantum physics is a flag bearer of an epochal paradigm shift currently taking place within human consciousness, deep within the collective unconscious, concerning the nature of reality itself. The discoveries of quantum physics are directly pointing to the hitherto unsuspected powers of the mind to cast reality in its image rather than the other way around. Quantum theory provides insight into how conscious entities, such as ourselves, can alter the course of the physically described aspects of reality through the decisions they make. The new physics is the beginning of the realization that the human psyche can intervene creatively in the physical and chemical processes of nature. Jeans writes, "We discover that the universe shows evidence of a designing or controlling power that has something in common with our own individual minds."[848] On an individual level, our awareness interacts with and affects the subatomic realm of our bodies, which then feeds back into and influences our awareness. In any case, though seemingly subtle in nature at the present moment, this shift in paradigms that quantum physics is initiating is an earthshaking affair, with ramifications beyond our present imagination.

The revelations of quantum physics can be used to destroy life or to enhance it beyond measure. These words from Banesh Hoffmann's book *The Strange Story of the Quantum*, published in 1947, are even truer today: "Now is the terrible crisis of our civilization. Now is the fateful hour of high decision. For better or worse, We, the People of the Earth, must choose our future."[849] Quantum physics tells us that the future is not written in stone, but instead is indeterminate and filled with infinite potential. How the world of the quantum manifests depends on how we dream it. As it says in the Bible (Deuteronomy 30:19), "I have set before you life and death, blessing and curse: therefore choose life, that both thou and thy seed may live." The choice is truly ours.

Wheeler writes, "Who could have imagined that the unbelievable burden of planning for the future of man as a race would some day be found loaded on man's own shoulders?"[850] As we see through the illusion that we exist separate from the universe, we naturally step into a more holistic and ecological mode of thinking in which we perceive ourselves as part of a greater ecosystem. The Dalai Lama adds, "Things depend entirely upon us; they rest on our shoulders. Therefore, the future of humanity is in the hands of humanity itself. We have the responsibility to create a better world."[851] Our future is literally in our own hands.

The energetic expression of realizing—not just intellectually, but in our hearts—that we do not exist separate from each other but are interconnected and interdependent at the deepest, most fundamental level of our being, is compassion. In a passage widely attributed to Einstein, it says:

> A human being is a part of the whole called by us universe, a part limited in time and space. He experiences himself, his thoughts and feeling as something separated from the rest, a kind of optical delusion of his consciousness. This delusion is a kind of prison for us, restricting us to our personal desires and to affection for a few nearest to us. Our task must be to free ourselves from this prison by widening our circle of compassion to embrace all living creatures and the whole of nature in its beauty.[852]

SYMBOLS

Just like the most awakened of alchemists realized, once we begin to recognize the dreamlike nature of reality, the opposites begin to merge: spirit becomes materialized (taking embodied form), and matter, which is recognized to be an unmediated revelation of spirit, becomes "divinized." The universe then assumes its revelatory and theophanic function of being a living oracle of and for itself, speaking to and from something within ourselves. The quantum universe is speaking symbolically in "dream speak" and once this is recognized, quantum physics reveals its heretofore hidden "hermetic" side.

Sounding like an alchemist, Pauli felt that "reality in itself" is "symbolic."[853] This sounds similar to the Tibetan Buddhist idea that the universe is a living symbol of itself.[854] Jung writes, "There are, and always have been, those who cannot help but see that the world and its experiences are in the nature of a symbol, and that it really reflects something that lies hidden in the subject himself, in his own transubjective reality."[855] Similarly, Nicolas Berdyaev writes, "Everything external, material, everything of the object, is only a symbol of what is taking place in the depth of the spirit, in man."[856] Historically, in the evolution of our species as well as in an individual's life, recognizing the symbolic dimension of life is a momentous event—a literal expansion of consciousness—that changes everything.

A symbol, by definition, is a conjoining of two factors or two levels of reality that are normally conceived of as being opposites. Synthesizing matter and spirit, for example, a symbol "always expresses the one through the other; it comprises both without being either," to quote Jung.[857] In other words, a symbol can express spirit through the medium of matter. In describing a symbol, Jung could just as well be describing a quantum entity when he writes, "This dual character of real and unreal is inherent in the symbol. . . . Only that can be symbolic which embraces both."[858] In describing quantum entities, which are said to "both exist and not exist simultaneously," physicists are describing the symbolic nature of reality. Quantum physics is a symbolic procedure used to render the symbolic nature of reality into form as well as into our minds.

Just like in a dream (sleeping or waking), there is a function in the unconscious that creates symbols while there is another function within us that "understands" the symbols. Both functions are aspects of ourselves—as if we have split ourselves in two so as to communicate with ourselves, the ultimate aim being unification on a higher, more integrated and conscious level, with greater degrees of freedom and deeper dimensions of wholeness. The symbolic dimension is a manifestation of the interplay between the dreamer and the seemingly external (sleeping or waking) dream. The dreamer and the dream are not separate, but indivisible, interrelated, and interactive parts of a unified quantum field. The dreamer and the dreamscape—just like the experimental quantum physicist and the world they are exploring—mutually reflect, inform, and affect each other, reciprocally co-arising together in a synchronistic, cybernetic feedback loop.

The unconscious takes the raw material (the alchemical "prima materia") of psychic life and shapes it into communicable symbolic form so as to reflect it back to ourselves. Once we are touched by and change in relation to what the dream is speaking to in us, the unconscious then, in turn, will respond and reflexively transform, changing its form relative to us. The integration that we've achieved as a result of metabolizing what the dream is reflecting back to us is instantaneously relayed back to the unconscious, which then reworks, rewords, and retransmits new mythopoetic reflections of itself and ourselves back to us. This process happens both over time and in no time at all. In other words, it not only happens instantaneously, but is happening at each and every moment, which is to say that this process is happening right now.

As if intimate partners in relationship, we are always having a dialogue with the unconscious (which being nonlocal, can manifest both within our mind/body as well as seemingly outside of ourselves). The unconscious and our (conscious) selves work in concert, co-inspiring (and conspiring with) each other to create symbolic meaning. To say this differently, we are always dreaming in, through, and as our life. Transmuting the world into symbols by definition transcends the distinction between the inner and outer, between the subject and object, between dreaming and waking, and between spirit and matter. Symbols create a bridge that connect the inner and outer reality.

Becoming more fluent in the symbolic nature of reality is to develop what Jung calls "symbolic awareness"—the language of dreaming. To see our life symbolically is the doorway which begins to unlock the never-ending magic of the quantum. Symbolic awareness and the symbols that are the objects of our symbolically attuned awareness are not separate things interacting but are interrelated parts of a whole, indivisible quantum system. The symbolic dimension is not "objective," existing separately from our own mind. The symbolic script of our universe is not something we are passively watching, rather it is a revelation that we are actively dreaming up and cocreating (knowingly or unknowingly) at each and every moment. The symbols precipitating out of the interplay between ourselves and the universe is the very language through which our consciousness communicates with itself/ourselves.

A primordial revelation that is speaking in the language of symbols, the universe is a living symbolic scripture, a literal and symbolic book of life that is thirsting to be interpreted as such. Realizing the symbolic dimension of reality liberates us from the curse of literal matter, quantitative space, and historical, linear time. "In the symbol," to quote Jung, "the *world itself* is speaking."[859] If the world is not recognized as the symbol of itself that it is, like prose that is written in stone, we demythologize the sacred dimension of the world and the world then solidifies as mute, immutable, faceless, forever collapsing into seemingly concretized objectivity.

Elementary particles are like living symbols, pregnant with a deeper meaning (in Jung's words, "an unsurpassed container of meaning"), manifestations of the dreamlike nature of reality that have crystallized out of and into our world, reflecting back to us the very dreamlike nature that they are expressions of. The creation of a symbol, Jung writes, "is like the becoming of human life in the womb . . . if the depths have conceived, then the symbol grows out of itself and is born from the mind, as befits a God."[860]

A symbol is an informational analogue of the experience that it is re-presenting; if decoded, the symbol can be converted back into the original experience, bringing us with it in the process. As if once again describing subatomic particles, Jung writes that for a symbol to be effective it must "be sufficiently remote from comprehension to resist all attempts of the critical intellect to break it down."[861] A symbol, like an inscrutable quantum entity, is always more than we can understand at first sight. Jung writes, "The symbol always says: in some such form as this a new manifestation of life will become possible, a release from bondage and world-weariness."[862] The quantum realm, when recognized as a foreshadowing of the dreamlike nature of reality that it is, truly portends "a new manifestation of life" and "a release from bondage." When we connect with the symbol, to quote Jung, "it is as if a door opens leading into a new room whose existence one previously did not know."[863]

QUANTUM VISUALIZATION

The quantum realm can't be visualized but its dreamlike nature can help us to use the practice of visualization itself to connect us with our ineffable yet always present true nature. Quantum entities don't objectively exist in the ordinary sense of the word but, as previously mentioned, if physicists treat them as if they do, then they manifest in our experience with all the resultant effects.

As a Tibetan Buddhist practitioner, I can't help but notice the similarities of certain features of quantum entities to the Buddhist practice of visualizing a deity. Students are instructed that the deity doesn't exist outside of ourselves in an objective way, while at the same time the visualized deity is not just the projection of our mind, not merely an unreal figment of our imagination. And yet if we relate to the deity as if it exists, the practitioner receives the blessings from the deity as if it really does.

In a sense the deity is brought into existence parallel with the percipient's awareness of it (just like in the quantum realm); previous to the perception of it, the deity cannot be said to "exist." The deity and its perception occur simultaneously; neither can be separated from the other at any time. They are part of one unified and dynamic quantum field overflowing with creativity.

Strangely enough, it could be said that the deity induces a potential percipient to perceive it. In other words, during its state of nonexistence the deity is able to cause its own existence. In a manner of speaking we could say the deity—just like the quantum universe—has a need to be perceived in order to complete and fulfill its existence, so out of itself it calls forth and creates an observer to bestow upon itself full existence. Instead of the deity being our projection, in a sense we are its projection.

The deity's self-generating nature brings to mind Wheeler's notion of the universe as a self-excited circuit. The percipient sees the deity because the deity causes the percipient to see it, but the deity did not come into existence until the percipient saw it. The effects of the deity are nonlocally felt before the deity exists. These effects are acausal, having no cause because their cause does not yet exist. Prior to its manifestation in the visualization, the deity exists as a nonlocalized quantum potential within the implicate order of the quantum plenum. Only retroactively will the effect have a cause; this turns our concept of linear time on its head.

The deity appears to come and go, just like the quantum, but in a sense it is always present. They are ubiquitous; there is no place or time where they are not. In these visualization practices, Buddhist wisdom is skillfully using the quantum nature of things—as well as the projective tendencies of our mind—so that, rather than distancing ourselves from reality (which projections are known to do), we become more deeply connected to the reality of our essential nature.

Quantum physics tells us that the universe is very fluid, malleable, and ever-changing in its nature. Quantum physics is revealing that in our most fundamental and deepest natures we are beings whose essences are not set in stone. We can create and step into new images of ourselves through our creative imagination. These images immediately get imprinted into the quantum field, establishing a new template that increases the probability of this new identity pattern becoming more natural and familiar to us, which increases the likelihood that we will more embody this new identity in our lives.

In other words, we have it in our power to change (or re-create) ourselves. Most of us, however, don't know this. But this power is not ours in the sense that we don't own, possess, or control it. It is a power that is beyond us. This power comes from and is itself the spontaneous creativity of the quantum realm that pervades the whole universe and our very being. This power is always flowing through us. It is merely a question of whether we have fallen into the habitual pattern of blocking it, or whether we let go and align ourselves with the creative power of the quantum, letting it flow through us like some form of grace. We then become a channel or instrument for the creativity of the quantum to manifest itself through us. All genuine spiritual practices are about deepening our familiarity with this power of quantum creativity, a power which is truly our birthright.

In Buddhist visualization practices, the practitioner is instructed to not just see the visualized deity (which symbolizes our enlightened nature) outside of and separate from ourselves, but at a certain point to merge with the essential nature of the deity, to identify with it to the point where we creatively imagine ourselves as becoming the deity itself. When we visualize, imagining ourselves, for example, as being the deity, we are tuning into and helping to manifest the part of ourselves that the deity symbolically represents, the part of us that is already healed, whole, and awake.

In our evocation, visualization, and identification with the deity, we are not merely creating a fabrication (imagining something that doesn't exist or isn't real), but through our act of creative imagination we are skillfully getting in touch with and becoming familiar with what the deity represents—our true nature. Our true nature has always been with us, but most of us have not been aware of it consciously. In consciously engaging with the creative quantum flux that comprises our being at its most fundamental level, we are simply taking advantage of our intrinsic, open-ended, and fluid quantum nature to reimagine ourselves in a way that is more in alignment with the truth of who we actually are.

The point is not that everyone should start doing Tibetan Buddhist visualization practices. I cite these practices as but one example to show how we can actually exploit and creatively engage our quantum nature for our own good. These Buddhist visualizations show how, for example, our quantum nature can empower our creative imagination to take advantage of our projective tendencies such that, instead of keeping us asleep, they help us to awaken. There are countless other ways that we can further take advantage of our intrinsic quantum nature that serve our evolutionary potential. We can all do our own unique creative experiments, be they in visualization or other ways. The quantum revelation gives us the power to creatively reengineer ourselves. The quantum universe, being a mirror, will simply reflect back our realization.[864]

• CHAPTER TWENTY-THREE •

GENIUSES WITH AMNESIA

Einstein famously said, "It is the theory which decides what we can observe."[865] Nature simply responds in accordance with the theory by which it is approached. The choice we make about what we observe makes a difference in what we find. If in a physicist's theory the universe is composed of separate parts, the physicist will ask questions, set up experiments, perceive, and interpret their results in a way that produces the very fragmentation they believe exists (which is the very fragmentation that characterizes their state of mind). Now having apparent "objective" proof of their presumed fragmentary worldview, they don't notice that they themselves, acting according to their unreflected on axiomatic sets, have brought about the seemingly real fragmentation that they are citing as evidence for the rightness of their fragmented viewpoint. Because reality is whole, when it is approached with a fragmentary perspective, it will invariably reflect back a correspondingly fragmented response. Through the choices we make, interpreting our world and placing meaning on our experience, "the corresponding world unrolls upon his [or her] screen," Wheeler writes in his journal. Associations of projecting onto an inkblot, which then simply reflects back our viewpoint, come to mind.

The intrinsic power to create our experience has boomeranged against us in a way that is not only not serving us, but is also limiting our creative brilliance. As if under a form of trauma (the aforementioned QPIT) many physicists seem to have put themselves under a self-created, self-impoverishing, and self-perpetuating hypnotic "spell." Like the perfect mirror of our minds that it is, quantum physics—not to mention the universe as a whole—is simply reflecting back this process. If this reflection is not recognized, the universe simply continues to reflect back the self-limiting consequences of our lack of recognition until, sooner or later, we recognize what is being revealed to us.

Because of the quantum, mirrorlike nature of reality, once we view the universe "as if" it independently, objectively exists, it will manifest in a way which confirms our viewpoint, appearing in an utterly convincing way to be independent and objective. One way to better understand this is to remember the dreamlike nature that quantum physics is continually reflecting. When we hold a viewpoint within a dream, the dreamscape, which is nothing other than a reflection of our mind, has no choice but to instantaneously shape-shift in such a way so as to supply perceptual evidence that justifies our viewpoint as being correct. Now having seemingly objective "proof" of the correctness of our viewpoint, we become even more firmly entrenched and fixed in our point of view, which in a seemingly endless feedback loop then dreams up the universe to supply more evidence of the truth of what we are seeing and imagining to be true, ad infinitum. This is a self-generated feedback loop originating in our own mind that happens over, in, through, and outside of time.

As if "bewitched," we entrance ourselves by our own innate, unrealized genius for cocreating reality. We are powerful wizards wielding a magic wand (the quantum).[866] But because we are disempowered and don't realize our own divine gift, we are using our power to create our world unconsciously, which is to say destructively. We have forgotten that we have reality-shaping powers at our disposal, thereby unwittingly placing ourselves in what William Blake calls "mind-forg'd manacles." It is like we are disoriented (and deranged) magicians who have created a world for ourselves that doesn't serve us, all the while thinking that we are just encountering—and being victimized by—an objective reality that we cannot change.

The truth of our situation, simply put, is that we are geniuses with amnesia. We have literally forgotten who we are and in so doing have disconnected from our vast creative powers for consciously shaping and cocreating reality. At any moment, to the extent we are aware of our true nature, we can help each other to remember. From the alchemical point of view, our true mission and vocation in this life is to become living philosopher's stones.

One of the greatest powers of the atom is to seize, captivate, and stimulate the human imagination. We think of atomic physics, one of the discoveries based on quantum physics, as unleashing the incredible power latent in the atom, and yet we have hardly begun to realize that quantum physics has likewise tapped into the vast world-transforming power of the psyche. As Jung never tired of pointing out, it is unconscious psychic forces which are the active "world powers" that rule over humanity. "The powers of the psyche" are so unimaginably vast, to quote Jung, that they "are far mightier than all the Great Powers of the earth."[867] The powers hidden within the psyche, as history shows, can transform entire civilizations in unforeseeable ways.

The riches hidden within quantum theory are far from fully mined; encoded within the quantum is a lode—the mother of all lodes—waiting to be exploited for the benefit of humanity. Imagine the immense energy accessed by atomic physics not being used to create new weapons of mass destruction, but instead channeled in a constructive, positive manner for the unlimited benefit of all beings. If we aren't able to successfully mine the revelatory treasure that is quantum physics, however, we might become "one of those poor souls without the critical power to save himself from pathological science."[868] This would be a tragic reflection of our inability to overcome our own self-destructive inner forces of psychopathology.

The revelations of quantum theory are a modern-day version of an occult-like secret—what theoretical computer scientist Scott Aaronson calls "the ultimate Secret of Secrets"—in need of being unlocked, liberated, and shared by all. To quote Aaronson, "For almost a century, quantum mechanics was like a Kabbalistic secret that God revealed to Bohr, Bohr revealed to the physicists, and the physicists revealed (clearly) to no one."[869] One of the problems in revealing the insights of quantum physics to the person on the street is that there are not enough words in the entire universe to do justice to the quantum state. The emergence of the quantum is like the discovery of the mythic Holy Grail, but this discovery does us, as a species, little good if we aren't able to effectively share it with each other.

Before the discovery of the quantum in the early twentieth century, no one would have taken seriously someone who was pointing it out. The ubiquitous quantum has been insistently giving us all the hints we need of its existence; we simply need to inwardly prepare ourselves to receive so priceless of a gift. Like a precious treasure thirsting to be brought forth, the mysterious place where the quantum abides waiting to be discovered is ultimately within our own minds. Wheeler writes, "But what agency selects the nourishment of our minds: the conversation we have, the thoughts we record, the books we read and the very issues we embrace? The richness of this nutrient stream can vary far more widely than the composition of the blood, and with far greater consequences. Man selects what makes the man."[870] Where we place our attention nourishes or starves our minds. Wheeler is rightfully espousing one of the major insights of quantum physics: Through the choices we make we create ourselves. Altering Descartes' famous principle, "I think therefore I am," quantum physics would instead say, "I choose therefore I am."

One of the greatest sovereign powers that we all wield as human beings, although often unknowingly or without awareness, is the power of choosing where to place our attention. As if we all have an unknown superhero power, the very power of creation lies invisibly enfolded within our field of attention. Quantum physics reveals to us that turning the gaze of our attention towards anything is a powerful creative act that alters, energizes, and potentiates whatever our gaze falls on. Focusing our attention is an act of creation in and of itself. Our beam of attention intersects and interacts with the multidimensional probability waves that hover in a ghostlike state of unrealized potentiality that comprise matter in its unobserved state. Once imbued with our attention, whatever we are looking at instantly materializes into a particular and perceivable appearance. The symbolic procedures of quantum physics awakens our attention, which in turn stimulates the development of consciousness.

We may pride ourselves in the cleverness of our mind in helping us understand the nature of the universe, but we so often fail to realize the nature of the very faculty through which we try to understand the universe—our minds. It can be profoundly powerful and self-transformative when, in introspective practice such as meditation, we put our attention on the contents of our own mind and our own awareness. Showing the power of the quantum observer effect turned inwards, by shifting our attention beam shimmering with the power of quantum creativity upon ourselves, we literally re-create and transform ourselves in the process.

MATRIX MAYA

To the extent that we are not awake to the dreamlike nature of our situation, we have fallen under what in Eastern traditions is called the power of "maya," the source of both our deepest illusions and our most exalted creativity. Etymologically, the word maya comes from the root "ma," which means "to measure." This reflects the fact that measuring something, though adding to our knowledge of the world on the one hand, can at the same time obscure us from the deeper, unbroken wholeness of reality on the other. It is noteworthy in this regard that the central enigma of quantum physics having to do with the role of the observer is called "the measurement problem." Maya has to do with the primal power that gives shape to the potentiality intrinsic to form. It has to do with the power to turn an idea into physical reality, and is related to the role of the imagination in creating the world. Maya refers to how the reality-creating power of our own mind can be unwittingly turned against us so as to entrance us. The creator of illusions, maya is that which makes the real appear unreal and the unreal appear real. The power of maya inspires delusions and false beliefs, such as when we believe the images in our minds are the external world. Through the wizardry of maya, we wield the cosmic creative power of shaping and giving meaning to appearances. Maya is related to the word "magic."

Schrödinger, in his book *Science and Humanism,* talks about the "problems" confronting modern physics. He is referring to maya when he cites "an evil godmother—if you please, like the thirteenth fairy in the tale of the Sleeping Beauty."[871] Likewise, he uses words such as "evil spell," "counter-spell," and "exorcise." The fact that one of the twentieth century's greatest scientists, in speaking about the discoveries of quantum physics, talks in such mythic, symbolic terms should give us pause. It should also help us gain insight into the archetypal energies of the psyche that modern physics has tapped into.

A contemporary mythic framework that can give meaningful insight into the significance of the discoveries of quantum physics is what is known by many as "the Matrix," the vast global, corporate, technocratic control and monitoring system. We live under greater surveillance than any civilization in all of history (think of the NSA and the Edward Snowden revelations). The Matrix is based on keeping people trapped within a paradigm of false and superficial knowledge of themselves and the universe known as materialism. It is fundamentally about centralizing power and control, as it enslaves people under deceptive lies of limitation and lack of options, keeping them disconnected from their own immense creative power.

The Matrix operates through the process of "compartmentalization," which prevents any one person from knowing too much. This is a reflection of a process of fragmentation going on within the human psyche that is being acted out and expressed in the outside world, which then serves to feed back into and reinforce to the psyche the very same fragmentation. Through a carefully orchestrated "need to know" basis, the Matrix keeps different groups of people who are serving its power structure partially informed and purposely disconnected from each other, so that no one but those at the top of the pyramid of power can know the overall big picture and hidden agenda in which they are unwittingly playing supportive roles. This isn't a paranoid conspiracy theory; the evidence is all around us for those who have eyes to see.

The Matrix control system has co-opted the powerful liberating knowledge of quantum physics to use instead for its own power-based agenda. There is a cultural bias, a subtle (and sometimes not so subtle) conspiracy that prevents us from knowing just how much we are contributing to the very world we are experiencing. Seen symbolically, by barring inquiry into quantum physics' metaphysical implications, the existing power structure has practically cast a materialist, nihilistic "evil spell" upon quantum physics itself. Like the evil godmother in "Sleeping Beauty," the Matrix keeps the liberating quantum gnosis asleep and under its control. It would be the powers that be's worst nightmare to have the spiritual dynamite that is the quantum revelation get out and be more widely accessible by the human population at large. This whole scenario is revealing a deeper archetypal process that exists in the collective unconscious, which is to say that the actual people who compose the Matrix (which is all of us, we are all just playing different interdependent roles) are merely *instruments* for this deeper process to potentially become conscious.

With the true power of the quantum spellbound, its liberating powers temporarily anesthetized, the Matrix is free to use the denatured knowledge of quantum physics to serve its own agenda of centralizing worldly power and extending its technocratic control over the material world. When it comes to the world's "body politic," it is crucial for us to realize the extent of the massive spell that is being woven all around us through the propaganda organs (such as the mainstream media) of the prevailing order, which is itself under the very spell it is casting. We are continually being conditioned, programmed, brainwashed, and hypnotized beyond belief so as to "buy into," to both "consume" and "be consumed by," an impoverished version of who we are. To the extent that we don't connect with our true essence and express our true creative selves we become "domesticated" like a trained animal.

Just like a dream (where the outer and inner are reflections of each other), what is playing out in the outside world is a reflection of an archetypal process happening deep within the human psyche. This is to say that what is referred to as the Matrix, though as real as real can be on the outer, relative level of reality, is at the same time an out-picturing of an inner dreaming process happening deep inside each one of us. The fact that our immense powers of quantum creativity are being used against us, when we get down to it, is *our own doing* (which is the meaning of the word "karma"), for we ultimately bear the responsibility if we subscribe to, identify with, and live out a false, disempowered and limited version of who we actually are.

Realizing that we have been unconsciously complicit in dreaming a self-limiting dream is the first step in taking our power back. Connecting with our true identity as sovereign creative agents who are learning to think and act independently invariably results in bringing an end to the Matrix's ability to control us. We can subsequently become proactively engaged with each other through our aforementioned "sacred power of dreaming" to reenvision ourselves into empowered creative agents who can further liberate the human spirit, allowing the latent genius intrinsic to this spirit to in-form and give shape to our world.

Quantum physics is showing us how we, both individually and collectively as a species, have been entrained, put under a self-created and self-limiting spell, and conditioned and programmed by our own mind. Its liberating insights are truly like some sort of miracle from on high, and yet they are the most natural thing of all. To quote Dr. Jeffrey Satinover, "Quantum Mechanics allows for the intangible phenomenon of freedom to be woven into human nature."[872] The revelations of quantum physics are offering us the keys to our intrinsic freedom. Heisenberg writes, "Classical physics seemed to bolt and bar the door leading to any sort of freedom of the will; the new physics hardly does this; it almost seems to suggest that the door may be unlocked—if only we could find the handle."[873] Whereas classical physics showed us a universe that, in Heisenberg's words, "looked more like a prison than a dwelling place," the new physics, he continues, "might conceivably form a suitable dwelling-place for free men."[874]

It is therefore up to us to cast a "counter-spell" and help to wake up the Sleeping Beauty of quantum physics by liberating it from the confines of an impoverished paradigm, freeing it from the ideological straitjacket of a power-hungry, utilitarian reductionism. The best way to do this is to partake in the most radical, subversive, and powerful form of activism that there is: recognizing and then engaging our own quantum nature. To quote Amit Goswami, "The central idea of quantum activism is not that we must advocate a dogma, but rather that we must learn how to be *free* of dogma."[875]

Schrödinger's cat is out of both the bag as well as the box, and we—each of us—have to come to terms with what the insights of quantum physics are revealing about ourselves and the world we live in. The biggest obstacle in bringing the liberating quantum gnosis into our world today is not the powers that be suppressing this knowledge, but rather the programmed, limited, and fear-based state of most people's minds. Many people, even in their wildest dreams, can't imagine the "good news" that the quantum is openly offering to everyone. To quote poet Ralph Waldo Emerson, "People see only what they are prepared to see."[876]

It is helpful to remember that due to the quantum's nonlocal nature, which allows it not to be bound by third-dimensional space and time, it can't actually or ultimately be hidden, kept down, bound, or imprisoned. This is to all of our advantages. Philip K. Dick refers to this as "the secret weapon of truth: it can't be suppressed, because of its nature; if it could be, it would be only opinion."[877] This state of affairs is typically portrayed in mythologies the world over, in which the highest, most sacred value is impossible to be destroyed.[878] The quantum is naturally buoyant, always ascending towards the light of consciousness, which is not separate from its—and our—nature.

PLATO'S CAVE

Just like the allegory of Plato's cave, wherein people looking at shadows reflected on the wall of the cave believed the shadows to be reality itself, physics is dealing with lower-dimensional shadow projections of reality, not reality itself.[879] "The greatest achievement of twentieth-century physics," to quote Sir James Jeans, "is the general recognition that we are not yet in contact with ultimate reality. We are still imprisoned in our cave, with our backs to the light, and can only watch the shadows on the wall."[880] It is as if the walls of the cave are representative of space and time, and matter is the shadows cast upon the screen of space and time, while the reality outside the cave which produces the shadows on the walls of the cave is outside of space and time. Using poetry to express the state of affairs, Jeans writes in his book *The Mysterious Universe* that events in time and space become "no other than a moving row / of Magic Shadow-shapes that come and go."[881]

The laws of physics are stencil-like descriptions of a lower-level cross section of a higher-dimensional, immaterial reality. Eddington writes, "In the world of physics we watch a shadowgraph performance of familiar life. The shadow of my elbow rests on the shadow table as the shadow ink flows over the shadow paper."[882] When physicists "look at" quantum reality, they are not seeing the unmediated "thing-in-itself," but rather, abstract mathematical symbols that represent reality. To quote Eddington, "We have learnt that the exploration of the external world by the methods of physical science leads not to a concrete reality but to a *shadow world of symbols*."[883] No one has ever directly seen the quantum world. Physicists "track" its ghostly footprints, inferring the world of the quantum through the results of their experiments. In doing physics, as Bohr points out, "It must be recognized that we are here dealing with a *purely symbolic procedure*."[884] The symbols utilized by physics represents the underlying reality, the nature of which remains hidden, mysterious, and inaccessible.

If physicists regard their theories as direct descriptions of reality as it is, there is a high possibility of falling into the error of confusing the map with the territory. One of the greatest advances of the new physics is the realization that it isn't dealing with reality per se, but with the projections of a deeper reality cast into the third dimension as it interfaces with our consciousness through the instruments of our brain and nervous system. Speaking about his own field of psychology, Jung writes, "I have not the faintest idea what 'psyche' is in itself, yet, when I come to think and speak of it, I must speak of my abstractions, concepts, views, figures, knowing that they are our specific illusions. . . . We have no idea of absolute reality, because 'reality' is always something 'observed.'"[885] Similarly, the elementary particles of the quantum realm are human abstractions, made up in our attempt to understand the mystery.

To quote Eddington, "The frank realization that physical science is concerned with a world of shadows is one of the most significant of recent advances."[886] Before the advent of quantum physics, physicists were under the delusion that they were dealing with reality itself, not realizing they were engaging with its mere shadows or projections. In studying physical reality, physicists are contemplating an echo of ultimate reality, not its source, as if they are seeing the reflection of God's countenance, not her face directly. When an echo is recognized for what it is, those who will not be satisfied with listening to anything but the source will begin to place their attention in the direction from which the echo emanates.

In some sense, the higher dimension contains its third-dimensional projections within itself. These projections exist only as abstractions, however, which is to say that the source of the projections—the higher dimension—is not the same as the projections, but is something else entirely, something of a nature beyond and transcendent to its lower-dimensional shadow projections. This is similar to how a reflection in the mirror is inseparable from the object that it is reflecting while also not being the same as the object, which is beyond and transcendent to its mere reflections.

Jung comments, "Science is the art of creating suitable illusions which the fool believes or argues against, but the wise man enjoys their beauty or their ingenuity, without being blind to the fact that they are human veils and curtains concealing the abysmal darkness of the Unknowable."[887] If we are in the cave of shadows and don't know it, the shadows appear to be the real world, as we have no point of comparison. And yet the shadows have no intrinsic, independent existence on their own, as they are merely derivative from, projections of, and inseparable from the light. After centuries of becoming habituated to a world of shadows, however, it is easy to become attached to the reality of the shadows, practically disbelieving in the light that informs them. As if mesmerized, enchanted, and spellbound by the display of the shadows, we have little suspicion of the light that is their source. Once we step out of the cave, however, we realize that there is and always has been only light. To recognize and see through the world of shadows is to be in the light.

The cutting edge of twenty-first century physics is beginning to wonder: What is this higher-dimensional form of light that is casting the shadows? How is it different from the electromagnetic light that the majority of scientists think of when they think of light? Could this higher-dimensional form of light be the light of awareness itself, thereby related to the mysterious factor of consciousness? Discovering (or are they creating?) a novel spiritual path, physicists are beginning to find within the world of shadows—in our case, physical reality—the very light of consciousness that is its source. This is the light *by which* we see, not the visible light *that* we see with our physical senses. The light that we see with our eyes is itself a shadow on the walls of the cave of the third-dimension, a lower-level shadow of the higher-dimensional uncreated, living light of consciousness itself.

This is an archetypal situation that wisdom traditions throughout the ages have been pointing at: encoded within the shadow is hidden the light. The greatest pioneers of modern physics are beginning to look beyond the shadows, beyond the cave, beyond physics, beyond the physical world altogether into the realm of metaphysics and their own minds.

This turning inwards and upwards, dimensionally speaking, is a watershed moment in the evolution of the science of the physical world. It is an epochal inflection point in the very nature of the scientific endeavor, in which the quest for a deeper understanding of the nature of the physical world has finally gone beyond the physical world per se. In our inquiry into nature, humanity is beginning to realize that we can't exclude the aspect of nature within ourselves that is investigating nature. Science finds itself in the unexpected position of being irresistibly drawn into the higher-dimensional worlds of consciousness, where, much to its surprise, it is finding the deeper source of what we call the physical world. This is big news indeed.

• CHAPTER TWENTY-FOUR •

LUCID-DREAMING QUANTUM PHYSICS

The observer effect, the central pillar of quantum physics, reveals that the act of observation is not merely a passive reception of information, but rather is a creative act that we are all—knowingly or unknowingly—participating in every moment of our lives. This process is tantamount to the same kind of dynamic creativity that we engage with in our sleeping dreams. In a dream the un-manifest potentialities (the wave function of dream possibilities) within the unconscious, depending upon the psyche of the dreamer, are actualized or "dreamed up" into specific appearances as the fabric of the dream.

In a typical nonlucid dream, the dreamer relates to the forms of the dream as if they exist objectively, separate from themselves (in the same way most of us relate to waking life), and then reacts to and becomes conditioned by the appearances within the dream *as if* they are other than their own mind. This further conjures up the dream to manifest *as if* it were objectively real and other than the dreamer in a self-perpetuating feedback loop whose generative source is the dreamer's own mind. To quote Buddhist teacher Khenpo Tsewang Dongyal Rinpoche, "The world of waking experience is similar to a dream in that it seems to exist objectively. But when you examine it closely, you realize it is an uninterrupted flow of sights, sounds, tastes, smells, tactile sensations, and thoughts. Like a dream, it is completely your own experience, and entirely within your mind."[888]

When we become lucid in a dream, the idea is not to control the dream, which would be an expression of our nonlucidity (i.e., the spell of the separate self), for in trying to control the dream, we are still relating to it as something other than ourselves. Instead, when we become lucid in a dream, we become naturally "in control" of ourselves and in touch with our own sovereign power of cocreating reality. When we become lucid, our *relationship* to the forms of the dream changes, for once we recognize that the dream is our own energy appearing seemingly outside of ourselves, we are able to creatively flow with the manifestations of the dream in a different way than when we were entranced by its forms.

When quantum physics was first formulated in the early part of the twentieth century, it was during a time of great collective somnambulism. Quantum physics' creation/discovery can be seen as an inkling of a burgeoning awakening in the collective unconscious being expressed through the realm of science. The process that quantum physics is articulating regarding how an observer/participant evokes or "dreams up" reality is generally unfolding *unconsciously* in most humans, which means most of us in the waking dream of life are dreaming nonlucidly. The central problem of today's theoretical physics, as well as so much of the insanity going on in our world, is, to quote Mindell, "the marginalization of the dreaming background of the universe."[889]

Quantum physics points out that the fundamental building blocks of reality begin in a virtual reality of open-ended potentiality that is undetectable to our senses and unfathomable to our mind. We understand the external world not through passively receiving its signals into our brain through our senses. Rather, the incoming signals from the world activate models that already exist in the neuronal structure of our brain that interact with these seemingly external signals so as to create our experience of both the world and ourselves. Prominent brain scientist Rodolfo Llinás comments, "The only reality that exists for us is already a virtual one—we are dreaming machines by nature."[890] Dreaming machines are a mechanistic sounding metaphor that I can get behind, as long as part of its programming includes the potentiality for becoming lucid in the waking dream and realizing the dreamlike nature of the world it is helping to create.

As more of us become increasingly aware of the dreamlike nature of the universe and the world-shaping creative power that we wield, how will things change? How would an amplification of our awareness regarding our powers of quantum creation affect these very powers as we are more and more consciously using them? As Wheeler reminds us, asking the right question is more important than finding the right answer.

How would the world itself change as more of us become familiar with what quantum physics is revealing to us about our incredible—but mostly untapped and unconscious—power to cocreate reality? I find myself imagining that the world would reflect back our inner realization of its quantum nature and would manifest in a much more fluid, malleable, and dreamlike way. And how would the realization of the dreamlike nature of our universe change our experience of ourselves and our understanding of who we truly are? I can only imagine.

How would applying the power of lucid observer/participancy to the theory of quantum physics itself transform the field of quantum physics, not to mention the very physics of existence? Could the very nature of quantum physics be expanded so as to register and reflect this awakening force of lucidity that is flowing through our species? Might quantum physics as a theory become transformed by its and our realizations into a new and more coherent version of itself? I find myself imagining that this realization would spawn a new awakening generation of lucid quantum physicists/dreamers. Instead of "Generation X," they would be "Generation QP," or "Generation LD," or maybe simply "Gen LDQP" (Lucid-Dreaming Quantum Physics). This new generation of quantum-physicized lucid dreamers would be able to creatively dream up an enriched, deeper, and more rigorous articulation of quantum physics which would go far beyond the present formulation, as it would more elegantly and completely reflect and codify the creative role that our consciousness plays in the very unfoldment of the universe. Or so I imagine.

A longtime student of quantum physics and a close friend of mine had a personal conversation years ago with David Bohm in which he shared that when he becomes lucid in his dreams, he does physics experiments. Bohm was very excited by this, saying that this was exactly the type of research that was needed for advancement in the field. He shared that he had done the same thing in a few of his lucid dreams. My friend was struck by how Bohm was taking seriously that these dreams had their own physics that was worthy of being studied, an idea that was summarily dismissed by most other physicists. Bohm was of the opinion that it is very important for physicists to carry on more careful physics experiments in their lucid dreams in order to compare how the physics of the dreamworld compares to the physics of the waking state, for the way consciousness intervenes in the waking dream seems "slower" and less obvious due to the apparent density of the waking dream. He was convinced that through this kind of physics research we could uncover the fact that there was a still yet to be discovered and articulated physics of the dream state that was connected to but possibly differed in some important ways from the physics of the physical world.

Bohm confided to my friend that he felt that lucid dreaming very likely held an important key to a deeper understanding of the connection between consciousness and the manifestation of our experience in the world—in both our sleeping and waking dreamworlds. Bohm humorously added that doing physics research in our lucid dreams would solve the ever-present challenge of obtaining funding, since in our lucid dreams we could potentially dream up our own laboratories, research assistants, and whatever other kinds of support we might need.

Though sleeping dreams and the waking dream appear to be different, the question arises: Are they, deep down in their essence, actually different, or are they made of the same "dream stuff"? All of the spiritual wisdom traditions on this planet from time immemorial (including quantum physics) have pointed out the dreamlike nature of reality. What if we were to take seriously what these converging wisdom streams are reflecting back to us about the dreamlike nature of our situation and step more into the dream, so to speak? What would happen if we continued our explorations into the nature of reality in this light? If we interpret experiences in our lives as though we were in a dream, how would this change our experience? "Excellent questions," I imagine Wheeler saying.

The idea is to not just do physics experiments in our lucid dreams, but to recognize that life itself is potentially the dream within which we can become lucid. The more we recognize the dreamlike nature of our waking experience, the more our waking life will reflect back this realization and manifest itself in a dreamlike way, thereby increasing our lucidity even further. In a positive feedback loop, our increasing lucidity, driven by consciousness, builds on itself and at a certain point becomes self-generating, reminiscent of Wheeler's idea of the universe as a self-excited circuit. Conjuring up some sort of over-unity device of the mind, we can help each other to awaken as we collectively dream ourselves awake (or so I imagine). Adding lucidity to our experience of life is a powerful spiritual practice, a form of "dream yoga." Becoming lucid in our waking dream changes everything. The question then becomes: "Who is the dreamer?" This is *the* question.

Could *lucidity* (a word which is etymologically related to the word "light") be the missing evolutionary ingredient that our species has been dreaming about? Could the process of transforming passive, semiconscious, nonlucid quantum physics into a more lucid quantum physics be the very shift that our species desperately needs in order to make the critical evolutionary transition from *Homo somnabulens* to *Homo lucidus*? Adding the light of lucidity to our collective human experience and thus to our current understanding of quantum physics may be the very factor that enables us to finally bring about and e*lucid*ate a comprehensive and long dreamed about spiritual/scientific synthesis. "It is no exaggeration to say," remarked Alfred North Whitehead, "that the future course of history depends on the decision of this generation as to the relations between religion and science."[891]

Could this bridging and blending of science and spirit provide humanity with a more refined and integral map of reality which could lead us back—both individually and collectively—to living, sharing, and having an enriched experience of the intrinsic wholeness that currently lies implicit but yet unlived and largely unfulfilled within every human being? This is a real potential and a very realizable outcome of Lucid-Dreaming Quantum Physics—a human world that is able to collectively embrace its power of open-ended lucid dreaming so as to dream into physical reality the many as yet unrealized yearnings that lie deep within the most sacred visionary chambers of the human heart.

SUMMARY OF KEY POINTS IN PART II

- Quantum physics is returning physics to its roots in metaphysics, thereby becoming a spiritual path.
- According to quantum physics, the quest for a single descriptive model of reality has to be given up.
- Quantum entities are continually being re-created anew, and hence have no continuous thread of identity from one moment to the next.
- In an example of the complementarity principle of quantum theory, the opposite of the above statement is equally true: The universe is never divided, for all division is only apparent division and everything is simply an expression of an indivisible wholeness.
- The apparently solid, self-existing forms of the world are in actuality self-perpetuating patterns within an endless flux.
- We can only recognize and experience the undivided wholeness of the quantum realm when we are in touch with the wholeness within ourselves, as the two are reflections of each other.
- The revelations of quantum physics radically change how we conceive of ourselves relative to the universe.
- Consciousness is inescapably playing a key role in physics as well as in the creation of the universe.
- Our ordinary, day-to-day universe is quantum through and through.
- Quantum physics is a revelation of the dreamlike nature of the universe.
- Encountering the quantum realm is like discovering the Holy Grail. Its revelations can be used for good or evil.

- It is our task to liberate and engage the power of quantum physics' revelations for the benefit of everyone. The best way to do this is to partake in the most radical, subversive, and powerful form of activism that there is—recognizing and then consciously engaging our own quantum nature.

• AFTERWORD •

How *The Quantum Revelation* Found the Light of Day

How this book came to be incarnated is a living example of the creative nature of the quantum in action. After a few rejections from various publishers and agents, I felt stuck and discouraged. The manuscript was literally sitting on my computer for a number of months without me doing anything in the way of trying to get it out. Getting a book published these days is quite challenging. My friends who had worked in publishing warned me that unless I spent months working on a book proposal (something I didn't want to do) and had all sorts of connections and a big-time agent (neither of which I had), my chances for getting the book picked up by a publisher weren't good. I decided to call a meeting with the dreaming community that has formed around my work for whoever was interested in helping me to "dream into" getting the book out. I remember feeling strong resistance and ambivalence about even going through with the meeting and I almost cancelled it.

However I decided to work through my resistance and go ahead with the meeting. Eight people from my various dreaming groups showed up. It was a very inspired, and inspiring meeting. Everyone there had one intention: to selflessly dream into and lend their support in helping to get the book out in whatever way they could. We threw lots of creative ideas around the room and imagined all sorts of possibilities. The dreaming field felt activated, alive, and filled with magic. The meeting ended with one of the participants spontaneously doing a short blessing, the energy of which was palpable. I think it is safe to say that all of us had the experience of something—energetically speaking—being set into motion as a result of the process.

Within days of the meeting, the process around the book began shifting dramatically. Within a week or so I had found an agent who felt like I was the next big thing and was eager to represent me. While I ultimately decided not to work with her, this agent added to and was an expression of the energy in the field that was emerging to manifest the publication of the book. Then I received a Facebook friend request from none other than the visionary best-selling author and teacher Jean Houston, whose work I've long admired. After accepting her request and becoming "friends," I asked her if she'd be willing to look at, so as to possibly endorse, my new manuscript on quantum physics. She happily agreed and a few weeks later sent me an enthusiastic email saying how much she loved the manuscript.

A little while later Jean called up her friend Kenzi Sugihara, who is the publisher and founder of SelectBooks, an independent publishing company in New York, and urged him to publish my book. As soon as he got off the phone with Jean, Kenzi picked up the phone and called me. As an author, you can imagine my surprise and delight to pick up the phone and have it be the head of a publishing company interested in publishing my work. Kenzi needed a week or so to look at the manuscript, and soon thereafter called me up and offered me a contract. He commented that it had never happened before that someone as esteemed as Jean Houston called him up out of the blue to strongly suggest he publish a book. Jean Houston soon thereafter kindly agreed to write the foreword.

I'm not sure how to "explain" what this experience has to do with quantum reality, but I, as well as everyone else at that meeting, have a strong intuitive sense that it was related. As I point out in the book, the quantum realm is not separate from our own mind but actually reflects back to us our beliefs, expectations, and intentions. When a group of people gets their intention into alignment with each other, magic can happen. Their shared intention gets imprinted into the very quantum field that in-forms and gives shape to our universe in a way that can really make a difference in the world.

That dreaming meeting was like a sacred ceremony in which everyone involved—whom I now refer to as "The Dreaming Eight" ("Nine," if I include myself)—put their energy together so as to clarify, express, and amplify to the universe the intention of getting the book published. The quantum universe, which just like a dream reflects the attitude of the dreamer, responded in kind. This is how *The Quantum Revelation* came to be.

ENDNOTES

1 What physicist John Archibald Wheeler refers to as "substance without substance." John Wheeler, *At Home in the Universe* (Woodbury, NY: The American Institute of Physics, 1994), 294.

2 Active imagination is the most powerful technique Jung ever encountered for metabolizing and assimilating the contents of the unconscious and hence, becoming conscious. Instead of passively watching the manifestations of the unconscious, in active imagination we fully engage with and participate in a creative dialogical relationship with its contents which facilitates their passage from an unconscious, potential state to a conscious, actual one. Notice the similarity to one of the central discoveries of quantum physics—through the act of interacting with quantum entities, we facilitate their transition from a state of potentiality to actuality.

3 For the purpose of this book, the terms quantum physics, quantum theory, and quantum mechanics are used interchangeably.

4 See Abraham Pais, *Niels Bohr's Times: In Physics, Philosophy, and Polity* (Oxford: Clarendon Press, 1991), 87.

5 To quote scientist George Wald, "There was a *monumental* generation of physicists in the first half of the [twentieth] century." Quoted in Denis Brian, *The Voice of Genius* (Cambridge, MA: Perseus Publishing, 1995), 138.

6 Pais, *Niels Bohr's Times*, 3.

7 J. S. Bell, *Speakable and Unspeakable in Quantum Mechanics* (Cambridge, UK: Cambridge University Press, 2004), 27.

8 Robert Oppenheimer, "Science and the Common Understanding," The Reith Lectures, BBC Radio 4, London, Dec 20, 1953.

9 Quoted in Nick Herbert, *Quantum Reality: Beyond the New Physics* (New York: Anchor Books, 1987), 15.

10 Richard Feynman, *The Character of Physical Law* (Cambridge, MA: MIT Press, 1995).

11 This is reminiscent of the alchemists who, in writing about *Mercurius*, the lapis, the philosopher's stone (all different ways of describing their "God-image"), in Jung's words, "did not know what they were writing about."

12 Quoted in L. Wolpert, *The Unnatural Nature of Science* (Cambridge, MA: Harvard University Press, 1993), 144.

13 Quoted in "A Quantum Sampler," *New York Times*, December 26, 2005.
14 Timothy Ferris, *The Whole Shebang* (New York: Simon & Schuster, 1997), 266.
15 Leon M. Lederman and Christopher T. Hill, *Quantum Physics for Poets* (Amherst, NY: Prometheus Books, 2011), 14.
16 Arnold Mindell, *Earth-Based Psychology* (Portland: Lao Tse Press, 1997), 6.
17 Ibid.
18 From a letter to Heinrich Zangger, quoted in Abraham Pais, *Subtle is the Lord: The Science and the Life of Albert Einstein* (Oxford: Clarendon, 1982), 399.
19 Quoted in Bruce Rosenblum and Fred Kuttner, *Quantum Enigma: Physics Encounters Consciousness* (New York: Oxford University Press, 2006), 81.
20 Albert Einstein, *Out of My Later Years* (New York: Philosophical Library, 1950), 229.
21 Henry P. Stapp, *Mindful Universe: Quantum Mechanics and the Participating Observer* (Berlin: Springer-Verlag, 2007), 4.
22 Murray Gell-Mann, "Questions for the Future," in *The Nature of Matter: Wolfson College Lectures*, ed. J. H. Mulvey (Oxford: Clarendon Press/ Oxford University Press, 1981), 170.
23 Werner Heisenberg, *Physics and Philosophy: The Revolution in Modern Science* (Amherst, NY: Prometheus Books, 1999), 102.
24 Richard Feynman, *The Quotable Feynman*, (Princeton, NJ: Princeton University Press, 2015), 76.
25 Remark made during "Mind and Matter," Oxford University seminar, 1991. Quoted in Danah Zohar and Ian Marshall, *The Quantum Society: Mind, Physics, and a New Social Vision* (New York: William Morrow, 1994), 40.
26 From a letter to Einstein's colleague Johann Laub. Quoted in John Wheeler, *Geons, Black Holes, and Quantum Foam* (New York: Norton & Company, 1998), 185.
27 Quoted in Jeremy Bernstein, *Quantum Profiles* (Princeton, NJ: Princeton University Press, 1991), vii.
28 Quoted in Dennis Overbye, "John A. Wheeler, Physicist Who Coined the Term 'Black Hole,' Is Dead at 96," *New York Times*, April 14, 2008.
29 Albert Einstein, *The Ultimate Quotable Einstein*, edited and collected by Alice Calaprice (Princeton, NJ: Princeton University Press, 2011), 157.
30 Wheeler, *At Home in the Universe*, 112.
31 John Wheeler, "Beyond the Black Hole," in *Some Strangeness in the Proportion*, ed. Harry Woolf (Reading, MA: Addison-Wesley, 1980), 376.
32 Freeman Dyson, "Preface," in *Science and Ultimate Reality: Quantum Theory, Cosmology,and Complexity*, eds. John D. Barrow, Paul C. W. Davies, and Charles L. Harper (Cambridge: Cambridge University Press, 2005), xviii.
33 Paul C. W. Davies et al., "John Archibald Wheeler and the Clash of Ideas," in *Science and Ultimate Reality: Quantum Theory, Cosmology, and Complexity*, 23.
34 John Wheeler, "From the Big Bang to the Big Crunch," interview by Mirjana R. Gearhart, *Cosmic Search Magazine*, vol. 1, no. 4, accessed October 10, 2017, www.bigear.org/vol1no4/wheeler.htm.
35 Heisenberg. *Physics and Philosophy: The Revolution in Modern Science*, 58.
36 This is the first of a handful of Wheeler quotes that are from his personal journals and notebooks. They are from Amanda Gefter's book, *Trespassing on Einstein's Lawn* (New York: Bantam Books, 2014). In her book she describes how she made a pilgrimage to the American Philosophical Society in Philadelphia where Wheeler's journals are kept. I am grateful for her efforts.

37 Jack Saffarti, "Wheeler's World: It from Bit?" in *Developments in Quantum Physics*, eds. Frank Columbus and Volodymyr Krasnoholovets (New York: Nova Science Publishers, 2004), 42.
38 To quote Einstein, "Curiosity has its own reason for existing. . . . Never lose a holy curiosity." From *The Expanded Quotable Einstein*, collected and edited by Alice Calaprice (Princeton, NJ: Princeton University Press, 2000), 281.
39 Quoted in Cheuk-Yin Wong, "Remembering John Wheeler, Physicist and Teacher," *Princeton Alumni Weekly*, July 16, 2008, http://paw.princeton.edu/article/remembering-john-wheeler-physicist-and-teacher.
40 John Wheeler, "Information, Physics, Quantum: The Search for Links," *Proceedings of the Third International Symposium on Foundations of Quantum Mechanics in the Light of New Technology* (Tokyo: Physical Society of Japan, 1990).
41 Werner Heisenberg, *Across the Frontiers* (Woodbridge, CT: Ox Bow Press, 1990), 227.
42 Jagdish Mehra, ed., *The Physicist's Conception of Nature* (Boston: D. Reidel, 1973), 244.
43 Wheeler, *Geons, Black Holes, and Quantum Foam*, 334.
44 C. G. Jung, *The Structure and Dynamics of the Psyche*, CW 8 (Princeton, NJ: Princeton University Press, 1969), para. 438.
45 Quoted in Wolfgang Hofkirchner, *The Quest for a Unified Theory of Information: Proceedings of the Second International Conference on the Foundations of Information Science* (Oxford: Taylor and Francis, 1999).
46 Arthur Zajonc, ed., *The New Physics and Cosmology: Dialogues with the Dalai Lama* (New York: Oxford University Press, 2004), 147.
47 To quote Walter Heitler, "The separation of the world into an 'objective outside reality' and 'us,' the self-conscious onlookers, can no longer be maintained. Object and subject have become inseparable from each other." Heitler, "The Departure from Classical Thought in Modern Physics," in *Albert Einstein, Philosopher-Scientist*, ed. P. A. Schlipp (Chicago: Open Court, 1970), 196.
48 Niels Bohr, *Atomic Physics and Human Knowledge* (Mineola, NY: Dover Publications, 2010), 81.
49 Paul Buckley and F. David Peat, *A Question of Physics: Conversations in Physics and Biology* (London: Routledge & Kegan Paul, 1979), 54.
50 Quoted in Michael Brooks, "The Second Quantum Revolution," *New Scientist*, issue 2609 (June 23, 2007). 30–33.
51 Wheeler, *At Home in the Universe*, 126.
52 Buckley and Peat, *A Question of Physics: Conversations in Physics and Biology*, 55.
53 Wheeler, *At Home in the Universe*, 75.
54 Richard Feynman, *Feynman Lectures on Physics*, vol. 3, (Reading, MA: Pearson/Addison-Wesley, 1963), 1.
55 Speaking of the double-slit experiment: "It has been a touchstone for discourse about the implications of quantum theory for three-quarters of a century," in Wheeler, *Geons, Black Holes, and Quantum Foam*, 334.
56 Feynman, *Feynman Lectures on Physics*, vol. 3, 1.
57 We could set up many identical versions of this experiment all over the world and emit just one subatomic particle through the slits in each of them at a prearranged moment. If we add together the results from all of these completely different experiments, we would find that the net result would look like the wave interference pattern. See John D. Barrow, *The World within the World* (New York: Oxford University Press, 1988), 133–136.

58 Wheeler, *Geons, Black Holes, and Quantum Foam*, 335.
59 Massimo Teodorani, *Entanglement: l'intrication quantique, des particules à la conscience* (Cesena: Macro Editions, 2007), 9.
60 Niels Bohr, "The Atomic Theory and the Fundamental Principles Underlying the Description of Nature," in *Atomic Theory and Description of Nature: Four Essays* (Cambridge: Cambridge University Press, 1961), 119.
61 Reminiscent of Wheeler's delayed-choice experiment, an observation made after the fact appears to reach backwards in time and change the experiment. We can turn on the detector at the slits after the photon, behaving like a spread out wave, has emerged from the slits but before it hits the photographic plate. Common sense indicates that it would be too late for the photon to suddenly decide to behave like a localized particle that has only passed through one of the slits, but this is exactly what happens. In such experiments, the interference pattern disappears.
62 A very creative, innovative, and unprecedented set of double-slit experiments have been carried out by Dean Radin (chief scientist at the Institute for Noetic Science) to investigate whether the effects of the experiment can be influenced by conscious intention. In these carefully designed and rigorously controlled series of experiments, no one was directly looking at the double-slit apparatus, which was sealed in a closed box. The only interaction was through a directed intention/visualization. The experiments repeatedly resulted in statistically significant results, which indicate that not just visually observing, but merely wielding our attention/intention in particular ways seems to have material effects. Here are two published papers in the physics journal *Physics Essays* describing the first nine of the seventeen experiments of this type that have been conducted. The subsequent experiments not mentioned in these two papers involve more careful controls and variations in experimental design to try to better understand the nature of this remarkable effect. See "Consciousness and the Double-slit Interference Pattern: Six Experiments," http://www.deanradin.com/evidence/Radin2012doubleslit.pdf, and "Psychophysical Interactions with a Double-slit Interference Pattern," http://www.deanradin.com/evidence/RadinPhysicsEssays2013.pdf.
63 Wheeler, "Beyond the Black Hole," in *Some Strangeness in the Proportion*, 359.
64 In a bit of quantum strangeness, if we ask whether the universe really existed before we started looking at it, the answer we get from the universe is that it *looks* as if it existed before we started looking at it.
65 John Wheeler and Wojciech Hubert Zurek, *Quantum Theory and Measurement* (Princeton, NJ: Princeton University Press, 1983), 169.
66 William James, *The Principles of Psychology* (New York: Dover, 1950), 290-291.
67 Wheeler, "Information, Physics, Quantum: The Search for Links."
68 Bohr, *Atomic Physics and Human Knowledge*, 92–93.
69 Brian Josephson, "Physics and Spirituality: the Next Grand Unification?" *Physics Education*, vol. 22, no. 1 (1987): 15–19.
70 Wheeler, *At Home in the Universe*, 226.
71 Albert Einstein and Leopold Infeld, *The Evolution of Physics* (New York: Simon & Schuster, 1966), 294.
72 Heisenberg, *Across the Frontiers*, 154.
73 F. David Peat, *Infinite Potential: The Life and Times of David Bohm* (New York: Basic Books, 1997), 205.

74 Richard Feynman, *QED: The Strange Theory of Light and Matter* (Princeton, NJ: Princeton University Press, 2006), 149.
75 Heisenberg, *Across the Frontiers*, 34.
76 Quoted in Herbert, *Quantum Reality*, 17.
77 Wheeler, "Information, Physics, Quantum: The Search for Links."
78 Henry P. Stapp, "Quantum Reality and Mind," in *Quantum Physics of Consciousness*, eds. Subhash Kak, Sir Roger Penrose, and Stuart Hameroff (Cambridge: Cosmology Science Publishers, 2011), 19.
79 Quoted in Gefter, *Trespassing on Einstein's Lawn*, 283.
80 Quoted in Sheila Jones, *The Quantum Ten* (New York: Oxford University Press, 2008), 269.
81 C. G. Jung, *The Symbolic Life*, CW 18 (Princeton, NJ: Princeton University Press, 1976), para. 826.
82 John von Neumann, *Mathematical Foundations of Quantum Mechanics* (Princeton, NJ: Princeton University Press, 1955), 420.
83 Wojciech Hubert Zurek, Alwyn van der Merwe, and Warner Allen Miller, eds., *Between Quantum and Cosmos: Studies and Essays in Honor of John Archibald Wheeler* (Princeton, NJ: Princeton University Press, 1988), 11.
84 Anton Zeilinger, "Why the Quantum? 'It' from 'Bit'? A Participatory Universe? Three Far-reaching Challenges from John Archibald Wheeler and Their Relation to Experiment," in *Science and Ultimate Reality: Quantum Theory, Cosmology, and Complexity*, eds. Barrow, et al., 201.
85 Heisenberg, *Physics and Philosophy: The Revolution in Modern Science*, 129.
86 Bernard d'Espagnat, "The Quantum Theory and Reality," *Scientific American*, vol. 241, no. 5 (November, 1979).
87 Zurek et al., eds., *Between Quantum and Cosmos: Studies and Essays in Honor of John Archibald Wheeler*, 414.
88 Albert Einstein, "Maxwell's Influence on the Evolution of the Idea of Physical Reality," in *James Clerk Maxwell: A Commemorative Volume* (Cambridge, UK, Cambridge University Press, 1931).
89 Wheeler, *At Home in the Universe*, 187–188.
90 Pais, *Subtle Is the Lord: The Science and the Life of Albert Einstein*, 9.
91 Letter from Einstein to D. Lipkin. Quoted in Arthur Fine, *The Shaky Game: Realism and the Quantum Theory* (Chicago: University of Chicago Press, 1996), 1.
92 Quoted in "A Quantum Sampler," *New York Times*, December 26, 2005, http://www.nytimes.com/2005/12/26/science/a-quantum-sampler.html.
93 In an example of active imagination, even after Einstein's death Bohr would argue with him as though he were still alive. Wheeler characterized their debates in this life as "friendly but deadly serious." Wheeler, *At Home in the Universe*, 114.
94 Niels Bohr, "Can Quantum Mechanical Description of Physical Reality Be Considered Complete?" in *Physical Review* 48 (1935), 696–702.
95 N. David Mermin, "Is the Moon There When Nobody Looks? Reality and the Quantum Theory," *Physics Today* (April 1985), 38–47.
96 Buckley and Peat, *A Question of Physics: Conversations in Physics and Biology*, 54.
97 Bernstein, *Quantum Profiles*, 89.
98 Wheeler, *Geons, Black Holes, and Quantum Foam*, 329.
99 Ibid., 262.
100 Ibid., 119.

101 Ibid., 182.
102 Ibid., 263.
103 Ibid., 148. Wheeler completes his thought by writing, "Yes, there is happiness to be found in the mere contemplation of the deepest mysteries."
104 Ibid., 153.
105 Ibid., 173.
106 Ibid., 223.
107 Quoted in Fritjof Capra, *Uncommon Wisdom: Conversations with Remarkable People* (New York: Bantam Books, 1989), 21.
108 Feynman, *The Quotable Feynman*, 75.
109 The word "psychic" is used as the adjective form of "psyche" and not with any parapsychological connotations.
110 Philip K. Dick, *The Selected Letters of Philip K. Dick: 1980–1982,* ed. Don Herron and Russell Galen (Nevada City, CA: Underwood Books, 2009), 146.
111 Pamela Jackson and Jonathan Lethem, eds., *The Exegesis of Philip K. Dick* (New York: Houghton Mifflin Harcourt, 2011), 294.
112 David Bohm, *Wholeness and the Implicate Order* (London: Routledge, 2007), 29.
113 David Bohm, *On Dialogue* (New York: Routledge, 2007), 58.
114 Richard Feynman, "What is Science?" National Science Teachers Association fourteenth convention lecture, April 1966.
115 Newton himself had a worldview that was far more spiritually oriented than the "Newtonian" worldview that is now attributed to him.
116 Quoted in Lothar Schafer, *Infinite Potential: What Quantum Physics Reveals About How We Should Live* (New York: Deepak Chopra Books, 2013), 127.
117 Ibid., 74.
118 Amit Goswami, *The Everything Answer Book: How Quantum Science Explains Love, Death, and the Meaning of Life* (Charlottesville, VA: Hampton Roads, 2017), 110.
119 Denis Postle, *Fabric of the Universe* (New York: Crown Publishers, 1976), 12.
120 Bernardo Kastrup, *Why Materialism is Baloney* (Winchester, UK: Iff Books, 2014), 9.
121 Quoted in Trish Pfeiffer and John Mack, eds., *Mind before Matter: Visions of a New Science of Consciousness* (Winchester, UK: O Books, 2007), 2.
122 Jung, *The Structure and Dynamics of the Psyche,* CW 8, para. 529.
123 This collapsing of the boundary between subject and object has myriad expressions in different forms of art, for example when the camera becomes part of the photograph or the movie, the author's reflections upon his or her writings become part of the novel, or the painter's own activities self-reflectively show up in the painting.
124 Every worldview, however, has a potential pathological expression. Scientific materialism, with its view that the world objectively exists separate from the mind that observes it, is clearly a perspective that is out of (and not connected to) its mind. The postmodern, integral-aperspectival approach is not without its own aberration, known as "aperspectival madness." This is the insane—and maddening—view that no one perspective is better than the other (except its own).
125 C. G. Jung, *Civilization in Transition*, CW 10 (Princeton, NJ: Princeton University Press, 1970), para. 315.
126 Albert Einstein, *Ideas and Opinions* (New York: Three Rivers Press, 1995).
127 Quoted in Herbert, *Quantum Reality*, 4.

128 Letter from Pauli to Markus Fierz, written on August 12, 1948. Quoted in Suzanne Gieser, *The Innermost Kernel: Depth Psychology and Quantum Physics. Wolfgang Pauli's Dialogue with C. G. Jung* (Berlin: Springer-Verlag, 2006).
129 Quoted in Schafer, *Infinite Potential: What Quantum Physics Reveals About How We Should Live*, 13.
130 Wheeler, *At Home in the Universe*, 126.
131 Quoted in Eldon Taylor, *What Does that Mean? Exploring Mind, Meaning, and Mysteries* (New York: Hay House, 2010).
132 Jackson and Lethem, *The Exegesis of Philip K. Dick*, 482–483.
133 This is also the message of the wisdom teaching in Tibetan Buddhism popularly known as *The Tibetan Book of the Dead*.
134 C. G. Jung, *Psychological Types*, CW 6 (Princeton, NJ: Princeton University Press, 1971), para. 60.
135 Wheeler and Zurek, *Quantum Theory and Measurement*, 210.
136 Quoted in Wheeler, "Information, Physics, Quantum: The Search for Links."
137 Bernstein, *Quantum Profiles*, 138.
138 John Wheeler, *Frontiers of Time* (Amsterdam: North-Holland Publishing Co., 1979), 5.
139 Wheeler, *At Home in the Universe*, 43.
140 Ilya Prigogine, and Isabelle Stengers, *Order Out of Chaos* (New York: Bantam Books, 1984), 293.
141 Wheeler, *At Home in the Universe*, 45.
142 Quoted in Rosenblum and Kuttner, *Quantum Enigma: Physics Encounters Consciousness*, 124.
143 Wheeler, *Cosmic Search Magazine*, vol. 1, no. 4.
144 Wheeler and Zurek, *Quantum Theory and Measurement*, 201.
145 Jackson and Lethem, *The Exegesis of Philip K. Dick*, 75.
146 David Bohm, *Quantum Theory* (New York: Dover Publications, 1979).
147 Quoted in Herbert, *Quantum Reality*, 29.
148 Buckley and Peat, *A Question of Physics: Conversations in Physics and Biology*, 58.
149 Wheeler, *Geons, Black Holes, and Quantum Foam.*
150 Wheeler, *At Home in the Universe*, 15.
151 Jack Sarfatti, "Wheeler's World: It from Bit?" in *Developments in Quantum Physics*, eds. Frank Columbus and Volodymyr Krasnoholovets (New York: Nova Science Publishers, 2004), 42.
152 Quoted in Heisenberg, *Across the Frontiers*, 182.
153 Ibid., 183.
154 Ibid., 179.
155 Ibid., 171.
156 Buckley and Peat, *A Question of Physics: Conversations in Physics and Biology*, 59.
157 Wheeler, *Geons, Black Holes, and Quantum Foam*, 103.
158 Davies, "John Archibald Wheeler and the Clash of Ideas," in *Science and Ultimate Reality: Quantum Theory, Cosmology, and Complexity*, eds. Barrow, et al. 3.
159 Wheeler, *At Home in the Universe*, 9.
160 C. G. Jung, *Psychology and Alchemy*, CW 12 (Princeton, NJ: Princeton University Press, 1968), para. 18.
161 Jung refers to this as the "transcendent function" or "reconciling symbol," whose function is to transcend and reconcile the seemingly contradictory and paradoxical opposites into a higher unity. In describing the reconciling symbol, Jung writes, "In it all paradoxes are abolished." Jung, *Letters*, vol. 1, 61.

162 Zurek et al., eds., *Between Quantum and Cosmos: Studies and Essays in Honor of John Archibald Wheeler*, 12.
163 Ibid., 10.
164 Quoted in Walter Isaacson, *Einstein: His Life and Universe* (New York: Simon & Schuster, 2007), 347.
165 Leon Lederman, introducing Carlo Rubbia and congratulating him on just having won the Nobel Prize in physics, Santa Fe, New Mexico, November 3, 1984. Quoted in Timothy Ferris, *The Whole Shebang*.
166 Sean Carroll, *The Particle at the End of the Universe* (New York: Plume, 2013), 33.
167 Quoted in Sonu Shamdasani, *C. G. Jung: A Biography in Books* (New York: W. W. Norton & Co., 2012), 58–59.
168 Wheeler, *Geons, Black Holes, and Quantum Foam*, 227.
169 Lawrence Sutin, ed., *The Shifting Realities of Philip K. Dick* (New York: Pantheon Books, 1995), 234.
170 Jung, *The Structure and Dynamics of the Psyche*, CW 8, para. 747.
171 Sutin, *The Shifting Realities of Philip K. Dick*, 233.
172 Ibid., 234.
173 Richard Feynman, "The Uncertainty of Science," John Danz Lecture Series, 1963.
174 Wheeler, *At Home in the Universe*, 18.
175 Dick, *The Selected Letters of Philip K. Dick: 1980 – 1982*, 153.
176 In Buddhist philosophy "mind-stream" is the moment-to-moment continuum of sense impressions and mental phenomena/awareness.
177 Sutin, *The Shifting Realities of Philip K. Dick*, 233.
178 In an interview, Wheeler repeats this exact phrase but replaces the word "blind" with "stupid," (Brian, *The Voice of Genius*, 135); Wheeler, *At Home in the Universe*, 310.
179 John Wheeler, *A Journey into Gravity and Spacetime* (New York: Scientific American Library, 1990), 15.
180 P. C. W. Davies and J. R. Brown, *The Ghost in the Atom* (New York: Cambridge University Press, 1986), 69.
181 F. C. Happold, *Mysticism* (Baltimore: Penguin Books, 1963), p. 28.
182 Freeman Dyson, *Dreams of Earth and Sky* (New York: New York Review Books, 2015), 192.
183 Albert Einstein, *The World as I See It* (New York: Citadel Press, 1984).
184 Feynman, *The Feynman Lectures on Physics*, vol. 3.
185 Wheeler, "Information, Physics, Quantum: The Search for Links."
186 C. F. von Weizsäcker, "Introduction," in *The Biological Basis of Religion and Genius*, Gopi Krishna (New York: Harper and Row, 1972), 35–36.
187 Feynman, *The Quotable Feynman*, 211.
188 Ibid., 87.
189 This makes me think of how in Tibetan Buddhism the practitioner is instructed to visualize him or herself—as well as everyone else—in the enlightened form, and it is pointed out that this isn't a mere fabrication, but is actually a way of getting in touch with our true nature; i.e., the visualization is picturing the true state of our situation.
190 C. G. Jung, *Letters*, vol. 1 (London: Routledge & Kegan Paul, 1973), 60.
191 Wheeler, *Geons, Black Holes, and Quantum Foam*, 84.
192 Ibid.

193 John Dewey, *The Quest for Certainty* (New York: Putnam, 1960).
194 Davies, "John Archibald Wheeler and the Clash of Ideas," in *Science and Ultimate Reality: Quantum Theory, Cosmology, and Complexity*, 6.
195 Wheeler, *Frontiers of Time*, 1.
196 Buckley and Peat, *A Question of Physics: Conversations in Physics and Biology*, 59.
197 Wheeler, "Beyond the Black Hole," in *Some Strangeness in the Proportion*, 350.
198 Zurek et al., eds., *Between Quantum and Cosmos: Studies and Essays in Honor of John Archibald Wheeler*, 9
199 Wheeler, *Frontiers of Time*, 20.
200 Wheeler, *At Home in the Universe*, 35.
201 Helen Dukas and Banesh Hoffman, eds., *Albert Einstein, the Human Side: New Glimpses from His Archives* (Princeton, NJ: Princeton University Press, 1981), 33.
202 Wheeler, *Frontiers of Time*, 7.
203 Wheeler, *At Home in the Universe*, 26.
204 Davies, "John Archibald Wheeler and the Clash of Ideas," in *Science and Ultimate Reality: Quantum Theory, Cosmology, and Complexity*, 10.
205 Wheeler, *Cosmic Search Magazine*, vol. 1, no. 4.
206 Erwin Schrödinger, "The Present Situation in Quantum Mechanics," in *Quantum Theory and Measurement*, eds. Wheeler and Zurek.
207 Wheeler, *Geons, Black Holes, and Quantum Foam*, 339.
208 Wheeler, *At Home in the Universe*, 45.
209 I have coined a term, "Aparticipatory Delusional Syndrome" (ADS for short), which is based on the deluded assumption that we are separate from and not participating in calling forth the very situation in the outside world to which we are reacting. ADS is the primary, underlying psychological "dynamic" or "engine" that fuels the "malignant" aspect of wetiko. The extent to which we feel ourselves the victim of circumstances and don't realize our complicity in what is playing out in our lives is the extent to which we have fallen prey to ADS. ADS effectively immobilizes and renders inoperative our ability to self-reflect, as it relates to the world through the fixed and nonnegotiable lens of assumptions that the world "object"-ively exists, independent of ourselves. When we are stricken with ADS, we react to our perceptions and interpretations as if they exist inherently and independently in the objects of the world, rather than realizing that they are automatic "reflex-ions" of the way we are looking, and are thus always revealing the subject (ourselves).
210 Wheeler, *At Home in the Universe*, 120.
211 Rosenblum and Kuttner, *Quantum Enigma: Physics Encounters Consciousness*, 201.
212 Freeman Dyson, "Comment on the Topic 'Beyond the Black Hole,'" in *Some Strangeness in the Proportion*, 380.
213 Quoted in John Horgan's book, *The End of Science: Facing the Limits of Knowledge in the Twilight of the Scientific Age* (Reading, MA: Helix Books, 1996), 88.
214 Albert Einstein, *The Ultimate Quotable Einstein*, collected and edited by Alice Calaprice (Princeton: Princeton University Press, 2013), 379.
215 Heisenberg, *Across the Frontiers*, 46.
216 Ibid., 53.
217 Quoted in Carl L. Jech, *Religion as Art Form: Reclaiming Spirituality without Supernatural Beliefs* (Eugene, OR: Resource Publications, 2013), 214.
218 Buckley and Peat, *A Question of Physics: Conversations in Physics and Biology*, 58.

219 Please see the section "Four-valued Logic" in my book *Dispelling Wetiko,* 40-44.

220 Four-valued logic was introduced in the second century CE by Nagarjuna, who is considered to be the most important Buddhist philosopher after the Buddha himself and who is one of the most original and influential thinkers in the history of Indian philosophy. Nagarjuna stated emphatically that the unnecessary and artificially distorting limitations of two-valued, either/or logic are the greatest cause of human suffering.

221 In reminiscing about his inspiring walks with Nobel Prize winner Hideki Yukawa at Princeton, Wheeler writes, "Never did I appreciate more than there the difference between the sharp 'this is true, that is not true' of Western thought and the 'maybe yes, maybe no' [four-valued logic] that Yukawa lived and breathed." Quoted in Wheeler, *At Home in the Universe,* 197.

222 Jung compared two-valued logic to four-valued logic by referring to them as "the niggardly either/or" and "the glorious both/and." Quoted in Tim Freke, *The Mystery Experience: A Revolutionary Approach to Spiritual Awakening* (London: Watkins Publishing, 2012), 43.

223 Abraham Pais, *The Genius of Science* (Oxford: Oxford University Press, 2000).

224 Richard Conn Henry, "The Mental Universe," *Nature,* vol. 436 (July 7, 2005): 29.

225 Ibid.

226 Ibid.

227 Brian, *The Voice of Genius,* 73.

228 Zurek et al., eds., *Between Quantum and Cosmos: Studies and Essays in Honor of John Archibald Wheeler,* 11.

229 Please see my discussion in the "Alchemy" section in this book on what is known as an "increatum."

230 Andrei Linde, "Inflation, Quantum Cosmology, and the Anthropic Principle," in *Science and Ultimate Reality: Quantum Theory, Cosmology, and Complexity,* eds. Barrow et al.

231 Wheeler, *Geons, Black Holes, and Quantum Foam,* 338.

232 This is reminiscent of Meister Eckhart's words, "God is a word that speaks itself."

233 Quoted in Owen Gingerich, *The Nature of Scientific Discovery: A Symposium Commemorating the 500th Anniversary of the Birth of Nicolaus Copernicus* (Washington, DC: Smithsonian Institution Press, 1975).

234 Wheeler's more precise definition was "a self-referential deductive axiomatic system." Quoted in Wheeler, *At Home in the Universe,* 300.

235 Wheeler, "Information, Physics, Quantum: The Search for Links."

236 Wheeler, *Geons, Black Holes, and Quantum Foam,* 338.

237 Wheeler, *At Home in the Universe,* 344, fn. #75.

238 Stephen Hawking, ed., *A Stubbornly Persistent Illusion: The Essential Scientific Works of Albert Einstein* (Philadelphia, PA: Running Press, 2009).

239 Einstein, although he intuited the illusory nature of time, was himself seemingly not able to break free from this ingrained assumption of clock-based time. The "time" that he refers to in both his special and general theories of relativity (the "time" part of his four-dimensional "space-time" continuum) can only be defined with reference to clocks. Perhaps this is partially why he was so uneasy with quantum physics, which is, in its essence, a timeless theory.

240 Wheeler, *At Home in the Universe*, 184.
241 This discussion is related to the shamanic perspective, which considers that the shaman "journeys," in the present moment, both backward and/or forward in time so as to effect changes in the past and/or future, thereby changing the present circumstance.
242 Wheeler, *At Home in the Universe*, 42.
243 Quoted in F. David Peat, *Synchronicity: The Bridge Between Matter and Mind* (New York: Bantam Books, 1987).
244 Wheeler, *At Home in the Universe*, 301.
245 N. David Mermin, *Boojums All the Way Through: Communicating Science in a Prosaic Age* (Cambridge: Cambridge Univesity Press, 1990), 111.
246 Wheeler, *Geons, Black Holes, and Quantum Foam*, 345.
247 Wheeler, *Frontiers of Time*, 31.
248 C. G. Jung, *Children's Dreams: Notes from the Seminar Given in 1936-1940* (Princeton, NJ: Princeton University Press, 2008), 360.
249 Louis de Broglie, "The Scientific Work of Albert Einstein," in *Albert Einstein: Philosopher-Scientist*, ed. Paul Arthur Schlipp (New York: MJF Books, 1970), 114.
250 Wheeler, *Geons, Black Holes, and Quantum Foam*, 346.
251 Ibid., 347.
252 Wheeler, *Frontiers of Time*, 74.
253 Ibid., 11.
254 Wheeler, *At Home in the Universe, 124.*
255 Ibid., 114.
256 Quoted in Anthony Peake, *The Labyrinth of Time: The Illusion of Past, Present, and Future* (London: Arcturus Publishing, 2012), 109.
257 Quoted in Rosenblum and Kuttner, *Quantum Enigma: Physics Encounters Consciousness*, 193.
258 Wheeler, *At Home in the Universe*, 126.
259 Ibid., 42.
260 What light is made of; from the Greek word *photos*, light. Also called "light quantum."
261 Wheeler, *At Home in the Universe*, 64.
262 Interestingly, Pauli describes the wave function as a "reconciling symbol." Like any symbol, it unites the opposites of, in this case, our conceptions of continuity and discontinuity characterized by particles and waves.
263 Stephen Hawking and Leonard Mlodinow, *The Grand Design* (New York: Bantam Books, 2010), 140.
264 Ibid., 82.
265 Wheeler, *Geons, Black Holes, and Quantum Foam*, 307.
266 Wheeler, *At Home in the Universe*, 124.
267 Graham Smetham, *Quantum Buddhist Wonders of the Universe* (Brighton, UK: Shunyata Press, 2012), 155.
268 Wheeler, *At Home in the Universe*, 26.
269 Wheeler, *Frontiers of Time*, 20.
270 Quoted in Vlatko Vedral, *Decoding Reality: The Universe as Quantum Information* (Oxford: Oxford University Press, 2010).
271 Wheeler and Zurek, *Quantum Theory and Measurement*, 189.
272 Banesh Hoffmann, *The Strange Story of the Quantum* (New York: Dover Publications, 1959), 152.

273 Wheeler, *At Home in the Universe*, 39.
274 Ibid., 39.
275 Ibid., 42–44.
276 Jackson and Lethem, *The Exegesis of Philip K. Dick*, 718.
277 The words of Christ on the cross come to mind: "Forgive them Father, for they know not what they do."
278 Wheeler and Zurek, *Quantum Theory and Measurement*, 200.
279 Ibid.
280 Bernstein, *Quantum Profiles*, 138.
281 Wheeler, *At Home in the Universe*, 115.
282 Note the similarity to Jung's description of a synchronistic phenomenon as "an act of creation in time."
283 Sir James Jeans, *The Mysterious Universe* (New York: The Macmillan Company, 1930), 148.
284 Wheeler, *At Home in the Universe*, 130.
285 Wheeler and Zurek, *Quantum Theory and Measurement*, 199.
286 Ibid., 196–7.
287 Ibid., 199.
288 Wheeler, *At Home in the Universe*, 123.
289 Ibid.
290 Ibid., 23.
291 Wheeler and Zurek, *Quantum Theory and Measurement*, 197.
292 Quoted in Karen Schroeder Sorenson, *Cosmos and the Rhetoric of Popular Science* (Lanham, MD: Lexington Books, 2017).
293 Wheeler, *At Home in the Universe*, 128.
294 Quoted in Colin Wilson, *The Outsider* (New York: J. P. Tarcher, 1982).
295 C. G. Jung, *Memories, Dreams, Reflections* (New York: Pantheon Books, 1973), 338.
296 Wheeler and Zurek, *Quantum Theory and Measurement*, 227.
297 Jung, *Memories, Dreams, Reflections*, 256.
298 Buckley and Peat, *A Question of Physics: Conversations in Physics and Biology*, 58.
299 Wheeler, *Geons, Black Holes, and Quantum Foam*, 337.
300 Quantum theory was developed early in the twentieth century to explain the "mechanics" governing the behavior of atoms.
301 Richard Feynman and Franck Wilczek, *The Character of Physical Law* (Cambridge, MA: MIT Press, 2017).
302 To quote Heisenberg, "Atoms are not things."
303 When asked, "Where can the electron be said to be?" Bohr's answer was: "To be? What does it mean, to be?" Wheeler comments, "The words 'to be' are tricky words when you get down to it in nuclear physics." Quoted in Brian, *The Voice of Genius*, 127.
304 Notice the similarity to Jung's idea of an archetype, which can never be seen directly. We can only build up a picture of an archetype based on its effects.
305 Zurek et al., eds., *Between Quantum and Cosmos: Studies and Essays in Honor of John Archibald Wheeler*, 11.
306 Wheeler, *Geons, Black Holes, and Quantum Foam*, 330.
307 As G. K. Chesterton put it: "Art is limitation; the essence of every picture is the frame." G. K. Chesterton, *The Collected Works of G. K. Chesterton, Volume 1* (San Francisco: Ignatius Press, 1994).
308 Brian, *The Voice of Genius*, 134.

309 Buckley and Peat, *A Question of Physics: Conversations in Physics and Biology*, 60.
310 Pais, *Subtle Is the Lord: The Science and the Life of Albert Einstein*, 9.
311 Quoted in Wheeler and Zurek, *Quantum Theory and Measurement*, 185.
312 Quoted in Zeilinger, "Why the Quantum?" in *Science and Ultimate Reality: Quantum Theory, Cosmology, and Complexity*, 218.
313 Heisenberg, *Physics and Philosophy: The Revolution in Modern Science.*
314 Quoted by Deepak Chopra in his foreword to Schafer, *Infinite Potential: What Quantum Physics Reveals About How We Should Live Life*, xiii.
315 Heisenberg, *Physics and Philosophy: The Revolution in Modern Science*, 186.
316 C. A. Meier, ed., *Atom and Archetype: The Pauli/Jung Letters 1932-1958* (Princeton: Princeton University Press, 2001), 87.
317 Ibid., 41.
318 Henry P. Stapp, *Mind, Matter and Quantum Mechanics* (Berlin Heidelberg: Springer-Verlag, 1993), 221.
319 Ibid.
320 Quoted in Arnold Mindell, *Quantum Mind: The Edge Between Physics and Psychology* (Portland, OR: Lao Tse Press, 2000), 19.
321 Meier, *Atom and Archetype: The Pauli/Jung Letters 1932-1958*, 95.
322 His Holiness the Dalai Lama, *The Universe in a Single Atom: The Convergence of Science and Spirituality* (New York: Morgan Road Books, 2005), 64.
323 Quoted in Robert Scholnick, ed., *American Literature and Science* (Lexington, KY: University Press of Kentucky, 2015).
324 Quoted in Paul Davies, *God and the New Physics* (New York: Simon and Schuster, 1983), 144.
325 Nagarjuna was considered to be the most important Buddhist philosopher after the Buddha himself and is one of the most original and influential thinkers in the history of Indian philosophy. To quote His Holiness the Dalai Lama, "About 15-20 years ago at some meeting, the Indian physicist Raja Ramanna told me that he had been reading Nagarjuna and that he'd been amazed to find that much of what he had to say corresponded to what he understood of quantum physics. A year ago at Presidency College in Kolkata the Vice-Chancellor Prof S Bhattacharya mentioned that according to quantum physics nothing exists objectively, which again struck me as corresponding to Chittamatrin and Madhyamaka views, particularly Nagarjuna's contention that things only exist by way of designation." Kalee Brown, "Dalai Lama: Spirituality without Quantum Physics Is an Incomplete Picture of Reality," Collective Evolution, April 26, 2017, http://collective-evolution.com/2017/04/26/dalai-lama-spirituality-without-quantum-physics-is-an-incomplete-picture-of-reality. Accessed November 9, 2017.
326 Quoted in Fritof Capra and Pier Luigi Luisi, *The Systems View of Life: A Unifying Vision* (New York: Cambridge University Press, 2014).
327 The Buddhist name (in Pali language) is *paticca samuppada*. In Sanskrit it is *pratitya samutpada*.
328 Nagarjuna, *The Fundamental Wisdom of the Middle Way: Nagarjuna's Mulamadhyamakakarika*, trans. Jay Garfield (Oxford: Oxford University Press, 1995).
329 Sir James Jeans, *Physics and Philosophy* (New York: Dutton Publications, 1981), 204.
330 Forty years ago during my freshman year in college, I was visiting my best friend at the time, who was a first-year physics student at Princeton. I accompanied him to his introductory physics class and right before the class started, he leaned over, and in a very impressed voice whispered in my ear,

"John Wheeler is sitting right behind you." Wheeler was close enough that I could have reached out and touched him. How cool is that? Wheeler, one of the century's greatest physicists, was evidently checking out what was going on in a freshman physics class. Unfortunately, the fact that John Wheeler was sitting behind me meant absolutely nothing to me at the time, for I literally had no idea who John Wheeler was. I now joke with my friends that Wheeler and my wave functions became phase-entangled at that moment.

331 Quoted in Frank Laloe, *Do We Really Understand Quantum Mechanics?* (Cambridge: Cambridge University Press, 2012).

332 Wheeler, *Geons, Black Holes, and Quantum Foam*, 339.

333 Wheeler, "Beyond the Black Hole," *Some Strangeness in the Proportion*, 350.

334 Renee Weber, ed., *Dialogues with Scientists and Sages* (New York: Routledge & Kegan Paul, 1986), 119.

335 In his book *Entangled Minds*, Dean Radin proposes that *psi* (psychic phenomena) is the human experience of the entangled universe, i.e., the human experience of quantum interconnectedness.

336 Commenting on this seemingly strange (from our perspective) nature of light, Bohm writes, "Not that we can actually reach that, but for the sake of imagination you can suppose it, and therefore you could say that something timeless seems to be involved. And spaceless—a fundamental relationship which is beyond time and space." David Bohm, *Unfolding Meaning* (London: Routledge, 1995), 121.

337 Huston Smith, "Foreword," in *Epiphanies: Where Science and Miracles Meet,* Ann Jauregui (New York: Atria, 2007), xiv.

338 Quoted in David Nicol, *Subtle Activism: The Inner Dimension of Social and Planetary Transformation* (New York: State University of New York Press, 2015).

339 See Vlatko Vedral, "Entanglement Hits the Big Time," *Nature* 425 (2003): 28–29.

340 Quoted in Mindell, *Quantum Mind: The Edge Between Physics and Psychology*, 351.

341 Jung, *Letters*, vol. 1, 58.

342 This image is to be found in *The Flower Garland Sutra* (*Avatamsaka Sutra*).

343 David Bohm, "On the Intuitive Understanding of Nonlocality as Implied by Quantum Theory," *Foundations of Physics*, vol. 5 (1975).

344 Weber, ed., *Dialogues with Scientists and Sages,* 41.

345 Bohm, *Wholeness and the Implicate Order*, 250.

346 Robert Nadeau and Menas Kafatos, *The Conscious Universe* (New York: Springer-Verlag, 1990), 118.

347 Wheeler, *Geons, Black Holes, and Quantum Foam*, 341.

348 Quoted in Mindell, *Quantum Mind: The Edge Between Physics and Psychology*, 237.

349 Wheeler, *Geons, Black Holes, and Quantum Foam*, 341.

350 Note how this is similar and obviously related to Jung's inquiries into synchronistic phenomena, in which the boundary between mind and matter dissolves.

351 Wheeler, *At Home in the Universe*, 296.

352 A qubit is a unit of quantum information; it is a quantum analogue of the classical bit. Many physicists now conceive of the entire space-time continuum (including all of the matter within it) as an emergent structure that is derived from a more fundamental substratum that is made out of qubits, i.e.,

pure quantum information. In other words, from this perspective the universe is built out of information with space-time and matter simply being virtual representations of underlying information.

353 Wheeler, *At Home in the Universe*, 299.

354 Dick, *The Selected Letters of Philip K. Dick: 1980 – 1982*, 255.

355 Wheeler, *Geons, Black Holes, and Quantum Foam*, 341.

356 Ibid.

357 Schafer, *Infinite Potential: What Quantum Physics Reveals About How We Should Live Our Life*, 105.

358 Wheeler, "Information, Physics, Quantum: The Search for Links."

359 Wheeler, *Geons, Black Holes, and Quantum Foam*, 64.

360 Quoted in Richard M. Martin, *Logical Semiotics and Mereology* (Philadelphia, PA: John Benjamins Publishing Company, 1992).

361 Weber, *Dialogues with Scientists and Sages*, 113.

362 Quoted in Massimo Teodorani, *The Hyperspace of Consciousness* (Sweden: Elementa, 2015).

363 Weber, *Dialogues with Scientists and Sages*, 113.

364 Jung, *Memories, Dreams, Reflections*, 338–339.

365 Quoted in Bernstein, *Quantum Profiles*, 94.

366 Quoted in Werner Heisenberg, *Encounters with Einstein: And Other Essays on People, Places, and Particles* (Princeton, NJ: Princeton University Press, 1983), 135.

367 Wheeler, *At Home in the Universe*, 186.

368 Jung, *Memories, Dreams, Reflections*, 340.

369 Weber, *Dialogues with Scientists and Sages*, 123.

370 Ibid., 119–120.

371 From his poem "Relativity." Quoted in Robert P. Crease and Alfred Scharff Goldhaber, *The Quantum Moment: How Planck, Bohr, Einstein, and Heidegger Taught Us to Love Uncertainty* (New York: W. W. Norton & Company, 2014).

372 Wheeler and Zurek, *Quantum Theory and Measurement*, 202.

373 Heisenberg, *Physics and Philosophy: The Revolution in Modern Science*, 52.

374 There is another interpretation, however, known as the "many worlds" interpretation that says that all of the other potentialities actually do occur but each in its own parallel universe.

375 Wheeler, *Geons, Black Holes, and Quantum Foam*, 338.

376 Ibid., 167–168.

377 Sir Roger Penrose, *The Road to Reality: A Complete Guide to the Laws of the Universe* (New York: Vintage Books, 2007), 1031.

378 Ken Wilber, ed., *Quantum Questions: Mystical Writings of the World's Greatest Physicists* (Boston: Shambhala Publications, 2001), 98.

379 Erwin Schrödinger, *Mind and Matter* (London: Cambridge UP, 1964), 30.

380 Ludwig Wittgenstein, *Tractatus Logico-Philosophicus* (Mineola, NY: Dover Publications, 1999).

381 Over the course of this book I have used numerous baseball analogies to elucidate quantum reality. Stealing a footnote from Wheeler's autobiography, "My apologies to readers not familiar with baseball. You may need to consult a baseball fan."

382 Hoffmann, *The Strange Story of the Quantum*, 217.

383 Quoted in Peter Russell, *From Science to God* (Novato, CA: New World Library, 2003), 68.

384 Jeans, *The Mysterious Universe,* 151.
385 Wheeler, "Information, Physics, Quantum: The Search for Links."
386 Brian, *The Voice of Genius,* 212.
387 Heisenberg, *Encounters with Einstein: And Other Essays on People, Places, and Particles,* 106.
388 Sir Arthur Stanley Eddington, *Space, Time, and Gravitation: An Outline of the General Relativity Theory* (Cambridge: Cambridge University Press, 1920).
389 Nicolas Berdyaev, *The Meaning of the Creative Act* (San Rafael, CA: Semantron Press, 2009), 57.
390 Quoted in Graham Smetham, *Quantum Buddhism: Dancing in Emptiness* (Brighton, UK: Sunyata Press, 2010), 37.
391 Jackson and Lethem, *The Exegesis of Philip K. Dick,* 588.
392 Sutin, *The Shifting Realities of Philip K. Dick,* 261.
393 Zajonc, *The New Physics and Cosmology: Dialogues with the Dalai Lama,* 145.
394 Wheeler, *Frontiers of Time,* 85.
395 Ibid., 8.
396 Wheeler, *Geons, Black Holes, and Quantum Foam,* 351.
397 Wheeler is the one who originally calculated the energy density of space to be 1.0 x 10 to the 94th power grams per cubic centimeter (that's 1 with 94 zeros after it). This is a practically inconceivable amount of potential energy enfolded within a small amount of seemingly empty space. He once described the proton, which is a very dense particle found at the center of atoms, as being vaporous, diaphanous, and cloudlike compared to the much greater energy density of the space surrounding it.
398 Along these lines, Wheeler developed a whole new branch of physics called "geometrodynamics" (he had a habit of starting whole new fields within physics and letting other people do the work of filling in the details), writing a very significant book of the same name in which he describes all forms of matter, particularly particles, as being dynamic geometric modifications of the plenum of space. He conceives of particles as being places where the geometry of space is highly curved (like microvortices of the plenum of space itself), to the point that they appear to have the property that we call mass. Wheeler is explicitly suggesting that mass (in all its forms) is simply highly geometrically curved emptiness, the curvature granting it the appearance of form, yet in its essence still nothing but formless emptiness.
399 Quoted in Heinz R. Pagels, *The Cosmic Code: Quantum Physics as the Language of Nature* (New York: Bantam Books, 1983), 243.
400 Quoted in Joseph Needham, *Science and Civilization in China,* vol. IV (Cambridge: Cambridge University Press, 1965), 33.
401 Khenchen Palden Sherab Rinpoche and Khenpo Tsewang Dongyal Rinpoche, *Ceaseless Echoes of the Great Silence: A Commentary on the Heart Sutra Prajnaparamita* (Sidney Center, NY: Dharma Samudra, 2012), 1.
402 Buckley and Peat, *A Question of Physics: Conversations in Physics and Biology,* 53.
403 Bohm, *Wholeness and the Implicate Order,* 192.
404 Jonathan Allday, *Quantum Reality: Theory and Philosophy* (Boca Raton, FL: CRC Press, 2009), 493.
405 Zajonc, *The New Physics and Cosmology: Dialogues with the Dalai Lama,* 89.
406 Ibid., 97.
407 Ibid., 160.
408 Quoted in Fritof Capra, *Uncommon Wisdom: Conversations with Remarkable People* (New York: Bantam Books, 1989), 40.

409 Robert Oppenheimer, *Science and the Common Understanding* (New York: Oxford University Press, 1954).
410 Quoted in Mindell, *Quantum Mind: The Edge Between Physics and Psychology*, 455.
411 E. F. Kelly, A. Crabtree, and P. Marshall, eds., *Beyond Physicalism: Toward Reconciliation of Science and Spirituality* (Lanham, MD: Rowman & Littlefield, 2015), xii.
412 Zajonc, *The New Physics and Cosmology: Dialogues with the Dalai Lama*, 209.
413 Kalee Brown, "Dalai Lama: Spirituality without Quantum Physics Is an Incomplete Picture of Reality," Collective Evolution, April 26, 2017, http://www.collective-evolution.com/2017/04/26/dalai-lama-spirituality-without-quantum-physics-is-an-incomplete-picture-of-reality.
414 C. G. Jung, *Alchemical Studies*, CW 13 (Princeton, NJ: Princeton University Press, 1970), para. 253.
415 C. G. Jung and Wolfgang Pauli, *The Interpretation of Nature and the Psyche*, Bollingen Series LI (Princeton: Princeton University Press, 1955), 175.
416 Henry P. Stapp, "S-Matrix Interpretation of Quantum Theory," *Physical Review D*, March 15, 1971.
417 Quoted in Heisenberg, *Across the Frontiers*, 33.
418 C. G. Jung, *The Archetypes and the Collective Unconscious*, CW 9i (Princeton, NJ: Princeton University Press, 1959), para. 291.
419 Henry Pierce Stapp, "Are Superluminal Connections Necessary?" *Nuovo Cimento*, vol. 40B (1977), 191.
420 Third-dimensional space-time can be thought of as being "the box" that we need to learn to think outside of.
421 Etymologically, the word "daemon" means the inner voice and guiding spirit.
422 Jung, *Alchemical Studies*, CW 13, para. 163.
423 Heisenberg, *Physics and Philosophy: The Revolution in Modern Science*, 197.
424 Quoted in Carolyn Merchant, *The Death of Nature*, 169.
425 Feynman, *The Quotable Feynman*, 153.
426 Wheeler, *At Home in the Universe*, 6.
427 Jung, *Psychology and Alchemy*, CW 12, para. 420.
428 C. G. Jung, *Mysterium Coniunctionis*, CW 14 (Princeton, NJ: Princeton University Press, 1989), para. 649.
429 In Peter 2:5: "Like living stones be yourselves built into a spiritual edifice."
430 William McGuire and R. F. C. Hull, eds., *C. G. Jung Speaking: Interviews and Encounters* (Princeton, NJ: Princeton University Press, 1977), 228.
431 Jung, *Psychology and Alchemy*, CW 12, par. 346.
432 C. G. Jung, Elizabeth Welsh, and Barbara Hannah, *Modern Psychology: Notes on Lectures Given at the Eidgenössische Technische Hochschule, Zürich* (Zurich: K. Schippert & Co., 1959), 14.
433 Speaking of the esoteric and secret substance of alchemy, Jung writes, "It is a characteristic of the arcane substance to have 'everything it needs;' it is a fully autonomous being, like the dragon [ouroboros] that begets, reproduces, slays and devours itself . . . a being without beginning or end, and in need of 'no second.' Such a thing can by definition only be God himself." (Jung, *Mysterium Coniunctionis*, CW 14, para. 143.)
434 Jackson and Lethem, *The Exegesis of Philip K. Dick*, 708.
435 C. G. Jung, *Psychology and Religion: West and East*, CW 11 (Princeton, NJ: Princeton University Press, 1973), para. 148.

436 Jung, *Mysterium Coniunctionis*, CW 14, para. 753.
437 N. David Mermin, "What is Quantum Mechanics Trying to Tell Us?" *American Journal of Physics* 66 (1998): 753-767.
438 Physicist Vlatko Vedral writes in a groundbreaking article in *Scientific American* in 2011, "Until the past decade, experimentalists had not confirmed that quantum behavior persists on a macroscopic scale. Today, however, they routinely do. These effects are more pervasive than anyone ever suspected." Vlatko, "Living in a Quantum World," *Scientific American* 304, no. 6 (June 2011): 38–43.
439 Wheeler, *Geons, Black Holes, and Quantum Foam*, 329.
440 Ibid., 292.
441 Hideo Mabuchi, "Quantum Feedback and the Quantum-classical Transition," in *Science and Ultimate Reality: Quantum Theory, Cosmology, and Complexity*, eds. Barrow et al., 329.
442 Wheeler, *Geons, Black Holes, and Quantum Foam*, 330.
443 Allday, *Quantum Reality: Theory and Philosophy*, 4.
444 It should be noted that as we stop solidifying the fluid, dreamlike nature of our universe and recognize its dreamlike quality, the more dreamlike and synchronistic (and hence, "weirdly") it will manifest. Once we stop superimposing our own fixed and limited ideas upon its dreamlike fabric, we allow it to reveal its dreamlike magic. Once we let go our grasp, the universe will exhibit more synchronistic phenomena, which literally reveal the underlying quantum. We can use our imagination to deepen our understanding of this. When we are in a dream while asleep, if we recognize that we are dreaming (instead of thinking the universe we are inhabiting is objectively real), we allow the dream to reveal even more of its dreamlike nature. For being like a dream, the universe is not separate from the consciousness that's observing it, a realization which it continually reflects back to itself through us.
445 Heisenberg, *Physics and Philosophy: The Revolution in Modern Science*, 54.
446 Wheeler, *Frontiers of Time*, 8–10.
447 Ibid., 1.
448 Ibid., 10.
449 Nathan Schwartz-Salant, ed., *Jung on Alchemy* (Princeton, NJ: Princeton University Press, 1995), 147.
450 Wheeler, *At Home in the Universe*, 307.
451 Wheeler, *Geons, Black Holes, and Quantum Foam*, 336.
452 Albert Einstein, *Sidelights on Relativity* (New York: Dover, 1983), 28.
453 Wheeler and Zurek, *Quantum Theory and Measurement*, xvi.
454 Feynman, *The Quotable Feynman*, 68.
455 Heisenberg, *Physics and Beyond: Encounters and Conversations*, 209.
456 Jung, *The Structure and Dynamics of the Psyche*, CW 8, para. 964.
457 Ibid., para. 417.
458 C. G. Jung, *Letters*, vol. 2 (London: Routledge & Kegan Paul, 1990), 54.
459 C. G. Jung, *Aion*, CW 9ii (Princeton, NJ: Princeton University Press, 1973), para. 412.
460 Jung, *Psychology and Alchemy*, CW 12, para. 394.
461 Wheeler, *Geons, Black Holes, and Quantum Foam*, 330.
462 Erwin Schrödinger, *Science and Humanism: Physics in Our Time* (Cambridge: Cambridge University Press, 1961), 27.
463 Quoted in Herbert, *Quantum Reality: Beyond the New Physics*, 131.

464 Quoted in Max Jammer, *Concepts of Space: the History of Theories of Space in Physics* (Mineola, NY: Dover Publications, 1993), 189.
465 Schrödinger, *Science and Humanism: Physics in Our Time*, 27.
466 Notice the similarity to dreaming. In the evening you go to sleep at Point A, and in the morning you wake up at Point B and remember a dream you had during your sleep. We can observe you at Point A before you went to bed and at Point B when you wake up, but where were you in between?
467 Jeans, *Physics and Philosophy*, 175.
468 John Wheeler, "Superspace and the Nature of Quantum Geometrodynamics," in *Battelle Rencontres: 1967 Lectures in Mathematics and Physics*, eds. C. DeWitt and John Wheeler (New York: W. A. Benjamin, 1968).
469 Quoted by A. Forsee in *Albert Einstein Theoretical Physicist* (New York: MacMillan, 1963), 81.
470 G. W. Leibniz, *Leibniz Selections*, trans. P. P. Wiener (New York: Scribners, 1951), 488.
471 C. G. Jung, *Synchronicity: An Acausal Connecting Principle*, trans. R. F. C. Hull (Princeton, NJ: Princeton University Press, 2010), 28.
472 Wheeler, *Geons, Black Holes, and Quantum Foam*, 256.
473 Quoted in Daniel Pinchbeck, *2012: The Return of Quetzalcoatl* (New York: Jeremy Tarcher, 2006), 206.
474 To Paul Ehrenfest, July 12, 1924, expressing his frustration over quantum theory. Albert Einstein, *The Ultimate Quotable Einstein*.
475 Buckley and Peat, *A Question of Physics: Conversations in Physics and Biology*, 61.
476 Ibid.
477 Bernstein, *Quantum Profiles*, 138.
478 Wheeler, *At Home in the Universe*, 113.
479 Wheeler, *Geons, Black Holes, and Quantum Foam*, 270.
480 Wheeler, *A Journey into Gravity and Spacetime*, xi.
481 Wheeler, *Geons, Black Holes, and Quantum Foam*, 90.
482 Dyson, "Preface," in *Science and Ultimate Reality: Quantum Theory, Cosmology, and Complexity*, eds. Barrow, et al., xviii.
483 John Wheeler, "Does the Universe Exist If We're Not Looking?" interview by Tim Folger, *Discover Magazine*, vol. 23, no. 6 (June 2002).
484 Quoted in Dyson, "Preface," in *Science and Ultimate Reality: Quantum Theory, Cosmology, and Complexity*, xviii.
485 Ibid.
486 This is why the practice of meditation, which trains the mind to recognize and then relax into the space between thoughts, is good training to better understand the quantum realm and directly experience our quantum nature.
487 Herbert, *Quantum Reality: Beyond the New*, 40.
488 To make this point, Wheeler, in his classic text *Quantum Theory and Measurement* (p. 203), quotes art historian E. H. Gombrich: "We can never neatly separate what we see from what we know. . . . What we call seeing is invariably colored and shaped by our knowledge (or belief) of what we see."
489 Smetham, *Quantum Buddhism: Dancing in Emptiness—Reality Revealed at the Interface of Quantum Physics and Buddhist Philosophy*, 244.
490 This is a characteristic of the psychic malady called wetiko that I write about in my recent book. When people are afflicted with wetiko, they react to their projections upon the world as if they are objectively existing, separate from themselves.

491 Wheeler, *Frontiers of Time.*

492 Wojciech Zurek, "The Evolution of Reality," The Foundational Questions Institute, November 10, 2009, www.fqxi.org/community/articles/display/122.

493 John Wheeler, "Does the Universe Exist If We're Not Looking?" interview by Tim Folger, *Discover*, vol. 23, no. 6 (June 2002): p. 44.

494 Quoted in Benjamin D. Koen, *Beyond the Roof of the World: Music, Prayer, and Healing in the Pamir Mountains* (Oxford, Oxford University Press, 2008).

495 Quoted in Frank H. Columbus and Volodymyr Krasnoholovets, eds., *Developments in Quantum Physics* (Hauppage: Nova Science Publishers, 2004).

496 Jung, *Psychology and Alchemy*, CW 12, para. 396.

497 Albert Einstein, "What Life Means to Einstein," interview by George Sylvester Viereck, *The Saturday Evening Post*, October 26, 1929.

498 Jung, *Psychology and Alchemy*, CW 12, para. 394.

499 Ibid., para. 399.

500 Jung, *Alchemical Studies*, CW 13, para. 286.

501 This makes me think of when Christ said in the Bible "Ye are Gods."

502 In my previous writings, I call this "intervening in the dimension of the dreaming."

503 Bernstein, *Quantum Profiles*, 132.

504 Ibid., Wheeler continues, "He [Leibniz] went on to say, 'this dream or phantasm to me is real enough if using reason well we are never deceived by it.'"

505 C. G. Jung, *The Psychology of Kundalini Yoga: Notes of the Seminar Given in 1932* (Princeton, NJ: Princeton University Press, 1996), 47.

506 Jung, *The Structure and Dynamics of the Psyche*, CW 8, para. 747.

507 Quoted in Carl Friedrich von Weizsäcker, *Die Einheit der Natur* (Munich: Hanser, 1971), 424.

508 Which from one sense is the same present moment that I am writing this (though not from the perspective of linear time, however); for quantum physics has revealed that there is only one singular "now" moment which encompasses all past and future moments in one undivided, all-embracing, and eternally-creative present moment within which everything that happens arises.

509 Wheeler, *At Home in the Universe*, 44.

510 In a letter to Schrödinger dated June 19, 1935, Einstein writes, "Physics is a kind of metaphysics." Quoted in Jim Baggott, *The Quantum Story: A History in 40 Moments* (New York: Oxford University Press, 2011).

511 Heisenberg, *Physics and Beyond: Encounters and Conversations*, 210.

512 "Mysticism" is another word that has a bad reputation in the physics community. Physicist David Bohm points out that it might be more appropriate to refer to our ordinary, consensus-reality mode of consciousness as "mysticism," in the sense that it mystifies us from our true nature. Our ordinary, classical way of thinking elaborately obscures its mode of functioning from itself as it engages in self-deception. Bohm used the term "obscurantism" so as to capture the self-obscuring quality of our everyday consciousness.

513 The origin of the word "mystic" is the Greek word "mystikos," which means "of mysteries." To quote Einstein, "The most beautiful experience we can have is the mysterious. It is the fundamental emotion which stands at the cradle of true art and true science." Quoted in Alan Lightman, *A Sense of the Mysterious*, p. 42.

514 Wheeler, *At Home in the Universe*, 7.
515 Quoted in T. A. Heppenheimer, "Bridging the Very Large and the Very Small," *Mosaic* (Fall 1990): 33.
516 Aldous Huxley, *Grey Eminence: A Study in Religion and Politics* (London: Vintage/Random House London, 2005), 98.
517 Brian Josephson, "Physics and Spirituality: The Next Grand Unification?"
518 Excerpt from a letter from Pauli to Huxley, August 10, 1956. Quoted in Nathan Field, Trudy Harvey, and Belinda Sharp, *Ten Lectures on Psychotherapy and Spirituality* (London: Karnac Books, 2005).
519 Wilber, *Quantum Questions: Mystical Writings of the World's Greatest Physicists*, 207.
520 Josephson, "Physics and Spirituality: The Next Grand Unification?"
521 We can also include non-ordinary and paranormal experiences such as near-death experiences, out-of-body experiences, lucid dreams, mystical experiences, and so on.
522 Albert Einstein, *The World as I See It.*
523 Jung, *Psychology and Alchemy*, CW 12, para. 13.
524 Feynman, *The Quotable Feynman*, 57.
525 Spoken by Wheeler at a conference celebrating his ninetieth birthday. Quoted in Gefter, *Trespassing on Einstein's Lawn*, 21.
526 Octavio Paz, *Alternating Current* (NY: Arcade Publishing, 2015).
527 C. G. Jung, *Visions: Notes of the Seminar Given in 1930-1934*, Vol. 2 (Princeton, NJ: Princeton University Press, 1997), 742.
528 Quoted in Arthur I. Miller, *137: Jung, Pauli, and the Pursuit of a Scientific Obsession* (New York: W. W. Norton & Co., 2009).
529 Heisenberg, *Across the Frontiers*, 227.
530 Davies and Brown, *The Ghost in the Atom,* 69.
531 Lederman and Hill, *Quantum Physics for Poets*, 277.
532 Dyson, "Comment on the Topic 'Beyond the Black Hole,'" in *Some Strangeness in the Proportion*, 376.
533 Quoted in Dennis Overbye, "John A. Wheeler, Physicist Who Coined the Term 'Black Hole,' Is Dead at 96," *New York Times,* Apr 14, 2008.
534 Wheeler, *Geons, Black Holes, and Quantum Foam*, 342–343.
535 Bernstein, *Quantum Profiles*, 134.
536 Wheeler, *At Home in the Universe*, 303.
537 Quoted in Barrow et al., eds., *Science and Ultimate Reality: Quantum Theory, Cosmology, and Complexity*, xiv.
538 Quoted in Bernstein, *Quantum Profiles*, 134.
539 Quoted by Banesh Hoffmann in "Working with Einstein," *Some Strangeness in the Proportion*, 476.
540 Wheeler, *Geons, Black Holes, and Quantum Foam*, 287.
541 Wheeler, *At Home in the Universe*, 272.
542 John Wheeler, "Foreword," in *Science and Ultimate Reality: Quantum Theory, Cosmology, and Complexity*, eds. Barrow et al., xi.
543 In an interview Wheeler says, "I have had two students who thought I was cuckoo. In fact, I glued a statement of one into one of my notebooks as a salutary warning to be aware of my shortcomings and failures. I thought it was healthy." Brian, *The Voice of Genius*, 122.
544 Wheeler, *Cosmic Search Magazine*, vol. 1, no. 4.
545 Quoted in Cheuk-Yin Wong, "Remembering John Wheeler, Physicist and Teacher," *Princeton Alumni Weekly*, July 16, 2008.

546 Wheeler, *Cosmic Search Magazine*, vol. 1, no. 4.

547 Richard Feynman, *The Pleasure of Finding Things Out: The Best Short Works of Richard P. Feynman* (New York: Basic Books, 1999), 115.

548 Feynman, *The Character of Physical Law.*

549 Feynman, *The Quotable Feynman*, 95.

550 Quoted in Dennis Overbye, "John A. Wheeler, Physicist Who Coined the Term 'Black Hole,' Is Dead at 96," *New York Times*, April 14, 2008.

551 Feynman, *The Quotable Feynman*, 173.

552 Quoted in Arnold Mindell, *The Quantum Mind and Healing* (Charlottesville, VA: Hampton Roads, 2004), 3. This is from a speech that Goldin gave in April 2001 at the Fermi National Laboratory in which he scolds physicists for forgetting how to dream.

553 Hoffmann, *The Strange Story of the Quantum*, 170.

554 Sean Carroll, *The Big Picture: On the Origins of Life, Meaning, and the Universe Itself* (New York: Dutton, 2017), 166.

555 Lederman and Hill, *Quantum Physics for Poets*, 15.

556 See Evelyn Fox Keller, "Cognitive Repression in Contemporary Physics," *American Journal of Physics*, vol. 47, (August 1979).

557 Feynman and Wilczek, *The Character of Physical Law.*

558 Quoted in "A Quantum Sampler," *New York Times*, December 26, 2005, http://www.nytimes.com/2005/12/26/science/a-quantum-sampler.html.

559 Quoted in Paul Davies, *About Time: Einstein's Unfinished Revolution* (New York: Simon and Schuster, 1996), 163.

560 Meier, *Atom and Archetype: The Pauli/Jung Letters 1932-1958*, 183.

561 Quoted in Jeffrey Satinover, *The Quantum Brain* (New York: John Wiley & Sons, 2001), 116.

562 Jung, *Civilization in Transition*, CW 10, para. 155.

563 Pais, *Subtle is the Lord: The Science and the Life of Albert Einstein*, 463.

564 These comments bear a striking resemblance to certain religious statements. In his book *Quantum Physics and Theology* (New Haven, CT: Yale University Press, 2007), John Polkinghorne, a theoretical physicist, theologian, and Anglican priest, quotes bishop William Temple: "If any man says he understands the relation of Deity to humanity in Christ, he only makes it clear that he does not at all understand what is meant by Incarnation." (Temple, *Christus Veritas*, 139). It certainly sounds as if the bishop and the physicist are pointing at a similar mystery. The similarity of articulations doesn't necessarily mean they are illumining the same mystery, however, but it should definitely get our attention. Polkinghorne wonders whether there is a connection between physicists embracing the wave/particle duality of light and—in a similar embracing of things that are normally thought of as being opposites and mutually exclusive—theologians embracing the divine/human duality in Christ. Polkinghorne wonders whether the Christ event (with its re-visioning of God, humanity, and their relationship) is a theological analogue to the radical paradigm shift that took place with the advent of quantum physics.

565 Quoted in Freke, *The Mystery Experience: A Revolutionary Approach to Spiritual Awakening.*

566 Referring to Ortega y Gassett's thoughts on "the specialized scientist," Schrödinger writes that he is "the typical representative of the brute ignorant rabble—the *hombre masa* (mass-man)—who endanger the survival of

true civilization. . . . Yet the awareness that specialization is not a virtue but an unavoidable evil is gaining ground." Schrödinger, *Science and Humanism*, 6-7.

567 Concerning the ever-increasing specialization in science, Jung writes that this leads to "detachment from the world and from life, as well as to a multiplication of specialized fields which gradually lose all connection with one another. The result is an impoverishment and desiccation not merely in the specialized fields but also in the psyche of every man who has differentiated himself up or sunk down to the specialist level." Jung, *Psychological Types*, CW 6, para. 84.

568 This brings to mind something that happened to me many years ago. I had written an article in which I quoted Jung many times. A Jungian analyst sent me a scathing email, saying how dare I quote Jung, that I had no right to quote Jung because I wasn't an officially trained and certified Jungian analyst.

569 Dyson, "Comment on the Topic 'Beyond the Black Hole,'" in *Some Strangeness in the Proportion*, 376.

570 "The Uncertainty of Science," John Danz Lecture Series, 1963.

571 Brian, *The Voice of Genius*, 130.

572 Buckley and Peat, *A Question of Physics: Conversations in Physics and Biology*, 57.

573 Jung, *The Structure and Dynamics of the Psyche*, CW 8, para. 962.

574 Lincoln Barnett, *The Universe and Dr. Einstein*, xviii.

575 This is reminiscent of when I refer, in my recent book *Dispelling Wetiko: Breaking the Curse of Evil,* to how the wetiko virus can usurp the healthy parts of the psyche into its service; this inner process of the psyche is playing out and revealing itself through the medium of the outside world.

576 The relationship between the corporatized physics establishment compared to the smaller, pure physics community that is committed to the search for truth above all else (as a reward unto itself) is analogous to the relationship between the allopathic mainstream medical establishment (which includes Big Pharma) and the more holistic approaches of, for example, naturopathic medicine and the alternative medicine communities.

577 Ravi Ravindra, *Science and the Sacred: Eternal Wisdom in a Changing World* (Wheaton, IL: Quest Books, 2002), 39.

578 Brian, *The Voice of Genius*, 142.

579 Quoted in K. C. Cole, *The Universe and the Teacup: The Mathematics of Truth and Beauty*, 1998.

580 Søren Kierkegaard, *Works of Love* (Princeton, NJ: Princeton University Press, 1998).

581 Josephson, "Physics and Spirituality: The Next Grand Unification?" After this quote, Josephson adds the following in parenthesis: "There has even been a suggestion, in the editorial pages of a prestigious scientific journal, that a particular book should be burnt because it propagated dangerous ideas."

582 J. C. Eccles, *Facing Reality: Philosophical Adventures by a Brain Scientist* (New York, NY: Springer-Verlag, 1970), 115.

583 Upton Sinclair, *I, Candidate for Governor: And How I Got Licked* (Oakland: University of California Press, 1994), 109.

584 Erwin Schrödinger, *What Is Life? With Mind and Matter and Autobiographical Sketches* (Cambridge: Cambridge University Press, 2004), 94.

585 William James, *Pragmatism—A New Name for Some Old Ways of Thinking* (Redditch, Worcestershire: Read Books, Ltd., 2015).
586 One of the chief features of the aforementioned wetiko disease is to be doing the very thing you are accusing others of doing.
587 Interestingly, in a science fiction convention in 1972 Philip K. Dick famously opined that any formulation "that attempts to act as an all-encompassing, all-explaining hypothesis about what the universe is about" (i.e., a "ToE") are "manifestations of paranoia." Philip K. Dick, "The Android and the Human," *Science Fiction Criticism: An Anthology of Essential Writings,* ed. Rob Latham (London: Bloomsbury Academy, 2017).
588 Wheeler, *Geons, Black Holes, and Quantum Foam,* 90.
589 Heisenberg, *Physics and Beyond: Encounters and Conversations,* 114.
590 Jean Baudrillard, *The Ecstacy of Communication* (Los Angeles, CA: Semiotext(e), 2012).
591 To cite a few examples: How can quantum entities be both waves and particles, which are as mutually exclusive as it gets? How do these quantum entities go from here to there in no time at all without traversing the path in between? How do these massless, intangible quantum entities, which have no weight and no substance and can't even be said to be real or even exist, give rise to the massive weight of the whole universe? And how do quantum entities at opposite ends of the universe seemingly communicate with each other faster than the speed of light?
592 His Holiness the Dalai Lama, *The Universe in a Single Atom: The Convergence of Science and Spirituality,* 66.
593 Interestingly, what indigenous people call wetiko, (which I write about in *Dispelling Wetiko*) is a psychic form of blindness, which in its self-deception believes itself to be sighted. It operates through our unconscious blind spots and feeds on our unconscious looking away and avoiding relating with parts of ourselves.
594 His Holiness the Dalai Lama, *The Universe in a Single Atom: The Convergence of Science and Spirituality,* 66.
595 Michael Talbot, *Mysticism and the New Physics* (London: Arkana, 1993), 140.
596 Schafer, *Infinite Potential: What Quantum Physics Reveals About How We Should Live Our Life,* 52.
597 Freeman Dyson, *The Scientist as Rebel* (New York: New York Review Books, 2008).
598 Quoted in Wolfgang Yourgrau and Allen D. Breck, eds., *Cosmology, History, and Theology* (New York: Plenum Press, 1977).
599 Schrödinger, *Science and Humanism,* 51.
600 Jung, *Memories, Dreams, Reflections,* 318.
601 Quoted in Deirdre Bair, *Jung: A Biography* (New York: Little, Brown, 2004), 553.
602 Jung, *Mysterium Coniunctionis,* CW 14, para. 768.
603 Jung and Pauli, *The Interpretation of Nature and the Psyche,* 208.
604 Ibid., 210.
605 Meier, *Atom and Archetype: The Pauli/Jung Letters 1932-1958,* 124.
606 Ibid., 99.
607 Ibid., 165.
608 Ibid., 127.
609 Richard I. Evans, *Jung on Elementary Psychology* (New York: E. P. Dutton and Co., 1976), 155–156.

610 Jung, *Mysterium Coniunctionis*, CW 14, para. 344.
611 Jung, *Visions*, vol. 2, 1015.
612 Jung, *Psychology and Religion: East and West*, CW 11, para. 448.
613 Sir Arthur Eddington, *The Nature of the Physical World* (London: Forgotten Books, 2017), 291.
614 Jung, *Aion*, CW 9ii, para. 413.
615 Jung, *The Structure and Dynamics of the Psyche*, CW 8, para. 418.
616 Quoted in David Lindorff, *Pauli and Jung: The Meeting of Two Great Minds* (Wheaton, IL: Quest Books, 2009), 97.
617 Meier, *Atom and Archetype: The Pauli/Jung Letters 1932-1958*, 127.
618 C. G. Jung, *Psychological Reflections* (Princeton, NJ: Princeton University Press, 1970), 14.
619 Jung, *Letters*, vol. 2, 71.
620 Jung, *The Structure and Dynamics of the Psyche*, CW 8, para. 429.
621 Quoted in Gieser, *The Innermost Kernel: Depth Psychology and Quantum Physics. Wolfgang Pauli's Dialogue with C. G. Jung.*
622 Helen Dukas and Banesh Hoffmann, eds., *Albert Einstein, the Human Side: New Glimpses from His Archives* (Princeton, NJ: Princeton University Press, 1981), 38.
623 Quoted in Lindorff, *Pauli and Jung: The Meeting of Two Great Minds*, 49.
624 Henry Corbin, *Spiritual Body and Celestial Earth: From Mazdean Iran to Shi'ite Iran* (Princeton, NJ: Princeton University Press, 1977), 81.
625 Mindell, *The Quantum Mind and Healing*, 12.
626 Schafer, *Infinite Potential: What Quantum Physics Reveals About How We Should Live Our Life*, 6.
627 Jung, *Letters*, vol. 1, 57.
628 *Aurora Consurgens* is a document attributed to Thomas Aquinas on the problem of opposites in alchemy. The literal translation is "rising dawn," and it refers to the soul rising up through the union of the opposites.
629 Quoted in Gieser, *The Innermost Kernel: Depth Psychology and Quantum Physics. Wolfgang Pauli's Dialogue with C. G. Jung.*
630 Marie-Louise von Franz, *Psyche and Matter* (Boston: Shambhala Publications, 1992), 72.
631 To quote Jung, "Synchronicity therefore means the simultaneous occurrence of a certain psychic state with one or more external events which appear as meaningful parallels to the momentary subjective state—and, in certain cases, vice versa." Jung, *The Structure and Dynamics of the Psyche*, CW 8, para. 850.
632 Meier, *Atom and Archetype: The Pauli/Jung Letters 1932-1958*, 128.
633 Written in a letter to Markus Fierz in 1949, *Letters*, vol. 1, p. 530.
634 Marie-Louise von Franz, *Number and Time* (Evanston, IL: Northwestern University Press, 1974), 292.
635 Jung makes the point that he felt this way provided that the psyche was understood to be the collective unconscious. Jung, "Ein Brief zur Frage der Synchronizitat," *Zeitschrift fur Parapsychologie und Grenzgebiete der Psychologie* (Bern & Munich, 1961), vol. 5, no. 1, pp. 1ff.
636 Jung, *Letters*, vol. 2, 344.
637 Wheeler, "Beyond the Black Hole," in *Some Strangeness in the Proportion*, 384.
638 Jung, *Civilization in Transition*, CW 10, para. 113.
639 Meier, *Atom and Archetype: The Pauli/Jung Letters 1932-1958*, xxix.
640 Jackson and Lethem, *The Exegesis of Philip K. Dick*, 706–707.

641 Wheeler, *At Home in the Universe*, 123.
642 This is Jung's phrase, what he refers to as "a modality without a cause." Jung, *The Structure and Dynamics of the Psyche*, CW 8, para. 965.
643 The "will to power" is one of the key concepts in the philosophy of Friedrich Nietzsche. Though he never systematically defined it, he considered it to be the primary underlying force that is driving humanity's behavior, both individually and collectively. In Nietzsche's conception, this drive is neither good nor bad, but has a myriad of expressions. Its positive aspect can be considered to be the instinct for growth, creativity and evolution. When this impulse becomes untethered from the heart of compassion, however, it can go rogue and feed into and strengthen the unconscious power drives of the human shadow, resulting in the urge to dominate others as well as the environment.
644 Please see my article "The Sacred Art of Alchemy" on my website www.awakeninthedream.com.
645 Jung, *The Structure and Dynamics of the Psyche*, CW 8, para. 967.
646 Ibid., para. 965.
647 Wheeler, *At Home in the Universe*, 126.
648 "False imagination" is a term used in Buddhism which means something that appears to the mind but doesn't exist in actuality.
649 This is reminiscent of the archetypal figure of the shaman/trickster, who oftentimes manifests as a "fool." This figure stands at the threshold of the community's unconscious in such a way that they oftentimes act out and give living form to the group's unconscious, thereby helping the community to assimilate the unconscious that informs the collective. In alchemy, this trickster-like figure is known as Mercurius (who, interestingly enough, is considered to be a conjunction of all opposites). Mercurius is a relative of wetiko (which contains both the poison and the medicine superposed in a quantum state of potentiality).
650 Mindell, *The Quantum Mind and Healing*, 173.
651 Autonomous complexes are what ancient indigenous cultures would call "demons."
652 Bernstein, *Quantum Profiles*, 138.
653 To quote physicist Hendrik Kramers, "The quantum theory has been like other victories: you smile for months and weep for years." Quoted in Sheila Jones, *The Quantum Ten*, 289.
654 Quoted in "A Quantum Sampler," *New York Times*, December 26, 2005, http://www.nytimes.com/2005/12/26/science/a-quantum-sampler.html.
655 Quoted in Lederman and Hill, *Quantum Physics for Poets*, 119.
656 Cited in A. Foelsing, *Albert Einstein, A Biography* (New York: Penguin Group, 1997), 574.
657 C. Enz and K. von Meyenn, eds., *Writings on Physics and Philosophy* (Berlin: Springer Verlag, 1994), 144.
658 Hoffmann, *The Strange Story of the Quantum*, 228–229.
659 Jung, *Psychology and Religion: West and East*, CW 11, para. 747.
660 Henry P. Stapp, "Quantum Interactive Dualism," 18.
661 Hoffmann, *The Strange Story of the Quantum*, 229.
662 Quoted in Max Jammer, *Einstein and Religion* (Princeton, NJ: Princeton University Press, 1999), 70.
663 Alfred North Whitehead, *Science and the Modern World, Lowell Lectures, 1925* (New York: The Free Press, 1967), 51.

664 His Holiness the Dalai Lama, *The Universe in a Single Atom: The Convergence of Science and Spirituality*, 51.
665 Ibid.
666 Schrödinger, *What Is Life? With Mind and Matter and Autobiographical Sketches*, 118.
667 Ibid.
668 The "not I" is a psychological term which has relevance in the development of self-consciousness, a process in which we differentiate our sense of identity (an "I") from what is other than ourselves (the "not I"). Please refer to the classic book *The I and the Not I: A Study in the Development of Consciousness* by Esther Harding (Princeton: Princeton University Press, 1965).
669 His Holiness the Dalai Lama, *The Universe in a Single Atom: The Convergence of Science and Spirituality*, 64.
670 Francisco Varela, Evan Thompson, and Eleanor Rosch, *The Embodied Mind: Cognitive Science and Human Experience* (Cambridge, MA: MIT Press, 2016), 143.
671 William James, *The Principles of Psychology*, vol. 1 (New York: Henry Holt and Co., 1918), 3.
672 Ibid.
673 George Spencer-Brown, *Laws of Form* (New York: Dutton, 1979).
674 Jung, *Mysterium Coniunctionis*, CW 14, para. 659.
675 Schrödinger, *What Is Life? With Mind and Matter and Autobiographical Sketches*, 119.
676 Lee Nichol, ed., *The Essential David Bohm* (London: Routledge, 2008), 232.
677 *The Philosophical Writings of Niels Bohr*, vol. 1 (Woodbridge, CT: Ox Bow Press, 1987), 54.
678 Jung, *The Practice of Psychotherapy*, CW 16, para. 543.
679 Wheeler, *At Home in the Universe*, 307.
680 Jung, *The Structure and Dynamics of the Psyche*, CW 8, para. 357.
681 Quoted in Russell, *From Science to God*, 84.
682 Jung, *The Psychology of Kundalini Yoga: Notes of the Seminar Given in 1932*, 57.
683 Jorge Luis Borges, *Other Inquisitions* (Austin: University of Texas, 2002).
684 To quote Eddington, "By 'mind' I do not here exactly mean mind and by 'stuff' I do not at all mean stuff." Quoted in Wilber, *Quantum Questions*, 199.
685 Schrödinger, *Science and Humanism: Physics in Our Time*, 49.
686 Ibid., 11.
687 Ibid., 52–53.
688 Wheeler, *Geons, Black Holes, and Quantum Foam*, 235.
689 Zajonc, *The New Physics and Cosmology: Dialogues with the Dalai Lama*, 108.
690 Wilber, *Quantum Questions: Mystical Writings of the World's Greatest Physicists*, 81.
691 Heisenberg, *Across the Frontiers*, 227.
692 Quoted in Jammer, *Einstein and Religion*, 68-69.
693 Wheeler, *A Journey into Gravity and Spacetime*, xi.
694 Wheeler, *Geons, Black Holes, and Quantum Foam*, 353.
695 Ibid., 262.
696 John Wheeler, *The Black Hole: An Imaginary Conversation with Albert Einstein*.
697 This is why another name I've given for wetiko, the psychospiritual disease that afflicts our species, is "ME disease"—it is a misidentification of who we think we are.
698 Synchronistically, as I was writing this, I got the following email from someone on January 15, 2017, who was quoting one of my Buddhist teachers,

Khenpo Tsewang Dongyal Rinpoche: "Even if you subdivide a particle a million times, there is ultimately nothing to find. When you see this, that is supreme seeing."

699 Stephen Hawking, ed., *The Dreams that Stuff is Made of* (Philadelphia, PA: Running Press, 2011), xi.

700 Jung, *The Structure and Dynamics of the Psyche*, CW 8, para. 623–624.

701 Schrödinger, *Science and Humanism: Physics in Our Time*, 48.

702 Quoted in Jeans, *Physics and Philosophy*, 199.

703 Schrödinger, *What Is Life? With Mind and Matter and Autobiographical Sketches*, 120.

704 Quoted in Mindell, *Quantum Mind: The Edge Between Physics and Psychology*, 519.

705 David Chalmers, "Facing up to the Problem of Consciousness," *The Journal of Consciousness Studies* 2, no. 3 (1995): 210.

706 Jackson and Lethem, *The Exegesis of Philip K. Dick*, 131.

707 Wheeler, *At Home in the Universe*, 188.

708 Schrödinger, *What Is Life? With Mind and Matter and Autobiographical Sketches*, 122.

709 Jung, *Psychology and Religion: East and West*, CW 11, para. 16.

710 Henry Corbin, *Alone with the Alone: Creative Imagination in the Sufism of Ibn 'Arabi* (Princeton, NJ: Princeton University Press, 1997), 359.

711 Wheeler, *Cosmic Search Magazine*, vol. 1 no. 4.

712 Richard Conn Henry, "The Mental Universe," *Nature*, vol. 436 (July 7, 2005): 29.

713 Ibid.

714 Andrei Linde, "Inflation, Quantum Cosmology, and the Anthropic Principle," in *Science and Ultimate Reality: Quantum Theory, Cosmology, and Complexity*, eds. Barrow et al., 451.

715 To read more about Jung's idea of "the reality of the psyche," see pp. 25–28 in my book *Dispelling Wetiko*.

716 Jung, *Psychology and Religion: West and East*, CW 11, para. 16–18.

717 C. G. Jung, *Symbols of Transformation*, CW 5 (Princeton, NJ: Princeton University Press, 1976), para. 296.

718 Jung, *The Structures and Dynamics of the Psyche*, CW 8, para. 683.

719 Ibid., para. 748.

720 Jung's "reality of the psyche" brings to mind Henry Corbin's "imaginal realm," also referred to by his term "mundus imaginalis." Please see Corbin, *Swedenborg and Esoteric Islam*, chapter 1.

721 W. Y. Evans-Wentz, *The Tibetan Book of the Great Liberation* (London: Oxford University Press, 1954), 232. The actual words are: "As a thing is viewed, so it appears." This text is a "terma," or hidden, rediscovered treasure, authored by Padmasambhava, considered to be the second Buddha and the Buddha of our age.

722 Erich Heller, *The Importance of Nietzsche* (Chicago: University of Chicago Press, 1988).

723 Jackson and Lethem, *The Exegesis of Philip K. Dick*, 376.

724 Hoffmann, *The Strange Story of the Quantum*, 42.

725 Feynman, *The Feynman Lectures on Physics*, vol. 3, 7.

726 Wheeler, *At Home in the Universe*, 3.

727 Ibid., 8.

728 Jean Houston, email to author, February 27, 2017.
729 Wheeler, "Information, Physics, Quantum: The Search for Links."
730 Notice the similarity to what I wrote in *Dispelling Wetiko* (p. 266) regarding the psychospiritual disease of the soul called wetiko: "There is no one definitive model for this disease, as each model has both its utility as well as its limits. When all of these models are combined and looked at together, it gives us a greater resolution and capacity to see what no one particular model by itself can reveal."
731 Hoffmann, *The Strange Story of the Quantum*, 171.
732 Ibid., 172.
733 Zajonc, *The New Physics and Cosmology: Dialogues with the Dalai Lama*, 208.
734 Richard Feynman, *Surely You're Joking Mr. Feynman* (New York: Bantam Books, 1989), 64.
735 Dyson, "Preface," in *Science and Ultimate Reality: Quantum Theory, Cosmology, and Complexity*, xix.
736 Wheeler, *At Home in the Universe*, 126.
737 Oppenheimer, *Science and the Common Understanding*.
738 The universe seems to work through two cross fields in the same way that the electromagnetic field described by Maxwell's equations (classical physics) involves an electric field and a magnetic field orthogonal to each other.
739 Quoted in Jung, *The Structure and Dynamics of the Psyche*, CW 8, 229, footnote #130.
740 Henry Corbin, *The Voyage and the Messenger: Iran and Philosophy* (Berkeley, CA: North Atlantic Books, 1998), LV.
741 In the Nyingma tradition of Tibetan Buddhism, these hidden treasures are called *terma*. Please see my article "Hidden Treasures," on my website www.awakeninthedream.com.
742 B. J. Hiley, and F. David Peat, eds., *Quantum Implications: Essays in Honor of David Bohm* (London: Routledge & Kegan Paul, 1988), 387.
743 Jung, *Psychological Types*, CW 6, para. 60.
744 Jung, *Alchemical Studies*, CW 13, para. 2.
745 Jung, *Psychological Types*, CW 6, para. 84.
746 Jung, *Mysterium Coniunctionis*, CW 14, para. 768.
747 Ibid., para. 760.
748 Jung compares this, psychologically speaking, to the self, which he refers to as the "ground and origin of the individual personality past, present and future." *Mysterium Coniunctionis*, CW 14, para. 760.
749 Jung, *Letters*, vol. 2, 342.
750 Schrödinger, *What Is Life? With Mind and Matter and Autobiographical Sketches*, 127.
751 Jung, *Mysterium Coniunctionis*, CW 14, para.767.
752 Ibid., para. 765.
753 Jung, *The Structure and Dynamics of the Psyche*, CW 8, para. 965.
754 This is why Christ—who symbolically represents the self—in addition to being referred to as the Son of God, refers to himself as the "Son of Man." Please see *Dispelling Wetiko*, 188.
755 Schrödinger, *Science and Humanism: Physics in our Time*, 11.
756 Ibid., 20.
757 Jackson and Lethem, *The Exegesis of Philip K. Dick*, 479.
758 Nick Herbert, "Scientists Explore Invisible Ocean of Glue," 2.

759 Schrödinger, *Science and Humanism: Physics in Our Time,* 17.

760 Quoted in Schafer, *Infinite Potential: What Quantum Physics Reveals About How We Should Live,* 180.

761 Ibid., 58.

762 Weber, *Dialogues with Scientists and Sages,* 93.

763 Schrödinger, *Science and Humanism: Physics in our Time,* 27.

764 Quoted in Ervin Laszlo, *Science and the Reenchantment of the Cosmos: The Rise of the Integral Vision of Reality* (Rochester, VT: Inner Traditions, 2006).

765 A standing wave is a pattern of dynamically moving waves that takes on the appearance of being solid, nonmoving, and stable. Standing waves appear to be "standing" still when in fact there is nothing but movement creating the very convincing illusion of a static object. Some physicists consider our universe as a whole to be one gigantic standing wave pattern.

766 Schrödinger, *Science and Humanism: Physics in Our Time,* 13.

767 Corbin, *Alone with the Alone: Creative Imagination in the Sufism of Ibn 'Arabi,* 201.

768 Ibid., 202.

769 Quoted in Pfeiffer and Mack, *Mind before Matter: Visions of a New Science of Consciousness,* 68–69.

770 Jackson and Lethem, *The Exegesis of Philip K. Dick,* 705–709.

771 Bohm, *Wholeness and the Implicate Order.*

772 Norbert Wiener, *The Human Use of Human Beings,* 96.

773 Quoted in H. Folse, *Philosophy of Niels Bohr,* 81.

774 Quoted in *The Cartoon Guide to Physics,* (New York: HarperCollins Interactive, 1995), CD-ROM.

775 Werner Heisenberg, *Philosophie Problems of Nuclear Science,* trans. F. C. Hayes (New York: Pantheon, 1952), 41.

776 Heisenberg, *Across the Frontiers,* 121.

777 Brian, *The Voice of Genius,* 124.

778 This brings to mind what Jung wrote to Zwi Werblowsky in 1952: "The language I speak must be equivocal, that is, *ambiguous,* to do justice to psychic nature with its double aspect. I strive consciously and deliberately for ambiguous expressions, because it is superior to unequivocalness and corresponds to the nature of being." C. G. Jung, *Letters,* vol. 2, 70-71.

779 Ludwig Wittgenstein, *Philosophical Investigations,* trans. G. Anscombe (New York: Macmillan, 1958), 48.

780 Quoted in Aage Peterson, *Quantum Physics and the Philosophical Tradition,* 1968.

781 Quoted in Stuart Chase's foreword to Whorf, *Language, Thought and Reality,* vi.

782 As Wheeler writes in his autobiography, the origin of this term was as follows: during a lecture, Wheeler remarked that one needed a shorter descriptive phrase other than "gravitationally completely collapsed object." An unidentified person in the audience then asked "How about black hole?" Wheeler writes, "I had been searching for the right term for months, mulling it over in bed, in the bathtub, in my car, whenever I had quiet moments. Suddenly this name seemed exactly right." A few weeks later at a more formal lecture, he started using this term, "dropping it into the lecture and the written version as if it were an old familiar friend." We could say that Wheeler was open to the universe helping him to find this term; in a sense, he merely popularized the expression. (Wheeler, *Geons, Black Holes, and Quantum Foam,* 296–297.)

783 Wheeler, *Geons, Black Holes, and Quantum Foam*, 297.
784 Ibid.
785 Quoted in Mindell, *Quantum Mind: The Edge Between Physics and Psychology*, 215.
786 Heisenberg, *Physics and Philosophy: The Revolution in Modern Science*, 174.
787 Harald Atmanspacher and Hans Primas, eds., *Recasting Reality: Wolfgang Pauli's Philosophical Ideas and Contemporary Science* (New York: Springer Science & Business Media, 2008). From the letter by Pauli to Rosenfeld of April 1, 1952.
788 Heisenberg, *Across the Frontiers*, 35–36.
789 Wheeler, *At Home in the Universe*, 294.
790 Quoted in Bryce DeWitt, "The Everett Interpretation of Quantum Mechanics," in *Science and Ultimate Reality: Quantum Theory, Cosmology, and Complexity*, eds. Barrow et al., 177.
791 Schrödinger, *Science and Humanism: Physics in Our Time*, 19.
792 Ibid., 20–21.
793 Heisenberg, *Across the Frontiers*, 170.
794 Jeans, *Physics and Philosophy*, 195.
795 Durr, *Physik and Transzendenz* (Bern: Verlag, 1988), 13.
796 Heisenberg, *Across the Frontiers*, 116.
797 Wilber, *Quantum Questions: Mystical Writings of the World's Greatest Physicists*, 121.
798 Jeans, *The Mysterious Universe*, 158.
799 Quoted in Schafer, *Infinite Potential: What Quantum Physics Reveals About How We Should Live*, 102.
800 Heisenberg, *Across the Frontiers*, 75.
801 Morris Kline, *Mathematics: The Loss of Certainty* (Oxford: Oxford University Press, 1980), 323.
802 Bohm, *Quantum Theory*, 171.
803 C. G. Jung, *The Development of Personality*, CW 17, para. 160.
804 Sutin, *The Shifting Realities of Philip K. Dick*, 277.
805 It is worth noting that Jung extrapolates this dynamic to a more cosmic scale when he pointed out in his work *Answer to Job* that whenever anyone becomes aware of "God," this realization has an effect on "God"—in the ultimate sense, evoking his incarnation.
806 Wolfgang Pauli, *Writings on Physics and Philosophy* (New York: Springer Verlag, 1992), 153.
807 This also can be thought of as the outer and inner aspects of reality. In theological terms, these can be conceived of as the earthly and heavenly realms.
808 C. G. Jung, *Modern Man in Search of a Soul*, 213.
809 Bohm, *Quantum Theory*, 172.
810 Max F. Muller, *Wisdom of the Buddha: The Unabridged Dhammapada* (Mineola, NY: Dover Publications, 2000), 1.
811 David Peat, *Infinite Potential: The Life and Times of David Bohm.*
812 Bohm, *Wholeness and the Implicate Order*, ix.
813 Quoted in Schafer, *Infinite Potential: What Quantum Physics Reveals About How We Should Live*, 75.
814 Henry P. Stapp, "Harnessing Science and Religion: Implications of the New Scientific Conception of Human Beings," *Science & Theology News*, vol. 1, no. 6 (Feb. 2001): 8.
815 See Lynne McTaggart, *The Intention Experiment* (New York: Free Press, 2007).
816 Heisenberg, *Physics and Philosophy: The Revolution in Modern Science*, 202.

817 Gerald Holton, ed., *Science and the Modern Mind: A Symposium* (Beacon Press, 1958), 84.
818 Quoted in Arthur Koestler and J. R. Smythies, eds., *Beyond Reductionism* (Boston: Beacon Press, 1971), 115.
819 Quoted in Arne A. Wyller, *The Planetary Mind* (Aspen, CO: MacMurray & Beck, 1996), 26.
820 Pfeiffer and Mack, eds., *Mind before Matter: Visions of a New Science of Consciousness.*
821 Zajonc, *The New Physics and Cosmology: Dialogues with the Dalai Lama*, 48.
822 David Bohm, *On Creativity* (London: Routledge, 1996), 93.
823 Jung, *Psychology and Religion: West and East*, CW 11, para. 380.
824 Jackson and Lethem, *The Exegesis of Philip K. Dick*, 886.
825 Schrödinger, *What Is Life? With Mind and Matter and Autobiographical Sketches*, 121.
826 Quoted in Walter Moore, *A Life of Erwin Schrödinger* (Cambridge, UK: Cambridge University Press, 1994), 181.
827 Quoted in John William Navin Sullivan, *Contemporary Mind: Some Modern Answers* (London: H. Toulmin, 1934).
828 Colin McGinn, "Consciousness and Cosmology: Hyperdualism Ventilated," in *Consciousness*, Martin Davies (Oxford: Blackwell, 1993), 160.
829 B. D'Espagnat, *On Physics and Philosophy*, 425.
830 Schrödinger, *What Is Life? With Mind and Matter and Autobiographical Sketches*, 121.
831 Jeans, *The Mysterious Universe*, 137.
832 Kak et al., eds., *Quantum Physics of Consciousness*, 79.
833 Jung, *Alchemical Studies*, CW 13, para. 229.
834 Jung, *Visions*, vol. 2, 1046.
835 Please see my article "The Light of Darkness" on my website www.awakeninthedream.com.
836 The *lumen naturae* is considered to be the inner, living light of primordial, ever-present, nondual awareness itself, which is the light of nature that literally lights up the whole of creation in this and every moment. To see the light that is hidden in the darkness is to become conscious, which, alchemically speaking, frees the spirit that is hidden and trapped inside the material world (which is the whole point of the alchemical opus). Being non-dual, the lumen naturae nonlocally and synchronistically configures co-arising events, both in the physical world and reciprocally within the landscape of our minds, as a way of revealing itself.
837 Wilber, *Quantum Questions: Mystical Writings of the World's Greatest Physicists*, 175.
838 Quoted in Diarmuid O'Murchu, *Quantum Theology: Spiritual Implications of the New Physics* (New York: Crossroad Publishing Co., 2004), 3.
839 Quoted in Richard Conn Henry, "The Mental Universe," *Nature*, vol. 436 (July 7, 2005): 29.
840 Brian, *The Voice of Genius*, 133.
841 I am using the word "religious" not in the dogmatic sense but in the true meaning of the word; that is, carefully considering, with a sense of awe and reverence, a living and dynamic agency that is conceived of and experienced as a numinous power greater than our own ego. Etymologically, the word *religio* derives from *religare*, which means to link back and reconnect to the

source. The term "religious" designates the attitude peculiar to a consciousness that has been changed by experience of "the numinous."

842 Quoted in The Conference on Science, Philosophy, and Religion in Their Relation to the Democratic Way of Life, *Science, Philosophy, and Religion: A Symposium* (New York: The Conference, 1941).

843 E. N. da Costa Andrade, *Listener* 37 (July 10, 1947): 134.

844 Cited by John Honner, "Niels Bohr and the Mysticism of Nature," *Zygon* 17 (1982): 246.

845 Quoted in Schafer, *Infinite Potential: What Quantum Physics Reveals About How We Should Live,* 29.

846 Carl Sagan, *The Demon-Haunted World* (New York: Ballantine Books, 1996).

847 Einstein, *Ideas and Opinions.*

848 Jeans, *The Mysterious Universe,* 159.

849 Hoffmann, *The Strange Story of the Quantum,* 231.

850 Wheeler, *At Home in the Universe*, 226.

851 Zajonc, *The New Physics and Cosmology: Dialogues with the Dalai Lama,* 221.

852 Quoted in Sampooran Singh, *A Scientific Guide to the Integration of Man, Society, and Mankind* (New Dehli: Siddharth Publications, 1987).

853 Pauli writes to Jung, "Any deeper form of reality—i.e., every 'thing as such'—is symbolic for me anyway." Meier, *Atom and Archetype,* 87.

854 One of the words for enlightenment, "mahamudra," translates as "the great symbol." Interestingly, another translation of this word is "the great attitude." This is pointing at, from the Tibetan Buddhist point of view, the state of enlightenment has to do with having the attitude that interprets this universe symbolically, as if we are in a dream.

855 Jung, *Psychology and Religion: West and East,* CW 11, para. 849.

856 Berdyaev, *The Meaning of the Creative Act,* 20.

857 Jung, *Civilization in Transition,* CW 10, para. 24.

858 Jung, *Psychological Types,* CW 6, para. 178.

859 Jung, *The Archetypes and the Collective Unconscious,* CW 9i, para. 291.

860 C. G. Jung, *The Red Book: A Reader's Edition* (New York: W. W. Norton & Co., 2009), 392–393.

861 Jung, *Psychological Types,* CW 6, para. 401.

862 Ibid., para. 435.

863 Jung, *The Red Book: A Reader's Edition,* 392.

864 Notice the similarity—in the apocryphal texts, Christ says, "I am a mirror to those that perceive me." It is as if Christ is the embodiment—the incarnation—of the quantum.

865 Quoted in Heisenberg, *Physics and Beyond: Encounters and Conversations,* 63.

866 Jung uses the word "Zauberlehrling," which translates as "sorcerer's apprentice."

867 Jung, *Civilization in Transition,* CW 10, para. 326.

868 Wheeler, *At Home in the Universe,* 307.

869 Scott Aaronson, "Are Quantum States Exponentially Long Vectors?" accessed September 27, 2017, http://www.scottaaronson.com/papers/are.pdf.

870 Wheeler, *At Home in the Universe,* 15–16.

871 Schrödinger, *Science and Humanism: Physics in our Time,* 57.

872 Quoted in *The Little Book of Bleeps* (London: Revolver, 2005).

873 Heisenberg, *Physics and Philosophy: The Revolution in Modern Science,* 216.

874 Ibid.

875 Goswami, *The Everything Answer Book: How Quantum Science Explains Love, Death and the Meaning of Life*, 225.
876 Ralph Waldo Emerson, *The Essential Writings of Ralph Waldo Emerson* (New York: Modern Library, 2000).
877 Jackson and Lethem, *The Exegesis of Philip K. Dick*, 76.
878 In a tradition with which I'm personally familiar, as the story goes, the king arrested Padmasambhava (we could think of him as representing the quantum) and burned him alive, but he transformed the pyre into a lake, and was found sitting, cool and fresh, on a lotus blossom in its center. The Christ event can be interpreted similarly. In killing Christ, the powers of darkness only further helped him reveal his true (quantum) nature via the resurrected body. These are but a couple of examples of how mythologies the world over have symbolized the miraculous nature of the quantum in various personified forms.
879 Note the similarity to what I wrote in *Dispelling Wetiko* regarding wetiko: "To awaken to the dreamlike nature of reality, the symbolic dimension of existence, is to realize that events in our world are lower-level shadows or reflections of a higher-dimensional reality. . . . The person momentarily afflicted with wetiko is like a shadow on a wall, cast from a globe hanging from the ceiling, relative to the globe. The shadow on the wall is a re-presentation and projection of the higher, three-dimensional entity of the globe into a lower dimension of space. Studying the shadow within its proper context relative to the globe is the way of understanding the object casting the shadow (the higher-dimensional entity of the "global" wetiko psychosis). By tracking the variety of shadows the object is casting, we are illuminating an unknown, mysterious object from as many different angles as we can imagine. From enough of these shadows it becomes possible to reconstruct the illumined object." 169–170.
880 Wilber, *Quantum Questions: Mystical Writings of the World's Greatest Physicists*, 8.
881 Jeans, *The Mysterious Universe*, 134.
882 Eddington, *The Nature of the Physical World.*
883 Ibid., 282.
884 Wilber, *Quantum Questions: Mystical Writings of the World's Greatest Physicists*, 6.
885 Jung, *Letters*, vol. 1, 57.
886 Eddington, *The Nature of the Physical World*, 7.
887 Jung, *Letters*, vol. 1, 57.
888 Khenpo Tsewang Dongyal Rinpoche, *Inborn Realization: A Commentary on His Holiness Dudjom Rinpoche's Mountain Retreat Instructions* (Sidney Center, NY: Dharma Samudra, 2016), 218.
889 Mindell, *Quantum Mind: The Edge Between Physics and Psychology*, 575.
890 Quoted in Schafer, *Infinite Potential: What Quantum Physics Reveals About How We Should Live*, 186.
891 Whitehead, *Science and the Modern World*, 181.

SELECTED BIBLIOGRAPHY

Barad, Karen. *Meeting the Universe Halfway: Quantum Physics and the Entanglement of Matter and Meaning.* Durham, NC: Duke University Press, 2007.

Barrow, John D., Paul C. W. Davies, and Charles, L. Harper, Jr., eds. *Science and Ultimate Reality: Quantum Theory, Cosmology, and Complexity.* Cambridge: Cambridge University Press, 2005.

Bohm, David. *Wholeness and the Implicate Order.* London: Routledge, 2000.

Capra, Fritof. *The Tao of Physics.* Berkeley: Shambhala, 1975.

Davies, P. C. W., and J. R. Brown, eds. *The Ghost in the Atom.* Cambridge: Cambridge University Press, 1993.

Friedman, Norman. *Bridging Science and Spirit.* St. Louis, MO: Living Lake Books, 1994.

Gefter, Amanda. *Trespassing on Einstein's Lawn.* New York: Bantam Books, 2014.

Gieser, Suzanne. *The Innermost Kernel: Depth Psychology and Quantum Physics. Wolfgang Pauli's Dialogue with C. G. Jung.* Berlin: Springer-Verlag, 2006.

Goswami, Amit. *The Self-Aware Universe: How Consciousness Creates the Material World.* New York: Jeremy P. Tarcher, 1993.

Gregory, Bruce. *Inventing Reality: Physics as Language.* New York: John Wiley and Sons, Inc., 1990.

Heisenberg, Werner. *Across the Frontiers.* Woodbridge, CT: Ox Bow Press, 1990.

Heisenberg, Werner. *Physics and Beyond: Encounters and Conversations.* New York: Harper and Row, 1972.

Heisenberg, Werner. *Physics and Philosophy: The Revolution in Modern Science.* Amherst, NY: Prometheus Books, 1958.

Herbert, Nick. *Quantum Reality: Beyond the New Physics.* New York: Anchor Books, 1987.

Hoffmann, Banesh. *The Strange Story of the Quantum.* New York: Dover Publications, 1959.

Jackson, Pamela, and Jonathan Lethem, eds. *The Exegesis of Philip K. Dick.* New York: Houghton Mifflin Harcourt, 2011.

Jeans, Sir James. *The Mysterious Universe.* New York: The MacMillan Company, 1930.

Jeans, Sir James. *Physics and Philosophy.* New York: Dover Publications, 1981.

Jung, C. G. *The Collected Works of C. G. Jung.* Princeton, NJ: Princeton University Press, 1979.

Kak, Subhash, Sir Roger Penrose, and Stuart Hameroff, eds. *Quantum Physics of Consciousness.* Cambridge: Cosmology Science Publishers, 2011.

Lindorff, David. *Pauli and Jung: The Meeting of Two Great Minds*. Wheaton, Illinois: Quest Books, 2009.

Malin, Shimon. *Nature Loves to Hide: Quantum Physics and the Nature of Reality, a Western Perspective*. Singapore: World Scientific Publishing Co., 2012.

Mansfield, Victor. *Synchronicity, Science, and Soul-Making*. Chicago: Open Court, 1995.

Meier, C. A. *Atom and Archetype: The Pauli/Jung Letters 1932-1958*. Princeton, NJ: Princeton University Press, 2001.

Mindell, Arnold. *Quantum Mind: The Edge Between Physics and Psychology*. Portland, OR: Lao Tse Press, 2000.

Nadeau, Robert and Menas Kafatos, *The Non-local Universe: The New Physics and Matters of the Mind*. New York: Oxford University Press, 1999.

Peat, F. David. *Einstein's Moon: Bell's Theorem and the Curious Quest for Quantum Reality*. Chicago: Contemporary Books, 1990.

Rosenblum, Bruce, and Fred Kuttner. *Quantum Enigma: Physics Encounters Consciousness*. New York: Oxford University Press, 2006.

Schafer, Lothar. *Infinite Potential: What Quantum Physics Reveals About How We Should Live Our Life*. New York: Deepak Chopra Books, 2013.

Schrödinger, Erwin. *Science and Humanism: Physics in Our Time*. Cambridge: Cambridge University Press, 1961.

Schrödinger, Erwin. *What Is Life? With Mind and Matter and Autobiographical Sketches*. Cambridge: Cambridge University Press, 2004.

Smetham, Graham. *Quantum Buddhism: Dancing in Emptiness—Reality Revealed at the Interface of Quantum Physics and Buddhist Philosophy*. Brighton, Sussex: Shunyata Press, 2010.

Stapp, Henry P. *Mind, Matter, and Quantum Mechanics*. Berlin: Springer-Verlag, 1993.

Stapp, Henry P. *Mindful Universe: Quantum Mechanics and the Participating Observer*. Berlin: Springer-Verlag, 2007.

Talbot, Michael. *Mysticism and the New Physics*. New York: Penguin Books, 1993.

Wallace, B. Alan, ed. *Buddhism and Science: Breaking New Ground*. New York: Columbia University Press, 2003.

Wallace, B. Alan. *Hidden Dimensions: The Unification of Physics and Consciousness*. New York: Columbia University Press, 2007.

Wheeler, John Archibald. *At Home in the Universe*. Woodbury, NY: The American Institute of Physics, 1994.

Wheeler, John Archibald, with Kenneth Ford. *Geons, Black Holes, and Quantum Foam: A Life in Physics*. New York: W. W. Norton & Company, 1998.

Wheeler, John Archibald. *Frontiers of Time*. Amsterdam: North-Holland for the Societa Italiana di Fisica, 1979.

Wheeler, John Archibald, and Wojciech Hubert Zurek, eds. *Quantum Theory and Measurement*. Princeton, NJ: Princeton University Press, 1983.

Woolf, Harry. *Some Strangeness in the Proportion: A Centennial Symposium to Celebrate the Achievements of Albert Einstein*. Reading, MA: Addison-Wesley Publishing Company, 1981.

Zajonc, Arthur, ed. *The New Physics and Cosmology: Dialogues with the Dalai Lama*. New York: Oxford University Press, 2004.

Zukov, Gary. *The Dancing Wu Li Masters: An Overview of the New Physics*. New York: William Morrow and Co., 1979.

Zurek, Wojciech Hubert, Alwyn van der Merwe, and Warner Allen Miller, eds., *Between Quantum and Cosmos: Studies and Essays in Honor of John Archibald Wheeler*. Princeton, NJ: Princeton University Press, 1988.

INDEX

ABOUT THE AUTHOR

PAUL LEVY was born in 1956 and grew up in Yonkers, New York. In the mid-seventies he attended the State University of New York at Binghamton (now called Binghamton University), receiving degrees in both economics and studio art. While an undergraduate, he was hired by Princeton University to do research in economics. Though not a physicist himself, Paul has been seriously studying and going down the quantum physics rabbit hole for decades.

In 1981, catalyzed by an intense trauma, Paul had a life-changing spiritual awakening in which he began to recognize the dreamlike nature of reality. During the first year of his spiritual emergence, he was hospitalized a number of times and was told he was having a severe psychotic break from reality and (mis)diagnosed as having a chemical imbalance. He was informed that he had what was then called manic depression (now called bipolar disorder) and that he would have to live with his illness for the rest of his life and would need to take medication until his dying breath. Little did the doctors realize, however, that he was taking part in a mystical-awakening/shamanic-initiation process, which at times mimicked psychosis but in actuality was a spiritual experience of a far different order that was completely off the map of the psychiatric system. Fortunately, over time he was able to extricate himself from the psychiatric establishment so that he could continue to unfold his inner process of awakening.

After the trauma of his shamanic breakdown/breakthrough, Paul became a certified art teacher and taught both painting and drawing for a handful of years to people of all ages. Intensely interested in the work of C. G. Jung, in 1988 he became the manager of the C. G. Jung Foundation Book Service in New York, as well as the advertising manager for the Jungian journal *Quadrant.*

In 1990 Paul moved to Portland, Oregon. In 1993, after many years of working on himself so as to integrate his non-ordinary experiences, he started to openly share his insights about the dreamlike nature of reality by giving talks and facilitating groups based on the way life is a shared waking dream that we are all cocreating and dreaming together. A pioneer in the field of spiritual emergence, Paul is a wounded healer in private practice, helping others who are also awakening to the dreamlike nature of reality. He is the founder of the Awakening in the Dream community in Portland, Oregon.

Paul is the author of *Dispelling Wetiko: Breaking the Curse of Evil, Awakened by Darkness: When Evil Becomes Your Father,* and *The Madness of George W. Bush: A Reflection of Our Collective Psychosis.* A Tibetan Buddhist practitioner for more than thirty years, Paul has intimately studied with some of the greatest spiritual masters of Tibet and Burma. He was the coordinator of the Portland chapter of the Padmasambhava Buddhist Center for over twenty years.

For more information on Paul's work, visit his website www.awakeninthedream.com; his email is paul@awakeninthedream.com.